Springer Series in
Experimental Entomology

Thomas A. Miller, Editor

Springer Series in Experimental Entomology

Editor: T. A. Miller

Insect Neurophysiological Techniques
By T. A. Miller (1979)

Neurohormonal Techniques in Insects
Edited by T. A. Miller (1980)

Sampling Methods in Soybean Entomology
By M. Kogan and D. C. Herzog (1980)

Cuticle Techniques in Arthropods
Edited by T. A. Miller (1980)

Neuroanatomical Techniques: Insect Nervous System
Edited by N. J. Strausfeld and T. A. Miller (1980)

Functional Neuroanatomy

Edited by
Nicholas J. Strausfeld

With Contributions by
M. E. Adams · J. S. Altman · J. P. Bacon · U. K. Bassemir
C. A. Bishop · E. Buchner · S. Buchner · H. Bülthoff · S. D. Carlson
Che Chi · F. Delcomyn · H. Duve · M. Eckert · T. R. J. Flanagan
G. Geiger · C. M. Hackney · B. Hengstenberg · R. Hengstenberg
N. Klemm · D. R. Nässel · M. O'Shea · W. A. Ribi · R. L. Saint Marie
H. S. Seyan · P. T. Speck · N. J. Strausfeld · A. Thorpe
J. Ude · D. W. Wohlers

With 208 Figures

Springer-Verlag
Berlin Heidelberg New York Tokyo
1983

Nicholas J. Strausfeld
European Molecular Biology Laboratory, D-6900 Heidelberg, F.R.G.

QL
494
.F87
1983

ISBN 3-540-12742-9 Springer-Verlag Berlin Heidelberg New York Tokyo
ISBN 0-387-12742-9 Springer-Verlag New York Heidelberg Berlin Tokyo

Library of Congress Cataloging in Publication Data. Main entry under title: Functional neuroanatomy. (Springer series in experimental entomology) Bibliography: p.   Includes index. 1. Nervous system—Insects. 2. Insects—Anatomy. I. Strausfeld, Nicholas James, 1942—   . II. Adams, M. E. (Michael E.) III. Series. [DNLM: 1. Nervous system—Anatomy. 2. Insects—Anatomy. WL 494 F979] QL494.F87   1983 595.7'048   83-14826

Typesetting, printing and bookbinding: Konrad Triltsch, Graphischer Betrieb,
D-8700 Würzburg

2131/3130-543210

Preface

The "functional" in the title of this book not only reflects my personal bias about neuroanatomy in brain research, it is also the gist of many chapters which describe sophisticated ways to resolve structures and interpret them as dynamic entities. Examples are: the visualization of functionally identified brain areas or neurons by activity staining or intracellular dye-iontophoresis; the resolution of synaptic connections between physiologically identified nerve cells; and the biochemical identification of specific neurons (their peptides and transmitters) by histo- and immunocytochemistry.

I personally view the nervous system as an organ whose parts, continuously exchanging messages, arrive at their decisions by the cooperative phenomenon of consensus and debate. This view is, admittedly, based on my own experience of looking at myriads of nerve cells and their connections rather than studying animal behaviour or theorizing. Numerous structural studies have demonstrated that interneurons in the brain must receive hundreds of thousands of synapses. Many neurons receive inputs from several different sensory areas: each input conveys a message about the external world and possibly also about past events which are stored within the central nervous system. Whether an interneuron responds to a certain combination of inputs may be, literally, a matter of debate whose outcome is decided at the postsynaptic membrane. A nerve cell responding to an overriding command is possibly a rare event.

To understand how the nervous system works we must be able to observe its parts. How can we visualize the integrative activity of the nervous system? Only a few cells can be simultaneously recorded by intracellular electrodes and filled with dyes. Can the data obtained be sufficient for us to understand how neurons interact to generate behaviour? How are we to recognize the identity, or even the existence, of all the other neurons involved in generating even a small subroutine of a behavioural pattern? This is the substance of this book and its predecessor.

Why *insect* neuroanatomy? Why do certain neurobiologists eschew mice, cats and monkeys? Insects are indisputably more convenient experimental animals. It matters not a whit that they have exoskeletons and compound eyes. What interests the neurobiologist is that insects have hundreds of thousands of nerve cells, synapses, axons and dendrites, organized in much the same way as those of vertebrates, and that insects have predictable patterns of behaviour. It is also mighty handy to put all the sections of a cricket or fly central nervous system on one slide and to tour through all its sensory and motor projections to and from recognizable regions of neuropil.

Alongside nematodes, gastropods, annelids and crustaceans, insects offer unique opportunities in neurobiology. One is the study of genetic control of CNS development and structure where neurological defects can be related to the chromosomal map and, eventually, to molecular genetics. Another is the study of cell lineage in early development of the nervous system (although arguably the leech is a length ahead): segmentation, neuronal differentiation and pathfinding by identified neurons at the cellular and molecular level. Yet another is the study of integration by uniquely identified neurons that are involved in specific behaviour patterns. I refer particularly to identified neurons in multimodal sensory pathways controlling locomotion – ranging from "simple" jump responses, through control of flight or walking, to the courtship of fruit flies or grasshoppers where we may expect to find a high level of pattern recognition by single neurons. Despite the use of the word, none of these behaviours is "simple" in terms of the sheer number and variety of neurons involved.

Many behaviours, sophisticated or otherwise, may also be modulated by peptidergic systems. To understand the role of these vertebrate-like neuropeptides it is necessary to identify the neurons that contain them and to see these in the context of sensory-motor pathways. Again, insects provide compact but behaviourally sophisticated nervous systems for this line of research.

The purpose of this volume is to describe strategies that hopefully will encourage more substantial descriptions of neuronal control mechanisms. This book is the promised follow-up of *Neuroanatomical Techniques: Insect Nervous System* edited by Thomas A. Miller and myself which is hereafter referred to as Vol. 1. That volume (published in 1980) dealt mainly with the application of more classical techniques to insect central nervous systems. It also included electron microscopy and intracellular dye injection methods as well as several chapters on cobalt techniques – one of the most powerful neuroanatomical methods for structural studies of invertebrate and amphibian nervous systems.

As in Vol. 1, certain subjects in this one are covered by several chapters, e.g. multiple cell marking. Many of us recognize the need to expand our tunnel vision, particularly when we look into the electron microscope. Thin sections restrict our view severely, unless we can distinguish landmarks belonging to something familiarly three-dimensional. Methods that allow us to recognize as many as three differently marked neurons that were selected earlier in

thick sections in the light microscope are a logical development from electron microscopy of the classical Golgi method (see Chaps. 1, 3 and 6).
As in Vol. 1, chapters are grouped together to cover specific fields. Chapters
1–5 include electron microscopy of marked neurons (Golgi, cobalt, nickel,
heme proteins) followed by a chapter about combined techniques (Chap. 6).
Chapter 7 describes the histology for Lucifer yellow marked neurons, and
Chaps. 8 and 9 outline computer reconstructions for displaying structural
data obtained by this and other methods. Direct anatomical-functional correlates are the subject of Chaps. 10 and 11 (both of which truly represent
functional anatomy). The next six chapters deal with the biochemical (peptide and transmitter) identity of neurons both in the light and electron microscope. The last two chapters focus on the detailed aspects of cell contacts and
synapses using freeze fracture and high-voltage electron microscopy.
I would like to thank all the authors for their contributions and some for
their patience while later manuscripts came in. My thanks also go to Springer-Verlag for their civilized custom of waiting almost sublimely, without
too frequent breathing down my neck. I am also grateful to Wendy Moses
and Heide Beschorner for typing some of the manuscripts, to Renate Weisskirchen for printing many of the photographs and to Susan Mottram for
researching incomplete references. Eduardo Macagno and Corey Goodman
kindly donated data and figures for Chap. 6 (Figs. 8, 9) and Chap. 16
(Fig. 8), respectively.
Special and personal thanks go to Sir John Kendrew, F. R. S. to whom this
book is dedicated. As founding Director-General of the European Molecular
Biology Laboratory (which was conceived as an institute for fundamental
biological research), he encouraged the individual scientist and provided the
wherewithal for a small group of insect neurobiologists, all of whom benefit-
ed greatly from his sponsorship. He in no small way contributed to the
achievement of this and the preceding volume.
Finally, I thank Camilla, my wife, for the effort she put into this book –
language editing and condensing many of its chapters – and for her help and
love.

Heidelberg, August 1983   NICHOLAS J. STRAUSFELD

Call to Authors

Springer Series in Experimental Entomology will be published in future volumes as contributed chapters. Subjects will be gathered in specific areas to keep volumes cohesive.
Correspondence concerning contributions to the series should be communicated to:

THOMAS A. MILLER, Editor
Springer Series in Experimental Entomology
Department of Entomology
University of California
Riverside, California 92521
USA

Contents

Chapter 1
Electron Microscopy of Golgi-Impregnated Neurons
WILLI A. RIBI. With 7 Figures

I.	Introduction	1
II.	General Description of the Procedure	3
III.	Results	10
IV.	Discussion	14
V.	Appendix: Schedules	15

Chapter 2
**Block Intensification and X-Ray Microanalysis of Cobalt-Filled
Neurons for Electron Microscopy**
URSULA K. BASSEMIR and NICHOLAS J. STRAUSFELD. With 10 Figures

I.	Introduction	19
II.	Method	20
III.	Ultrastructure	28
IV.	Identification of Size and Nature of Precipitate	37
V.	Discussion	37
VI.	Applications	42

Chapter 3
Horseradish Peroxidase and Other Heme Proteins as Neuronal Markers
DICK R. NÄSSEL. With 26 Figures

I.	Introduction	44
II.	The Versatility of Exogenous Heme Proteins as Neuronal Markers	45
III.	Chemistry of Peroxidase-Active Proteins and Reactions for their Demonstration	46
IV.	Methods	51

V.   Cytology of Peroxidase-Labelled Neurons   78
VI.   Mechanisms of Neuronal Uptake and Transport
      of Heme Proteins   85
VII. Concluding Remarks   88
VIII. Addendum   89
IX.   Appendix 1: Method   90
X.   Appendix 2: Solutions   91

Chapter 4
Intracellular Staining with Nickel Chloride
FRED DELCOMYN. With 1 Figure

I.   Introduction   92
II.   Method   93

Chapter 5
Rubeanic Acid and X-Ray Microanalysis for Demonstrating Metal Ions in Filled Neurons
CAROLE M. HACKNEY and JENNIFER S. ALTMAN. With 5 Figures

I.   Introduction   96
II.   Use of Different Metal Ions   96
III. Rubeanic Acid Development   98
IV.   Applications of Rubeanic Acid Development   100
V.   X-Ray Microanalysis for Detection of Metal Ions   105
VI.   Conclusion   111

Chapter 6
Double Marking for Light and Electron Microscopy
HARJIT SINGH SEYAN, URSULA K. BASSEMIR and NICHOLAS J. STRAUSFELD. With 9 Figures

I.   Introduction   112
II.   Double Marking for Light Microscopy   113
III. Double Marking for Light and Electron Microscopy   117
IV.   Alternative Strategies   128

Chapter 7
Lucifer Yellow Histology
NICHOLAS J. STRAUSFELD, HARJIT SINGH SEYAN, DAVID WOHLERS and JONATHAN P. BACON. With 10 Figures

I.   Introduction   132
II.   Filling from Electrodes   134
III. Passive Back- or Forwardfilling   135

IV.   Fixing   135
V.   Buffers, Ringers   137
VI.   Whole-Mount Viewing   138
VII.   Embedding and Sectioning   141
VIII. Microscopy   143
IX.   Photography   145
X.   Fading   149
XI.   Reconstructions   151
XII.   Geography   151
XIII. Storage   151
XIV.   Artefacts   154
XV.   Conclusions   155

Chapter 8
Portraying the Third Dimension in Neuroanatomy
PETER T. SPECK and NICHOLAS J. STRAUSFELD. With 22 Figures

I.   Introduction   156
II.   Why Computer Graphics in Neuroanatomy?   156
III.   Designing the System   157
IV.   The NEU System   159
V.   Alignment of Sections   161
VI.   Interactive Profile Acquisition   163
VII.   Noninteractive Operations   167
VIII. Final Remarks   177

Chapter 9
**Three-Dimensional Reconstruction and Stereoscopic Display of Neurons
in the Fly Visual System**
ROLAND HENGSTENBERG, HEINRICH BÜLTHOFF and
BÄRBEL HENGSTENBERG. With 12 Figures

I.   Introduction   183
II.   Procedure   183
III.   Hardware Configuration   185
IV.   The Data Acquisition Program HISDIG   185
V.   The Reproduction Program HISTRA   188
VI.   Stereoscopic Vision   188
VII.   Examples of Displays and Stereopairs   190
VIII. Further Applications   203
IX.   Concluding Remarks   205

Chapter 10
**Laser Microsurgery for the Study of Behaviour and Neural Development
of Flies**
GAD GEIGER, DICK R. NÄSSEL and HARJIT SINGH SEYAN. With 14 Figures

I.   Introduction   206
II.   The Laser Microbeam Unit   207
III. Procedure of Laser Surgery   210
IV.   Histological Analysis   215
V.   Anatomical-Behavioural Correlations of Laser-Eliminated
      Lobula Plate Neurons   219
VI.   Aspects of Neuronal Development   220
VII. Discussion   224

Chapter 11
**Anatomical Localization of Functional Activity in Flies Using
³H-2-Deoxy-D-Glucose**
ERICH BUCHNER and SIGRID BUCHNER. With 6 Figures

I.   Introduction   225
II.   Essentials of the Technique   227
III. Results   230
IV.   Concluding Remarks   237

Chapter 12
Strategies for the Identification of Amine- and Peptide-Containing Neurons
MICHAEL E. ADAMS, CYNTHIA A. BISHOP and MICHAEL O'SHEA.
With 3 Figures

I.   Introduction   239
II.   Neutral Red: A Nonspecific Stain for Amine-Containing
       Neurons   239
III.   Neutral Red: An Indicator of Peptidergic Neurons   241
IV.   Permanent Preparations of Vital Staining with Neutral Red ·   242
V.   Use of Immunohistochemical Approaches to Neuron
       Identification   243
VI.   Immunohistochemical Screening: Whole-Mount Method . . .   243
VII.   Identification: Immunohistochemistry and Dye Injection   . .   245
VIII. Confirmation of Immunohistochemistry: Cell Isolation,
       Extraction and Assay   247
IX.   Concluding Remarks   248

Chapter 13
**Immunochemical Identification of Vertebrate-Type Brain-Gut Peptides
in Insect Nerve Cells**
HANNE DUVE and ALAN THORPE. With 10 Figures

I.	Introduction	250
II.	Immunocytochemistry: Basic Principles	251
III.	Techniques of Immunocytochemistry	252
IV.	Problems of Specificity	255
V.	Brain-Gut Peptides in Insects	259
VI.	Extraction and Purification	259
VII.	Conclusions	264
VIII.	Appendix 1: Immunofluorescence: The Indirect Method . . .	265
IX.	Appendix 2: Immunoperoxidase: The PAP Method	266

Chapter 14
**Immunocytochemical Techniques for the Identification of Peptidergic
Neurons**
MANFRED ECKERT and JOACHIM UDE. With 31 Figures

I.	Introduction and Survey	267
II.	Preparation of Antigens	271
III.	Production and Isolation of Antibodies	275
IV.	Absorption of the Antisera Before Use for Immuno-cytochemistry.	277
V.	Isolation of Hapten-Specific Antibodies	278
VI.	Methods for Antibody Isolation	278
VII.	Immunocytochemical Techniques	281
VIII.	Immunocytochemical Staining Methods	288
IX.	$CoCl_2$-Iontophoresis and Indirect Immunofluorescence Method	292
X.	Supplementary Methods	292
XI.	Electron Microscopy	293
XII.	Conclusions	301

Chapter 15
**Detection of Serotonin-Containing Neurons in the Insect Nervous System
by Antibodies to 5-HT**
NIKOLAI KLEMM. With 8 Figures

I.	Introduction	302
II.	General Considerations of Antibody Staining	303
III.	The Immunofluorescence Technique	304
IV.	Fluorescence Microscopy and Photography	310
V.	The Unlabelled Antibody Enzyme Method for Sections	311
VI.	A Whole-Mount Method for Antibody Staining	313
VII.	Specificity of Anti-5-HT Labelling	315

Chapter 16
Monoaminergic Innervation in a Hemipteran Nervous System:
A Whole-Mount Histofluorescence Survey
THOMAS R. J. FLANAGAN. With 10 Figures

I.   Introduction   317
II.   Materials and Methods   317
III.   Results   320
IV.   Discussion   329

Chapter 17
Identification of Neurons Containing Vertebrate-Type
Brain-Gut Peptides by Antibody and Cobalt Labelling
HANNE DUVE, ALAN THORPE and NICHOLAS J. STRAUSFELD.
With 3 Figures

I.   Introduction   331
II.   Method   333
III.   Interpretation of the Results   336
IV.   Conclusions   338

Chapter 18
Interpretation of Freeze-Fracture Replicas of Insect Nervous Tissue
STANLEY D. CARLSON, RICHARD L. SAINT MARIE and CHE CHI.
With 15 Figures

I.   Introduction   339
II.   Procedure   342
III.   Interpretation of Replicas   350
IV.   The Cleaved Cell: A Survey   351
V.   Recent Advances and Future Prospects   374

Chapter 19
High-Voltage Electron Microscopy for Insect Neuroanatomy
CHE CHI. With 6 Figures

I.   Introduction   376
II.   Rationale for HVEM for Biological Research   377
III.   Method   379
IV.   HVEM of Insect Neurons   380

References   386

Subject Index   420

List of Contributors

MICHAEL E. ADAMS, Zoecon Corporation, 974 California Avenue, Palo Alto, California 94305, USA

JENNIFER S. ALTMAN, University of Manchester Institute of Science and Technology, P.O. Box 88, Manchester M6O 1QD, England

JONATHAN P. BACON, Universität Basel, Zoologisches Institut, Rheinsprung 9, CH-4051 Basel, Switzerland

URSULA K. BASSEMIR, European Molecular Biology Laboratory, Meyerhofstrasse 1, D-6900 Heidelberg, Federal Republic of Germany

CYNTHIA A. BISHOP, Department of Psychology, Stanford University, Building 420, Jordan Hall, Stanford, California 94305, USA

ERICH BUCHNER, Institut für Genetik und Mikrobiologie der Universität Würzburg, Röntgenring 11, D-8700 Würzburg, Federal Republic of Germany

SIGGI BUCHNER, Institut für Genetik und Mikrobiologie der Universität Würzburg, Röntgenring 11, D-8700 Würzburg, Federal Republic of Germany

HEINRICH BÜLTHOFF, Max-Planck-Institut für biologische Kybernetik, Spemannstrasse 38, D-7400 Tübingen, Federal Republic of Germany

STANLEY D. CARLSON, Department of Entomology, University of Wisconsin, Madison, Wisconsin 53706, USA

CHE CHI, Nicolet Instrument Corporation, Biomedical Division, Madison, Wisconsin 53706, USA

FRED DELCOMYN, Department of Entomology and Program in Neural and Behavioral Biology, University of Illinois, Urbana, Illinois 61801, USA

HANNE DUVE, School of Biological Sciences, Queen Mary College, University of London, Mile End Road, London E1 4NS, England

MANFRED ECKERT, Wissenschaftsbereich Tierphysiologie der Sektion Biologie, Friedrich-Schiller-Universität, DDR-6900 Jena, German Democratic Republic

THOMAS R. J. FLANAGAN, Cold Spring Harbor Laboratories, P.O. Box 100, Cold Spring Harbor, New York 11724, USA

GAD GEIGER, Research Laboratories of Electronics, Massachusetts Institute of Technology, Cambridge, Massachusetts 02139, USA

CAROL M. HACKNEY, E. M. Unit, Departments of Botany and Zoology, University of Manchester, Oxford Road, Manchester M13 9PL, England

BÄRBEL HENGSTENBERG, Max-Planck-Institut für biologische Kybernetik, Spemannstrasse 38, D-7400 Tübingen, Federal Republic of Germany

ROLAND HENGSTENBERG, Max-Planck-Institut für biologische Kybernetik, Spemannstrasse 38, D-7400 Tübingen, Federal Republic of Germany

NIKOLAI KLEMM, Fakultät für Biologie, Universität Konstanz, Postfach 5560, D-7750 Konstanz, Federal Republic of Germany

DICK R. NÄSSEL, Department of Zoology, University of Lund, Helgonavägen 3, S-22362 Lund, Sweden

MICHAEL O'SHEA, Department of Pharmacological and Physiological Sciences, University of Chicago, 947 E. 58th Street, Chicago, Illinois 60637, USA

WILLI A. RIBI, Quaderstrasse 22, CH-7000 Chur, Switzerland

RICHARD L. SAINT MARIE, Department of Anatomy, Boston University School of Medicine, 80 E. Concord St., Boston, Massachusetts 02118, USA

HARJIT S. SEYAN, European Molecular Biology Laboratory, Meyerhofstrasse 1, D-6900 Heidelberg, Federal Republic of Germany

PETER T. SPECK, European Molecular Biology Laboratory, Meyerhofstrasse 1, D-6900 Heidelberg, Federal Republic of Germany

NICHOLAS J. STRAUSFELD, European Molecular Biology Laboratory, Meyerhofstrasse 1, D-6900 Heidelberg, Federal Republic of Germany

ALAN THORPE, School of Biological Sciences, Queen Mary College, University of London, Mile End Road, London E1 4NS, England

JOACHIM UDE, Wissenschaftsbereich Tierphysiologie der Sektion Biologie, Friedrich-Schiller-Universität, DDR-6900 Jena, German Democratic Republic

DAVID W. WOHLERS, Max-Planck-Institut für Verhaltensphysiologie, Abteilung Huber, D-8131 Seewiesen, Federal Republic of Germany

Chapter 1

Electron Microscopy of Golgi-Impregnated Neurons

Willi A. Ribi

Max-Planck-Institut für biologische Kybernetik
Tübingen, F.R.G

Introduction

This chapter will describe and discuss a combined Golgi–electron microscopy (EM) technique that successfully preserves gross morphology and ultrastructure, especially the synaptology, of identified neurons. In addition, various combinations of fixation, chromation and impregnation useful for insect and vertebrate nervous tissue are described. The Appendix lists schedules for fixation and chromation, giving impregnation times.

The most pleasant experience for a neuroanatomist is to know that certain profiles seen in the EM belong to a neuron already seen in the light microscope (LM), particularly in the case of intricate cells whose perikarya and largest processes only are identifiable in thin sections. The present Golgi–EM modification permits synaptic and morphometric analysis of all parts of a nerve cell.

Golgi-stained neurons have been observed in LM and EM by many investigators working on vertebrates (Stell 1964; Blackstad 1965), and insects (Trujillo-Cenóz and Melamed 1970; Strausfeld 1973; Strausfeld and Campos-Ortega 1973; Campos-Ortega and Strausfeld 1973; Ribi 1975), all of whom devised methods for re-embedding Golgi-stained neurons. Other examples of Golgi–EM studies are: Kolb (1970), LeVay (1973), Pinching and Brooke (1973), West (1976), Parnavelas et al. (1977), Peters and Fairén (1978), Peters et al. (1979) and Fairén and Valverde (1980).

Ideal sections containing neurons of special interest, impregnated against a clear background, were removed for EM examination before the embedding medium (usually Permount) had set. The sections were washed in xylene and re-embedded in araldite for ultrathin sectioning. Additional contrast was rarely (or sparingly) applied, since aqueous treatment risks removing the silver chromate precipitate that fills impregnated cells. It is not surprising that the general quality of ultrathin sections has not been satisfactory. In particular, tissue was poorly preserved after rough and unphysiological treatment and suitable only for LM. The omission of OsO_4, uranyl acetate and lead citrate was generally disadvantageous, so that a specific modification of existing Golgi–EM techniques was called for.

The development of aldehyde-based fixatives with physiologic osmolarities and ionic concentrations has refined the method so as to allow combined LM and EM examination of impregnated neurons in the context of their immediate surroundings of unimpregnated neighbours.

Attempts have also been made, especially by vertebrate neuroanatomists, to preserve the cytological details of the impregnated neurons by de-impregnation of normal Golgi-stained tissue prior to embedding for EM. Blackstad (1970) used either sodium sulfite or ammonium nitrate to remove silver chromate deposits, while Ramón-Moliner and Ferrari (1972) substituted silver chromate by lead nitrate. More recently, Ramón-Moliner and Ferrari (1976) have improved their procedure by using lead lactate to produce lead chromate. The initial results were not encouraging, as the preservation of the tissue was poor and the substitution difficult to control.

Another method proposed by Blackstad (1975a, b) consisted of reducing silver chromate to metallic silver by putting thick sections (20–30 μm) containing silver-impregnated cells in glycerol and exposing them to UV-irradiation. Afterwards excess silver chromate was removed with sodium thiosulfate. Only scattered silver particles remained in the neuron originally impregnated with silver chromate. Apart from being difficult to control, the usefulness of this method is limited: sections thicker than 30 μm limit penetration of UV-irradiation.

Fairén et al. (1977) successfully substituted part of the silver chromate deposit with metallic gold prior to removing the remaining silver chromate with sodium thiosulfate. This technique leaves only a minimum amount of gold deposit in the previously impregnated structures. The fine structure of the impregnated fibre is well-preserved, allowing recognition of synapses in the impregnated neuron.

To overcome the complexity of de-impregnation and substitution methods, a sparser precipitation can be obtained using short impregnation times (Scott and Guillery 1974).

The procedure described here is a Golgi–EM technique initially employed for insects (Ribi 1976a) and modified for vertebrate nervous tissue (Ribi and Berg 1980). In this method, silver chromate precipitation is stopped before the cell processes are completely filled, resulting in an incomplete impregnation. This can be achieved by weak chromate and silver solutions and shorter impregnation times, providing a simple, rapid, and reliable Golgi–EM method. Impregnated neurons can be studied in thick sections; they have improved preservation of ultrastructure and are easier to thin-section.

General Description of the Procedure

Insects

Golgi–EM investigations were carried out on the optic lobes of bees and flies. Insects were immobilized by cooling to 4 °C for a few minutes before the head capsules were opened to allow the fixative to directly contact the brains. Opening certain parts of the head capsule helps local impregnation. Removing the cornea results in impregnated neurons in the first optic neuropil. Removing ommatidia results in impregnated neurons in the first to third optic neuropils and in the midbrain. To resolve neurons extending from the head to the body, openings are made ventrally in the thorax to allow diffusion of silver nitrate (Strausfeld 1980).

Vertebrates

Golgi–EM investigations were mainly carried out on the cerebellar and cerebral cortex of mice and rats. Animals, aged 4 to 5 months, were anaesthetized by intraperitoneal injection of 35% chloral hydrate at a dosage of 0.1 ml/100 g body weight (Palay and Chan-Palay 1974), artificially respired with 95% oxygen and 5% carbon dioxide, and then perfused through the aorta using a two-step aldehyde solution [in 0.12 M phosphate buffer with 0.02 mM $CaCl_2$ (pH 7.35) at 20 °C as recommended by Palay and Chan-Palay (1974)]. Initially 150 ml (1% formaldehyde, 1.25% glutaraldehyde) was used for each animal. After perfusion the heads were removed and stored overnight at 4 °C in a fresh perfusion solution.

The following day the brains were removed from the skulls and divided into pieces not larger than 125 mm^3 and rinsed in three changes of buffer before further treatment.

Fixation

In general, any aldehyde fixation suitable for EM can be used before chromation. The use of aldehyde fixative combined with phosphate buffer (Millonig 1961) gives good results on the above-mentioned species. Modifications of fixatives and buffers may favour impregnation of certain brain areas in different species (see Appendix).

Chromation and silver impregnation can be carried out in the dark or light without noticeable differences in quality of impregnation. Insect and vertebrate nervous tissue shown here was, however, processed in the dark at 4 °C.

The fixation time depends mainly on the tissue size and temperature. Insects the size of *Drosophila* are fixed for at least 3–4 h at 4 °C, whereas after perfusion vertebrate tissue was left in the final fixative overnight at 4 °C. The pH of the fixative is less critical than is its osmolarity, which should be physiologically appropriate. Fixatives were buffered to a pH value of between 7.2 and 7.4. Tissue was later washed in several changes of buffer to remove excess fixative and then stored in buffer at room temperature prior to osmication and chromation.

Osmication and Chromation

Since osmium tetroxide penetrates slowly, only a small zone of tissue near the surface will be secondarily fixed unless tissue is cut into small pieces. The normal concentration is 1%−2% aqueous or buffered OsO_4. Fixation is for 1−2 h, depending on the volume of the tissue. The duration and concentration of chromation can, however, be varied between 2 and 5 days without resulting in ultrastructural change or differences in the quality of silver impregnation.

Four different procedures gave equally good results:

1. Chromation is incorporated into initial fixation.
2. Chromation follows initial fixation, preceding osmication.
3. Fixation is followed by OsO_4 before chromation.
4. Chromation is incorporated with osmication (the "Golgi-rapid" method).

The following describes the third procedure. Chromation for 2−12 h (at 4 °C) in 1% $K_2Cr_2O_7$ solution (buffered or unbuffered) follows osmication (buffered 2% OsO_4). Tissue is then rinsed in phosphate buffer and, finally, rinsed in double-distilled water. Prolonged chromation does not increase silver impregnation (Strausfeld 1980).

Silver Impregnation

Following chromation, tissue is briefly rinsed in distilled water and then transferred through several changes of 0.5% silver nitrate until there is no precipitate from the tissue surface. To obtain incomplete impregnation, the immersion time is short and the silver concentration low.

Silver impregnation in 0.5% $AgNO_3$ (pH 6.9) is at 4 °C for 0.5 h to (maximally) 4 h in the dark. The tissue blocks were lightly agitated approximately every 15 min to facilitate access of the solution to all sides of the tissue.

After silver impregnation the tissue is washed several times in double-distilled water, dehydrated through ethanols, from 30% to absolute, and washed twice in propylene oxide and finally embedded in epoxy resin in the usual way for routine EM.

Thick Sectioning

The purpose of a planned Golgi−EM study is to analyse individual neurons, axons or dendrites. In such cases it is essential to first cut 20−80 μm thick sections, since searching for the desired structures by EM would require hundreds of ultrathin sections. Thick sections are inspected in the LM before re-embedding and proceeding further for EM.

The Golgi−EM technique uses a mixture of "soft araldite" as the embedding medium. This allows easy cutting of up to 100-μm-thick sections. In the early days of Golgi−EM, thick sections were first mounted in Permount between glass slide and coverslip. Removing a suitable section meant immersing the whole slide in xylene to dissolve the Permount. This procedure (contact between araldite, Permount and xylene) changed the cutting quality of the resin, making ultrathin sectioning difficult.

The following technique considers both requirements: easy handling and cutting of thick, semithin and ultrathin sections. Hard plastic sections (normal EM epoxy mixture) 20−80 μm thick are cut on a sliding microtome equipped with a heavy steel knife or with sturdy razor blades (Schick Injector blades) mounted in a special holder (Jung). The block face is softened during sectioning by illuminating it with an infrared lamp (Philips IR 230−250 V/250 W). Alternatively, a small soldering iron mounted a few millimeters in front of the block face can be used to warm up its surface. The correct degree of softening can be regulated by the cutting speed, or the distance and temperature of the soldering iron or IR bulb.

Instead of thick sections being mounted under Permount between glass slide and coverslip, they are placed in liquid resin between acetate sheets. The advantage of this is that single sections can easily be cut out and stripped of acetate foil without damaging the tissue, or loosening neighbouring sections (Ribi 1978). The cutting quality of the section also remains stable. Thick sections are mounted as follows:

A piece of 0.1-mm-thick acetate foil is attached to a glass slide with double-sided Scotch tape. Thick sections are glued in rows on the acetate foil with a little resin. To flatten the sections the slide is placed on a hot plate (60 °C) for a few minutes. A second piece of acetate foil is then put on the top of the sections. Another glass slide is then weighted down on top of it. After 48 h polymerization of 60 °C the top slide is removed and the top acetate foil peeled off from the sections (Fig. 1).

Re-embedding

Thick sections mounted on the lower acetate sheet are examined under the LM, and appropriately impregnated cells photographed, dissected out with a microscalpel, and remounted in a flat silicon mould (Fig. 1) for an ultrathin cross- or longitudinal-section series.

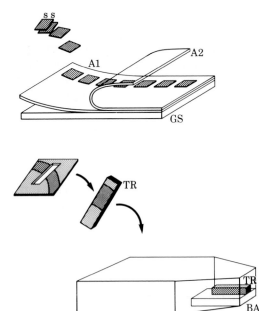

Fig. 1. Re-embedding procedure. $25-80\,\mu m$ thick sections (*S*) are mounted with a drop of araldite between two sheets of a 0.1 mm thick acetate foil (*A1* and *A2*) on a glass slide *GS*. After polymerization in the oven for 24 h at 60 °C the top acetate foil (*A2*) can be removed. Sections containing neurons of interest are dissected out and the target neuron (*TR*) re-embedded in a flat silicon mould. Small pieces of hardened araldite (*BA*) are first put beneath the section to prevent it from tilting or sinking

Semithin Sectioning

The conventional technique of cutting, staining and examining semithin sections (0.5−2 μm thick) is used on Golgi−EM material to control trimming and to aid orientation before making ultrathin sections.

Semithin sections are stained with 1% toluidine blue in 1% borax. Mounting sections in high viscosity immersion oil under a coverslip prevents rapid bleaching.

Ultrathin Sectioning

Trimming and ultramicrotomy of Golgi−EM treated material is carried out conventionally after the block face has been trimmed as much as possible to minimize the search area.

Ultrathin sections are collected on slot grids, coated with 0.25% Formvar and stained for 30 min with a saturated uranyl acetate solution made up in methanol. They are then stained by a modified lead citrate procedure (Venable and Coggeshall 1965) for 20 min.

Fig. 2. A A Purkinje cell of the mouse cerebellum showing the entire inner membrane ensheathed by a thin layer of precipitate. Even small dendritic spines (*small arrows*) contain the precipitate. **B** Pyramidal cell of the mouse cortex. Dendrites with numerous spines and the single axon are lightly impregnated. The cell body appears darker. *Small arrows* indicate dendritic spines, *Ax* axon; *D* dendrite. Bars **A, B** = 20 μm

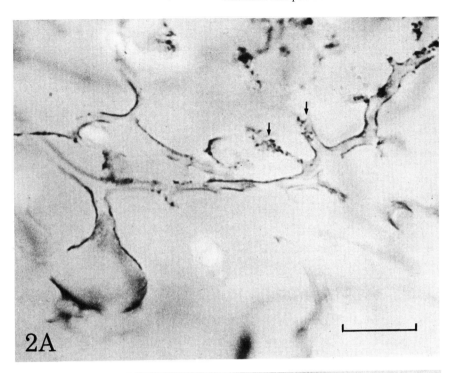

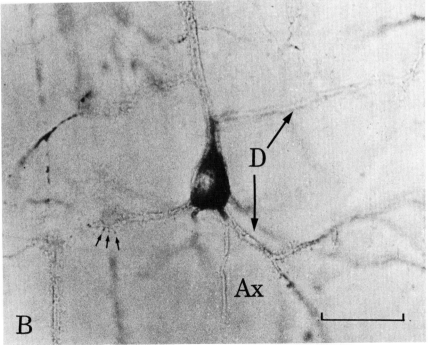

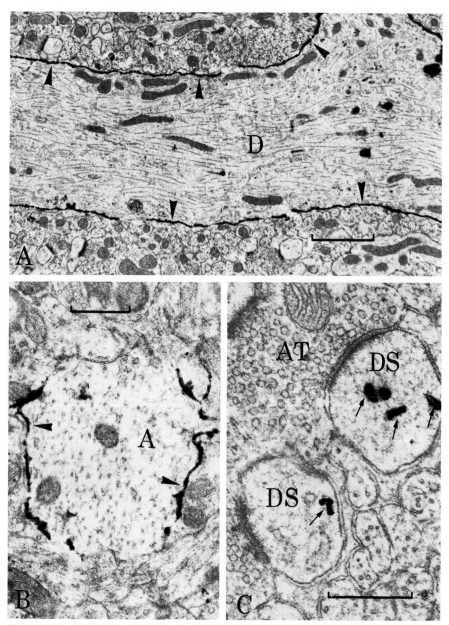

Fig. 3. A Longitudinal section of a dendritic branch of a Purkinje cell delineated by fine silver chromate precipitates along the cell membrane. *D* dendrite; *arrowheads* silver precipitate. **B** Cross section through an axonal (*A*) branch of a pyramidal cell. Like the dendrites of cerebellar neurons, the cortical pyramidal cells are delineated by a fine silver precipitate (*arrowhead*). **C** Two dendritic spines in cross section with silver chromate particles located on the spine apparatus (*arrows*). Pre- and postsynaptic sites are well-preserved. *AT* axon terminal; *DS* dendritic spine. Bar **A** = 2 μm, **B, C** = 0.5 μm

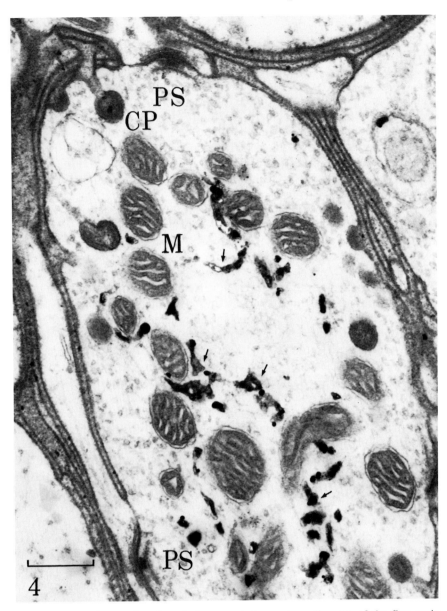

Fig. 4. Electron micrographs of partly impregnated receptor cell axons of the first optic ganglion of the fly (*Calliphora*). *Arrows* indicate the silver-impregnated smooth endoplasmic reticulum of the axon. The rest of the cytoplasm is free of precipitate. Presynaptic site *PS* and capitate projections *CP* appear unaffected. *M* mitochondria. Bar = 0.2 μm

Results

The present method produces silver-impregnated neurons that are (1) clearly recognizable by LM, (2) structurally intact and (3) well-delineated by EM. The technique overcomes limitations of previously described techniques, such as preparation time and limited quality of the stained material.

In LM, impregnated perikarya and their processes show a variety of silver-impregnated densities, from lightly tinted (giving them a somewhat transparent appearance) to black. High magnification resolves a fine silver chromate precipitate inside the cell membrane, as seen in vertebrate nervous tissue (Figs. 2 and 3), or silver chromate particles located mainly within the cytoplasm, as in insects (Figs. 4 and 5).

The staining intensity in both vertebrate and insect cells often varies within and between individual neurons. In most cases, cell bodies are more intensively stained than dendrites or axons.

In electron micrographs silver chromate precipitates appear as electron-dense particles of various shapes. Variable impregnation densities seen by LM can be resolved at the EM level. In vertebrate neurons almost the entire plasma membrane is delineated by a fine layer of precipitate (Figs. 2 and 3 A, B). Cell bodies, axons, and dendritic branches are outlined by a thin layer of precipitate which mainly occurs along the cell inner membrane. The majority of dendritic spines are marked unambiguously either by precipitate covering the inner plasma or by electron-dense silver particles located in the cytoplasm (Fig. 3 C). As expected, cells appearing darker in LM were found to have a more extensive precipitate electron microscopically, covering microtubules and smooth endoplasmic reticulum, as well as some dendritic spine apparati. However, no tendency was observed for precipitates to diffuse extracellularly.

In insects, densities of impregnation also differ. Whereas in mice and rats the silver chromate precipitate mainly delineates the entire inner plasma membrane of the cell, in insects silver chromate can be preferentially deposited on the smooth endoplasmic reticulum (Figs. 4, 5). Under LM, in insects some types of nerve cells appear spotted or banded (Fig. 6 B, C). Electron microscopic observations resolve microtubules and mitochondria with silver chromate precipitates (Fig. 7), whereas in vertebrates mitochondria are not stained. It seems, therefore, that the origin of silver impregnation may differ in various nerve cells of vertebrates and insects. In the bee the silver chro-

---

Fig. 5. A Partly impregnated receptor cell axon profile showing impregnated smooth endoplasmic reticulum as parallel lines (*arrows*). No other cytoplasmic organelles appear impregnated. **B** Light micrograph of a frontal section through the bee lamina. *Arrow* indicates a fully impregnated *L 1* monopolar cell; *R* retina; *L* lamina. **C** Cross section of a bee photoreceptor axon at the third proximal stratum of the lamina. The heavy silver chromate impregnation completely obscures the cytoplasm of the axon. Bar **A** = 0.5 μm; **B** = 50 μm; **C** = 2 μm

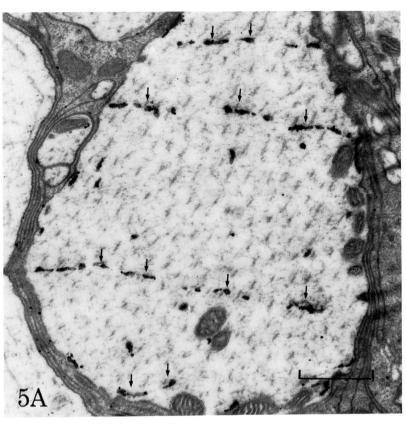

5A

R

L

B

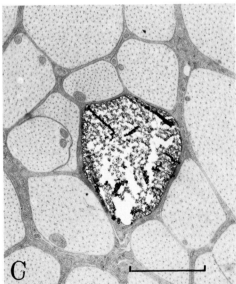

C

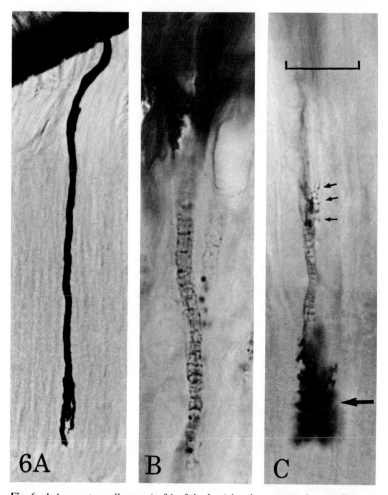

Fig. 6. A A receptor cell axon (svf 1 of the bee) has been treated according to the original Golgi–Kopsch technique. The cytoplasm appears completely stained by the silver chromate precipitation (see Fig. 5C). **B** Partly impregnated receptor cell axon of the fly shown at high magnification. The tangentially arranged silver impregnation pattern has its origin in impregnated, horizontally arranged, smooth endoplasmic reticulum. **C** Partly impregnated ending of a second-order neuron (L-fibre) in the bee's medulla. The terminal arborization is indicated by *large arrow. Small arrows* indicate the impregnated spines from its axon. Bar **A, C** = 20 μm; **B** = 10 μm

mate precipitate also delineates cell membranes in a fashion comparable to vertebrate central nervous system neurons. In fly photoreceptors, however, precipitates can first be seen on the smooth endoplasmic reticulum and microtubules (Figs. 4, 6B and 7B) using extremely short impregnation times. However, precipitates may also form at the inner cell membrane

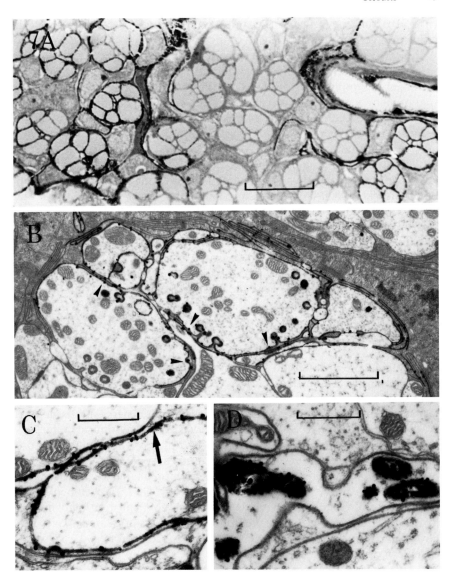

Fig. 7. A Cross section through the distal part of the bee's lamina showing heavily stained (light micrograph) multilayered glial cells. **B** Cross section through an "optic cartridge" of the fly *Calliphora*. Smooth endoplasmic reticulum of some receptor cell axons is filled with silver chromate (see Fig. 4). However, other axons show initial impregnation at cell membranes (*arrowheads*). **C** Cross section through the neuropil of the bee's lamina showing heavily impregnated cell membranes (*arrow*). **D** Impregnated mitochondria, mainly seen in glial cells here shown from a cross section through the bee's lamina. Bar **A** = 10 μm; **B** = 2 μm; **C, D** = 0.5 μm

(Fig. 7 B). Long impregnations at room temperature can completely fill the cell with silver chromate (Figs. 5 B, C and 6 A).

In vertebrates selective silver precipitation on the plasma membrane, smooth endoplasmic reticulum and microtubules might be explained by an increased argyrophilia of these cell structures (Fig. 3). Possibly impregnation of many vertebrate and insect neurons starts preferentially at the inner membrane proceeding into the cytoplasm, conforming to the model of impregnation recently proposed by Strausfeld (1980; see Appendix I). However, even in the most heavily stained cells, precipitate was never observed in the rough endoplasmic reticulum, the Golgi apparatus, the nuclear envelope or synaptic vesicles. This finding is in agreement with those of many other authors (Blackstad 1965; Chan-Palay and Palay 1972; Pinching and Brook 1973; Ramón-Moliner and Ferrari 1972; LeVay 1973), except that we also observed precipitate in smooth endoplasmic reticulum, insect mitochondria and microtubules. Such variations do not, however, limit the usefulness of the procedure. Silver-impregnated neurons appearing light brown, spotted, tinted or "hollow" in the LM (with just parts of the cytoplasm or the cell membrane impregnated) were the most useful for EM investigation.

Discussion

The Golgi−UV−EM method developed by Blackstad (1975 a, b) is a relatively simple procedure that removes enough silver chromate to allow the examination of ultrastructure of the impregnated neurons. However, the effectiveness of this technique is limited by section thickness, and tissue preservation is not optimal. Scott and Guillery (1974) reported a Golgi−EM method producing partially impregnated structures by interrupting the procedure in statu nascendi, resulting in a certain transparency of parts of some neurons whose dendritic spines remained unimpregnated even when other cells were fully impregnated nearby. Fairén et al. (1977) developed one of the most useful Golgi−EM methods employing gold substitution, which allows cytological analysis of the neuron in quite well-preserved tissue. It is, however, relatively tedious to perform. De-impregnation and ion substitution can cause distortion of cytoplasmic fine structure between organelles. The more gentle treatment of tissue by the present technique, employing short periods in low concentration solutions, usually results in good tissue preservation and impregnation.

A further advantage of the present method lies in the embedding and re-embedding procedure which follows immediately after silver impregnation and dehydration, thus eliminating artefacts previously reported from sectioning unembedded tissue (Blackstad 1975 b) or tissue embedded in agar (Fairén et al. 1977).

Golgi−EM techniques have also been used for high-voltage EM studies (Chan-Palay and Palay 1972; Scott and Guillery 1974). It is conceivable that the more transparent neurons, such as those produced by the present Golgi−EM method, may be more readily studied by high-voltage than conventional Golgi preparations, which contain more solid silver chromate precipitates.

With the present technique it is possible to examine the fine cytology of identified neurons and to ascertain the distribution and types of synapses in them; it is a valuable tool for identifying functional contacts between identified neurons. A further improvement is to combine Golgi−EM with either degeneration (Peters et al. 1979), HRP (Somogyi et al. 1979), or dye-injection techniques (Ribi, in prep.) for labelling adjacent neurons at the EM level.

Acknowledgements. I am indebted to N. J. Strausfeld who initiated me in this fascinating technique and let me experience all its mysteries. I also wish to thank T. W. Blackstad for fruitful discussion and comments on early stages of the Golgi−EM method; to G. Berg, V. Braitenberg, K. Kirschfeld and K. Hausen for helpful advice and encouragement and R. Cook for reading the manuscript.
The work was supported by the M.P.I. für biol. Kybernetik, Tübingen, FRG and a grant No. 3 102-0.81 by the Fonds National Suisse de la Recherche Scientifique.

Appendix: Schedules

Buffer and Fixative for Invertebrate Procedures

Phosphate Buffer After Millonig (1961)

Solution A: 0.164 M monosodium phosphate $= 2.26\%$ $NaH_2PO_4 \cdot H_2O$
$= 2.56\%$ $NaH_2PO_4 \cdot 2\,H_2O$
(store in refrigerator; keeps several weeks)

Solution B: 0.63 M sodium hydroxide $= 2.52\%$ NaOH in pellets

pH	6.0	6.2	6.4	6.8	7.0	7.2	7.3	7.4	7.6	7.8
A ml	96.2	94.7	92.5	87.9	85.8	83.9	83.0	82.5	81.6	80.8
B ml	3.8	5.3	7.5	12.1	14.2	16.1	17.0	17.5	18.4	19.2

Add to 100 ml buffer 0.4 ml 0.6% $MgCl_2$ and 1.3 g sucrose to adjust ionic balance and osmolarity.

Phosphate Buffer After Sörensen (see Dawson et al. 1969)

Solution A: 0.066 M potassium dihydrogen phosphate $= 0.908\%$ KH_2PO_4

Solution B: 0.066 M disodium phosphate $= 1.188\%$ $Na_2HPO_4 \cdot 2\,H_2O$
$= 1.786\%$ $Na_2HPO_4 \cdot 7\,H_2O$
$= 2.387\%$ $Na_2HPO_4 \cdot 12\,H_2O$

pH	6.0	6.8	7.0	7.2	7.4	7.6	7.8	8.0
A ml	87.7	50.8	39.2	28.5	19.6	13.2	8.6	5.5
B ml	12.3	49.2	60.8	71.5	80.4	86.8	91.4	94.5

$MgCl_2$ and sucrose are added as above.

Piperazine (PIPES) Buffer (Baur and Stacey 1977)

0.3 M PIPES (piperazine N-N bis 2 ethanol sulfonic acid) – Dissolve 30.24 g PIPES in 1 l double-distilled H_2O by addition of concentrated NaOH until solution clears (add NaOH drop by drop). The pH is adjusted to between 7.0 and 7.4 by addition of excess 1 N NaOH.

Fixatives; Paraformaldehyde – Glutaraldehyde:

In Combination with Millonig's Phosphate Buffer

1. Dissolve 2 g paraformaldehyde in a certain amount of solution B according to the desired pH (see above) by swirling in an Erlenmeyer flask in hot water bath; do not let temperature of solution exceed 60 °C as this causes the aldehyde to break down.
2. When completely dissolved, add solution A to make 100 ml.
3. Add contents of one vial 25% glutaraldehyde.
4. Adjust pH if necessary.

In Combination with Sörensen's Buffer

1. Dissolve 2 g paraformaldehyde in 25 ml double-distilled H_2O.
2. Heat in Erlenmeyer flask under hood to 60 °C – stir – after 5 min add 1 to 3 drops 1 N NaOH (= 4%) until suspension clears; cool.
3. Add contents of a 10 ml vial of 25% glutaraldehyde.
4. Add buffer to make 100 ml.

In Combination with PIPES Buffer

1. Add 2 g paraformaldehyde and a pellet of NaOH to 80 ml double-distilled H_2O and warm up to 60 °C in a hot water bath.
2. When completely dissolved and cooled down add the contents of one 10 ml vial of 25% glutaraldehyde.
3. Add 3.024 g PIPES and dissolve using concentrated NaOH.
4. Adjust the pH to between 7.0 and 7.4.

5. Bring the final volume to 100 ml double-distilled H_2O.

6. Adjust pH if necessary.

Buffer Used for Vertebrate Procedure (Palay and Chan-Palay 1974)

Standard Phosphate Buffer (0.4 M):

5.3 g $NaH_2PO_4 \cdot H_2O$
28.0 g K_2HPO_4
add double-distilled water to make 500 ml.

Rinse Solution:

8 g dextrose
30 ml 0.4 M phosphate buffer
0.4 ml 0.5% $CaCl_2$
add double-distilled water to make 100 ml.

Procedure. Add dextrose and buffer and make up to almost final concentration with H_2O. Slowly add the calcium and complete the dilution.

Double-strength Buffer for Osmium Tetroxide (OsO₄):

7 g dextrose
30 ml 0.4 M phosphate buffer
add double-distilled water to make 50 ml.

Procedure. Add to 15 ml double-strength buffer, 15 ml 4% OsO_4 (in double-distilled water) and 0.15 ml 0.5% $CaCl_2$.

Golgi−EM Procedure for Vertebrate Nervous Tissue

1. Perfuse with a solution containing 1% paraformaldehyde and 1.25% glutaraldehyde made up in a 0.12 M phosphate buffer with 0.02 mM $CaCl_2$.
   − Perfuse with a solution containing 4% paraformaldehyde and 5% glutaraldehyde. Both solutions should have a pH of 7.35. Perfuse at room temperature. Leave the decapitated head overnight in the final concentrated solution at 4 °C.
2. Divide the tissue into pieces of no more than 125 mm³.
3. Wash three times in buffer (3 min each).
4. Transfer the tissue to a buffered 2% OsO_4 solution and leave for 2 h at room temperature.

5. Wash three times in buffer (3 min each).
6. Bring the tissue into a 1% $K_2Cr_2O_7$ solution for 12 h at 4 °C in the dark (refrigerator).
7. Wash briefly in distilled water and then in a 0.5% $AgNO_3$ solution (made up in double-distilled water) until the washing solution remains clear (no precipitate forms). Leave sample in 0.5% $AgNO_3$ solution for 2 to (maximally) 4 h in the dark at 4 °C (refrigerator).
8. Wash several times in distilled water before dehydration in an ethanol series 30%, 50%, 70%, 80%, 95%, (5−10 min each step) twice in 100% for a total of 15 min and twice in propylene oxide for a total of 30 min.
9. Embed in araldite (hard EM mixture) as for routine EM procedure.
10. Cut thick sections (10−80 μm) on a rotary or sledge microtome using solid razor blades. Preheat the block surface using an IR heating bulb or a soldering iron to make cutting easier.
11. Collect the sections on an acetate foil (0.1 mm thick) and cover them with fresh araldite and a second acetate foil.
12. Examine the polymerized sections under the LM. Remove and reembed well-impregnated fibre elements of interest for EM.

Golgi−EM Procedure for Invertebrate Nervous Tissue

1. Fix small pieces of tissue (3 mm each side) in 2% paraformaldehyde and 2.5% glutaraldehyde in Millonig's phosphate buffer. Add D-glucose (1 g per 100 ml solution) and 1% $CaCl_2$ (0.9 ml per 100 ml solution) as desired; pH should be 7.2−7.3.
2. Fix for up to 4 h in the refrigerator at 4 °C.
3. Wash three times in buffer (3 min each).
4. Transfer the tissue to a 1%−2% OsO_4 solution (containing $CaCl_2$ and D-glucose) and leave for 1−2 h (depending on the size of the tissue) at room temperature.
5. Wash three times in buffer (3 min each).
6. Put the tissue in a 1% $K_2Cr_2O_7$ solution (made up with double-distilled water) and leave for 4 h in the dark and at 4 °C (refrigerator).
7. Wash briefly in distilled water and then in a 0.5% $AgNO_3$ solution (made up with double-distilled water) until the washing solution remains clear, and then leave the tissue in the solution for 0.5−4 h (depending on the size of the tissue and the impregnation grade desired) in the dark at 4 °C (refrigerator). Further steps 8−9 given on p. 18 (above).

Chapter 2

Block Intensification and X-Ray Microanalysis of Cobalt-Filled Neurons for Electron Microscopy

Ursula K. Bassemir and Nicholas J. Strausfeld

European Molecular Biology Laboratory
Heidelberg, F.R.G.

Introduction

Many special methods visualize a nerve cell in its entirety. Modifications of Golgi's (1873) silver impregnation (Strausfeld 1980) stochastically select single nerve cells. The more selective modern neuronal markers include labelling with various metals (Kerkut and Walker 1962; Lux and Globus 1968), injection of dyes and fluorescent markers (Thomas and Wilson 1966; Kato et al. 1968; Stretton and Kravitz 1968, 1973; Kater and Nicholson 1973; Stewart 1978), application of enzyme proteins (see Chap. 3) and specific antibody staining (see Chaps. 12–16). All these methods give invaluable information concerning function, shape and chemical identity of neurons within a neuropil. However, the identification of small structures that imply functional connections, such as synapses, can only be done at high resolution by electron microscopy. This requires good tissue preservation, a feature that is lacking in all labelling methods developed for light microscopy alone. Furthermore, a marker substance is required that allows identification of the neuron at low magnification by light microscopy and produces sufficient electron scattering at the fine-structural level for its own unambiguous detection.

Although there is a vast variety of cellular markers available, only a few of them provide sufficient contrast at low and high resolution. Excellent candidates for this are cobalt solutions, intracellular stains which delineate the finest details of neurons. Their ionic nature and high mobility within neurons make them especially suitable for electrophysiological and structural studies.

Pitman et al. (1972, 1973) introduced the cobalt staining technique to resolve cockroach motor neurons by light and electron microscopy. Since then several attempts have been made to overcome specific difficulties of this method, one of them being poor fine-structural preservation of the filled neurons and unambiguous visualization of cobalt (precipitated as sulphide) within the marked cell. Better preservation of the fine structure was achieved by aldehyde fixation prior to sulphidation as shown by Rade-makers (1977). A combined fixation and precipitation procedure, together with the use of minimal amounts of cobalt, has been previously recom-mended by Tyrer and Bell (1974). However, in our experience, fixation be-fore or during sulphidation can result in displacement of cobalt from filled neurons.

It is difficult to see small amounts of cobalt sulphide in both thick and thin plastic sections. Timm's (1958) sulphide-silver method was first used by Tyrer and Bell (1974) to intensify the cobalt sulphide precipitate in thick paraffin sections. They also used it on ultrathin sections for electron mi-croscopy. Their cobalt intensification method shows good tissue preser-vation due to aldehyde fixation and osmication and allows identification of labelled profiles at both levels of resolution by means of intensely staining silver precipitates. Since then, the method has been successfully applied to the vertebrate and invertebrate nervous system (Rademakers 1977; Székely and Kosaras 1976, 1977; Altman et al. 1979; Hausen et al. 1980; Phillips 1980). There remain two disadvantages, however, with the procedure. First, it requires tedious handling of ultrathin sections during the intensification stage. Second, because silver precipitation is in tissue that has already been processed for conventional transmission electron microscopy, which includes osmication, osmium may cause unspecific reduction of silver from the physical developer solution (Mobbs 1976). This may increase background staining and give rise to difficulties in interpretation, especially of small in-tensified profiles.

In this chapter we describe three block-intensification methods for elec-tron microscopy illustrating their fine-structural features. All methods have been applied to the nervous system of the fly, *Musca domestica*, but this de-scription should act as a useful guide for using them on other species. In-cluded are schedules for embedding and processing of thick sections. Some advice about ultrathin sectioning and electron microscopy should help the investigator keep problems to a minimum.

Method

1. Cobalt Filling

We have tested the method on the peripheral and central nervous system of the housefly. Either primary antennal sensory cells or descending neurons in the brain were allowed to take up cobalt through their cut axons. Living flies were mounted with wax onto glass slides. Filling was performed by putting a cobalt-filled (5% cobalt chloride in distilled water) glass electrode over the cut stump of the second antennal segment or by immersing the cut thoracic ganglion in a 5% cobalt chloride solution contained in a leak-proof vaseline trough. After an uptake time of 1 h for the antennal sensory cells and 2 h from the thoracic ganglion, the cobalt solution was removed and the preparation was kept for an additional 1-h diffusion period at 4 °C.

2. Cobalt-Sulphide Precipitation and Processing up to Block Intensification

The posterior cuticle is removed from the head capsules, and the heads are processed as follows:

a) Immerse head capsules in ammonium sulphide in 0.05 M cacodylate buffer, pH 7.4, containing 5%−6% sucrose [4 drops concentrated $(NH_4)_2S$ solution per 10 ml buffer], and rotate specimen rapidly in closed vials for 2 min on a rotator.
b) Transfer head capsules to fresh buffer containing vials and rotate for additional 2 min.
c) Fix tissue for 2 h in 2.5% glutaraldehyde (in the same buffer) containing 5%−6% sucrose, at room temperature.
d) Transfer specimen to fresh buffer solution and dissect brains out under buffer. Wash several times in fresh buffer and store tissue overnight in buffer at 4 °C.
e) Wash tissue carefully in several changes of fresh 5%−6% sucrose solution to remove excess of buffer, total 0.5 h.
f) Process brains through intensification procedure P, G or H.

The vehicle for glutaraldehyde and osmium fixation is either (0.1 M, pH 7.4) Pipes buffer [Piperazine N,N′-bis(2-ethane sulfonic acid)] or sodium cacodylate buffer (0.05 M, pH 7.4). Tyrer et al. (1980) suggested that cacodylate should be avoided because of its possible interaction with the silver developer. From our experience, however, careful washing and presoaking the tissue in sucrose solution before silver intensification does not give rise to undue silver precipitation, and ultrastructure was generally better after using cacodylate-buffered fixatives.

3. Mild Block-Intensification Procedures for Electron Microscopy

All solutions are freshly prepared. For solution A heat water to 60 °C, dissolve the gum arabic, and afterwards add the remaining ingredients. Keep all glassware and solutions at the same temperature as that used for intensification. Avoid vigorous shaking, especially of the mixed A + B solutions.

Pyrogallol Method (P)

Soln. A: 0.8% pyrogallol
           0.6% citric acid
           1.0% gum arabic   in 100 ml distilled water
           0.5% DMSO
           5.0% sucrose
Soln. B: 1% $AgNO_3$ in distilled water.

The brains are processed as follows:

a) Incubation at 4 °C for 1 h in solution A.
b) Incubation at 4 °C overnight in 9 vol. A + 1 vol. B.

All steps are carried out in the dark.

Glycine Method (G)

Soln. A: 5% glycine
           0.75% citric acid
           1.0% gum arabic   in 100 ml distilled water
           4.0% sucrose

Soln. B: 1% $AgNO_3$ in distilled water.

The brains are incubated at 24 °C for 24 h in 10 vol. A + 1 vol. B in the dark.

Hydroquinone Method (H)

Soln. A: 0.05% hydroquinone
           0.7% citric acid
           2.0% gum arabic   in 100 ml distilled water
           1.0% DMSO
           5.0% sucrose

Soln. B: 0.5% $AgNO_3$ in distilled water.

The brains are processed as follows:

a) Incubation at 20 °C for 1 h in Soln. A.
b) Incubation at 37 °C for 1 h in 9.9 vol. A + 0.1 vol. B.
c) Incubation at 37 °C for 1 h in 9.75 vol. A + 0.25 vol. B.
d) Incubation at 37 °C for 1.5 h in 9.5 vol. A + 0.5 vol. B.
e) Incubation at 37 °C for 1 h or less in 9 vol. A + 1 vol. B.

All incubation steps are carried out in the dark.

4. Processing After Block Intensification. Osmication, Dehydration and Embedding Schedule

a) Wash tissue well in 5% sucrose solution for 1 h in the dark, with four changes. If a wash follows intensification at 37 °C, then begin the wash at that temperature and let tissue cool to room temperature. Likewise warm slowly from 4 °C to room temperature.
b) Transfer tissue into appropriate buffer solution, wash for 0.5 h with three changes (or store overnight at 4 °C in the refrigerator).
c) Incubate tissue in 1% OsO_4 in buffer. Fix for 1 h in an ice bath in the dark.

d) Transfer tissue to appropriate buffer, wash well for 1 h with four changes.

Dehydration

e) Dehydrate through 70%, 80%, 90%, 96% and 100% (dried) ethanol, each 2 × 10 min. Then soak 2 × 10 min in propylene oxide.

Infiltration and embedding

f) Transfer tissue to "soft" araldite mixture/propylene oxide: 1 vol. araldite + 2 vol. propylene oxide and incubate for 2 h at room temperature (closed vials, at low speed on a rotator).
g) Transfer to "soft" araldite mixture/propylene oxide: 2 vol. araldite + 1 vol. propylene oxide, incubate for 4−5 h at room temperature (closed vials, at low speed on a rotator).
h) Transfer tissue into pure "soft" araldite mixture and incubate for 1 h in open vials, then continue infiltration overnight in closed vials at room temperature in the dark (rotator, low speed).
i) Transfer tissue to pure "soft" araldite mixture and infiltrate for 5−6 h in closed vials in the dark (rotator, low speed).
j) Transfer tissue in a drop of "soft" araldite onto silicone rubber plates, orient tissue and put an araldite-filled BEEM or gelatine capsule upside down over the tissue.
k) Polymerize overnight at 60 °C.

5. Section Re-embedding Procedure for Electron Microscopy

As cobalt-filled neurons cannot be visualized in whole-mount preparations that have been processed through osmication procedures, the tissue has to be sectioned prior to selection of intensified areas.

Prepare:

a) Clean pieces of acetate sheet (translucent, colourless sheets for overhead projection) are cut to approximately the size of a glass slide.
b) Stick the ends of the sheet onto a glass slide with adhesive tape.
c) Distribute a thin layer (0.3−0.5 mm) of unpolymerized soft araldite mixture over the acetate foil. Avoid producing air bubbles.

Perform:

1. Cut 20-μm plastic sections with a sliding microtome.
2. Mount sections with Permount and cover with coverslip.
3. Select appropriate sections, take photographs if necessary.
4. Scratch a window into the coverslip above the desired section with a diamond and remove the glass.

5. Remove the section from Permount with fine forceps.
6. Swirl the section for a few seconds in fresh xylol to remove excess of Permount and glass splinters.
7. Dry the section by touching it briefly with filter paper.
8. Transfer the section into the "soft" araldite mixture prepared on the acetate sheet (steps a−c). Make sure that the section is completely covered by plastic and that there are no air bubbles by checking with the dissection microscope.
9. Cover with a second clean piece of acetate foil of about the same size as the one underneath.
10. Polymerize at 60 °C overnight.
11. Remove carefully the upper acetate sheet. The section remains enclosed in the thin layer of araldite. To avoid contamination do not touch the surface.
12. Fill a BEEM or gelatine capsule with "soft" araldite mixture and put it upside down onto the already embedded section.
13. Polymerize again at 60 °C overnight.
14. First separate the acetate foil from the glass slide. Then separate the foil from the re-embedded section on the BEEM capsule.
15. Remove the capsule and trim for ultramicrotomy.

Comment: It is possible to leave out the Permount mounting (steps 2−7) and immediately embed the section in araldite (step 8). However, the light-diffraction properties of the acetate sheet make selection and photography of suitable sections more difficult.

6. Light Microscopy

Filled profiles are identified by light microscopy and a suitable area is chosen for further electron microscopic study. The relevant section is photographed and then re-embedded for ultramicrotomy.

The contrast between intensified profiles and the osmicated background varies greatly, the highest contrast being obtained by the pyrogallol method (method P) (Fig. 1 A, B) and the hydroquinone method when this is run for 2−3 h at 37 °C at step(e) (method H; see Fig. 1 E−G). The finest precipitates for electron microscopic identification of synaptic areas are achieved by the glycine method (G) or by the hydroquinone method running the final step(e) for 1 h or less (Fig. 2). We routinely use method H. Although offset by its excellence as an electron microscopic marker one obvious disadvantage of method H is its poor contrast for light microscopy. Black and white photography is difficult (Fig. 2 C−E), and we recommend serial step-focussed colour transparencies for reconstructing identifiable neurons (Fig. 2 A, B). A general feature of these methods is the occurrence of various densities of unspecific coarse background silver grains. However,

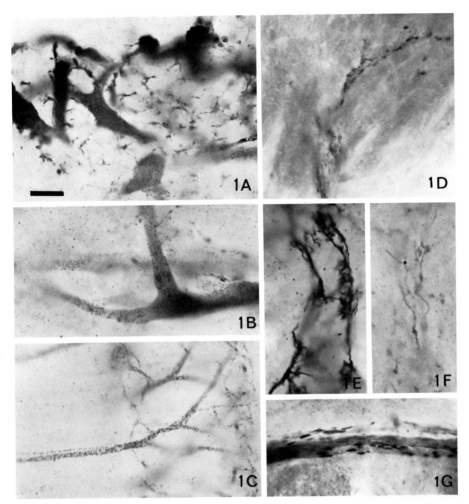

Fig. 1. Light microscopy of block-intensified CoS-filled profiles (All figures in this chapter are from *Musca domestica*). Except for **C,** tissue was fixed in cacodylate-buffered glutaraldehyde, intensified, and then osmicated. **A** and **B** Various degrees of intensification in descending neurons in the brain (method P). **D** A male unique neuron "transsynaptically" filled in the lobula, intensified by method **G** (note the darkly staining mitochondria). **E, F** and **G** Various strengths of intensification of midbrain neurons treated by method H. These three preparations had a prolonged wash in sucrose solution before intensification and the background of unspecific silver grains is consequently reduced (cf. Fig. 2 C−E). **C** Gallyas method of intensification, used for double marking, showing a "transsynaptically" filled Col A cell of the lobula. Precipitates are coarse and do not selectively show loci of mitochondria. Bar **A−G** = 10 μm

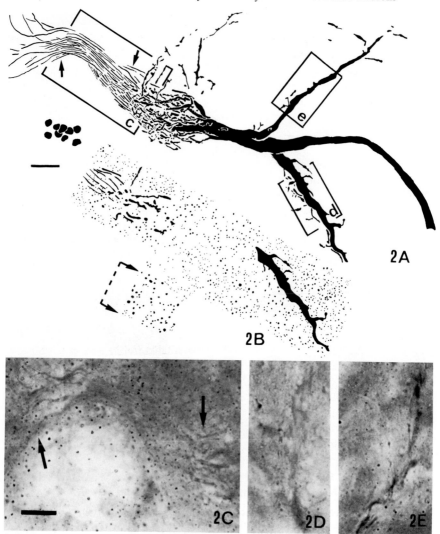

Fig. 2. Hydroquinone intensification (method H) can give rise to the finest-grained precipitates which allow exact electron microscopic identification of pre- and postsynaptic sites (see Figs. 5B; 6C–F; 7A, B). However, for this, optimal intensification of CoS-filled profiles should be pale. Also, intensification sometimes gives rise to many unspecific coarse-grained deposits within the tissue. Despite low contrast, entire neurons can be reconstructed. **A** shows the Giant Descending Neuron contiguous with Col A cells "transsynaptically" filled. Col A axons (*arrowed*) and perikarya are resolved, but cell body fibers are too pale to see. **B** shows the distribution of grains in a 35-μm-thick plastic section containing part of the depicted cell. The density of grains varies according to the composition of brain tissue, being largest and most widely spaced in the cell body layer (*arrowed* in **B**). Very few grains are encountered in ultrathin sections, whereas cobalt-silver intensified profiles are unambiguously recognized. **C, D** and **E** illustrate the rather poor contrast between intensified profiles (boxed *c, d* and *e* in **A**) and the heavily osmicated background with unspecific silver grains. Bars **A, B** = 20 μm; **C, D, E** = 10 μm

even when these occur in abundance, as shown in Fig. 2B, they are seldom encountered in an 80-nm-thick section even though we examine areas of up to 0.06 mm², where an average of six large grains is encountered per section. These unspecific precipitates may shatter during cutting. However, in our experience this happens rarely, and when it does, the resulting fragments are large enough to be easily recognized as background artefacts. The number of unspecific coarse silver grains can be minimized by removing excess cacody-late buffer before intensification by washing several times in a 0.1 M sucrose solution (Fig. 1E−G).

Another technique that we occasionally use is the Gallyas (1971) intensi-fication (Fig. 1C) which we find useful as a double marker in conjunction with the Golgi technique. The method provides unambiguous identification of cobalt-silver profiles by light microscopy, where they are clearly dis-tinguishable from Golgi-impregnated elements. The method provides a rela-tively large-grained marker for electron microscopic identification of pro-files, with little or no unspecific coarse-grained background.

7. Ultramicrotomy, Section Staining and Electron Microscopy

Cutting silver-intensified tissue may give rise to some problems. If possible one should avoid using glass knives. Cobalt-silver precipitate grains are ex-tremely hard and damage the "soft" glass edge. A diamond knife, although more expensive, saves time and will result in better quality sections. We have obtained good results by cutting with an angle of 4° to 4.5° at low speed (0.3−0.5 mm s⁻¹).

There are extreme differences in hardness between intensified profiles and softer surrounding tissue. In heavily intensified material, big silver grains are often pushed through the tissue, causing scratches. Positioning the intensified areas furthest away from the knife edge may help avoid this.

Another serious problem is secondary section contamination by silver grains. This may be the case in material containing very fine granular de-posits. Precipitate grains become loosened during sectioning and either ad-here to the knife's edge or float on the water surface within the trough. In the first case they damage the sections; in the second, they give rise to secondary unspecific section contamination. Unless adequate precautions are taken, the latter may complicate the identification of labelled profiles. Careful and fre-quent cleaning of the knife edge and trough is therefore essential for obtain-ing unambiguous identification of filled neurons.

Ultrathin sections should be analysed either unstained or stained with uranyl acetate alone. Good results are obtained with 2% uranyl acetate in 50% ethanol (20 min). Single or secondary lead staining to enhance contrast has to be avoided. The grains of lead stains have the same size and electron density as fine cobalt-silver precipitate (see Fig. 10A) and are indistinguish-able from them by conventional transmission electron microscopy.

Ultrastructural analysis must be carried out in a liquid nitrogen cooled specimen stage. Local heating of the examined areas by the electron beam is enormous and results in evaporation of silver. Without the cooling unit surrounding the specimen, silver vapours recondense immediately onto the section. This causes secondary silver contamination of the areas to be analysed and hampers interpretation.

Ultrastructure

a) Unlabelled Neurons

Considering that brain tissue is processed through at least 4 min of sulphiding and washing prior to fixation, the ultrastructural preservation of the tissue is satisfactory in that cytological details are recognizable. These include preservation of mitochondria, microtubules, synaptic vesicles, postsynaptic membrane specialization and presynaptic ribbons, which are unambiguously resolved (Figs. 3A, 5B, 6E). However, some degree of vacuolization and mitochondrial swelling, as well as occasional formation of myelinlike figures in the tissue or shrinkage of profiles, cannot be avoided under the processing procedures described. For comparison, Fig. 10B shows brain tissue which has been prepared for conventional ultrastructural observations. Fixation was in 2.5% glutaraldehyde in 0.05 M cacodylate buffer, pH 7.4 with 5% glucose, for 2 h. A 1-h wash in the same buffer followed, and subsequently the tissue was postfixed in 1% OsO_4 in the same buffer for 1 h. Thereafter the brains were processed as after block intensification.

b) Primarily Labelled Neurons

At primary magnifications between 10 000 × and 25 000 ×, cobalt-silver labelled profiles are detectable by their content of electron-dense deposits which are clearly restricted to filled profiles. There is a varying composition of small 4−6 nm granules and larger, more irregularly shaped, 30−90 nm precipitate grains. In brains treated by method P (Fig. 3A−C), the larger grains predominate, whereas material processed by method G (Fig. 4A, B)

→

Fig. 3A−C. Electron micrographs of method P-intensified neuronal profiles, demonstrating a mixed population of cobalt-silver grains. **A** and **B** Unstained sections. **C** has been double-stained for 2 min with uranyl acetate and 1 min with lead citrate. Note the enhanced contrast in **C** at the expense of fine silver precipitate resolution: it is not possible to determine if the mitochondrion-containing profile carries fine silver deposits. *m* mitochondrion; *arrow* presynaptic T-shaped ribbon, the apposing postsynaptic membrane thickening is hidden by a large silver grain; *arrowheads* cytoskeletal microtubular elements. Bars = 1 μm

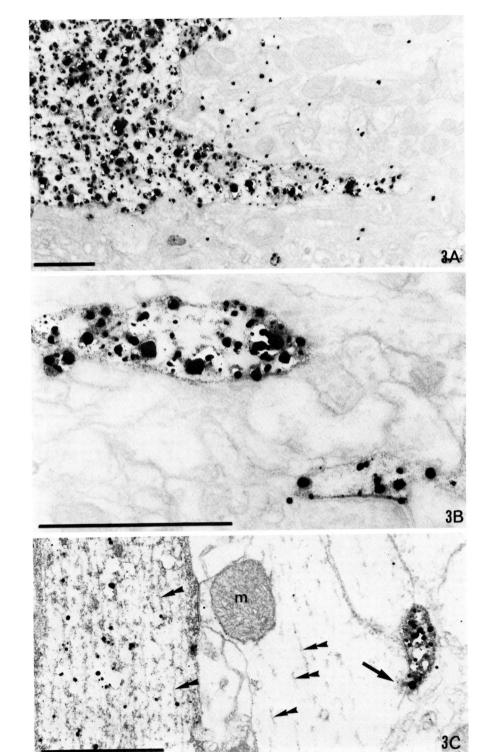

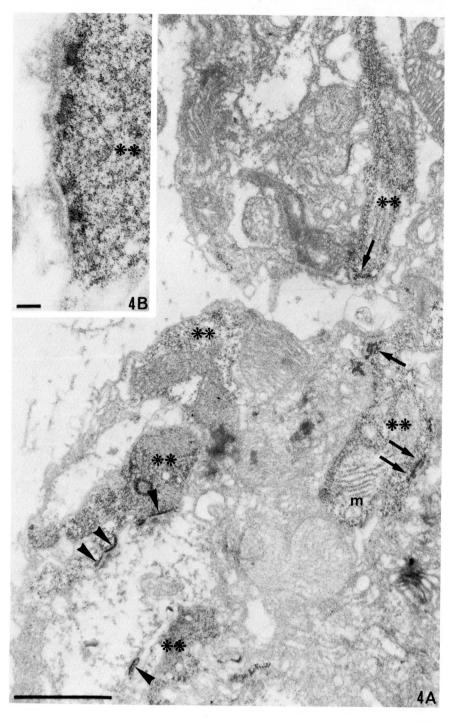

and method H contains mostly small grain deposits (Figs. 5A, B; 6C–F; 7A, B).

The abundance of grains depends on the degree of filling. The distribution of grains appears more homogeneous in smaller profiles (Figs. 4A, B; 5A), whereas precipitates in larger cell regions of bigger profiles are frequently aligned along microtubular elements (Fig. 5B).

Mitochondria contain various amounts of silver precipitate grains within the matrix and in the intracrista space. Although a variety of differently labelled mitochondria may exist within one filled cell, most are heavily labelled (cf. Figs. 4A; 5B; 7A; 8A, B).

All labelled cells are characterized by one constant and prominent feature: membrane specializations, such as postsynaptic thickenings (Figs. 3C; 4A; 5B; 6E, F; 7A, B), presynaptic ribbons (Figs. 4A, B) and

---

◀ **Fig. 4A, B.** Electron micrographs of method G-intensified cobalt-filled antennal mechanosensory axons. Sections are stained with uranyl acetate. **A** Homogeneously distributed Co-Ag precipitate of the small category is visible in the filled profiles (*double asterisks*). *m* labelled mitochondrion; *arrows* presynaptic T-bars enhanced in contrast by the Co-Ag precipitate; *arrowheads* postsynaptic membrane specialisations outlined by Co-Ag precipitate grains. Bar = 1 μm. *Inset* **B** enlargement of an intensified (*double asterisk*) cobalt-filled axon terminal. Note the grain of the label and its distribution throughout the cytoplasm and the presynaptic ribbons. Bar = 0.1 μm

Fig. 5A, B. Electron micrographs of primarily cobalt-filled (method H-intensified) antennal mechanosensory axons (**A**) and a DNVS descending neuron postsynaptic to Vertical cells from the lobula plate (**B**; see Chap. 7. Fig. 9A, B). **A** Unstained section, note distribution of silver precipitate within the mechanosensory axons. **B** Uranyl acetate-stained section. *double asterisk* cytoplasm of the primarily filled descending neuron; *m* mitochondrion of an unlabelled presynaptic profile; *m** labelled mitochondrion in the primarily filled descending neuron, note the different distribution of label in the neighbouring mitochondria in this cell; *arrow* accumulation of cobalt-silver grains at postsynaptic sites in the labelled cell; *arrowheads* silver grains aligned with microtubules. Bars = 1 μm

Fig. 6. A, B Extracellular cobalt-silver grains (*arrow*) in a leaky preparation. *c* cytoplasm of an unfilled profile in the dorsal deutocerebrum; *arrowheads facing each other* cytoplasmic membranes of two unlabelled profiles; *asterisk* labelled profile. (method H-processed material, uranyl acetate stain). Bar = 0.1 μm. **C, D** Precipitate accumulations (*arrows*) at gap junctionlike close membrane appositions between a primarily (*double asterisk*) labelled neuron (Giant Descending Neuron) and a "transsynaptically" filled (*single asterisk*) presynaptic columnar relay cell (type Col A). Precipitates of cobalt-silver grains are also present in the mitochondrion (*m*) of the transneuronal filled cell. (method H-processed material, uranyl acetate stain) Bar = 0.1 μm. **E** Postsynaptic cobalt-silver accumulations (*arrowheads*) in a primarily filled cell (*double asterisk*) (descending neuron of V-cell system). The presynaptic ribbon of an unfilled, unidentified profile is not outlined by precipitate. (method H intensification, uranyl acetate stain). Bar = 0.1 μm. **F** Postsynaptic cobalt-silver grains (*arrowheads*) in a primarily filled cell (*double asterisk*) (Giant Descending Neuron). A presynaptic ribbon of one "transsynaptically" filled profile (*single asterisk*) (Col A cell) is outlined by precipitate grains, enhancing the structure in contrast. *Arrows* cobalt-silver grains aligned at gap-junctionlike close membrane appositions between the Giant Descending Neuron and another transneuronally filled Col A cell. (method H-intensified, uranyl acetate stain). Bar = 0.1 μm

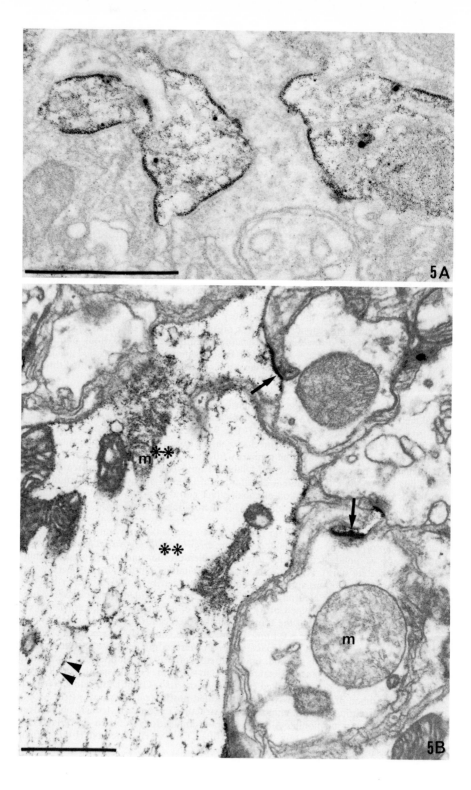

5A

5B

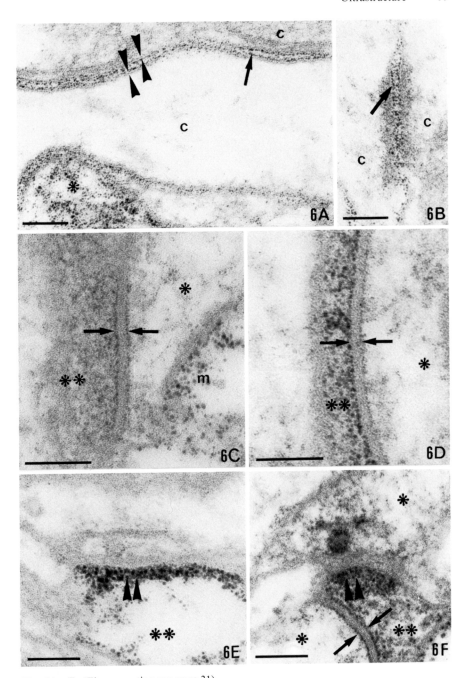

Fig. 6A – F. (Figure caption see page 31)

◀ **Fig. 5A, B.** (Figure caption see page 31)

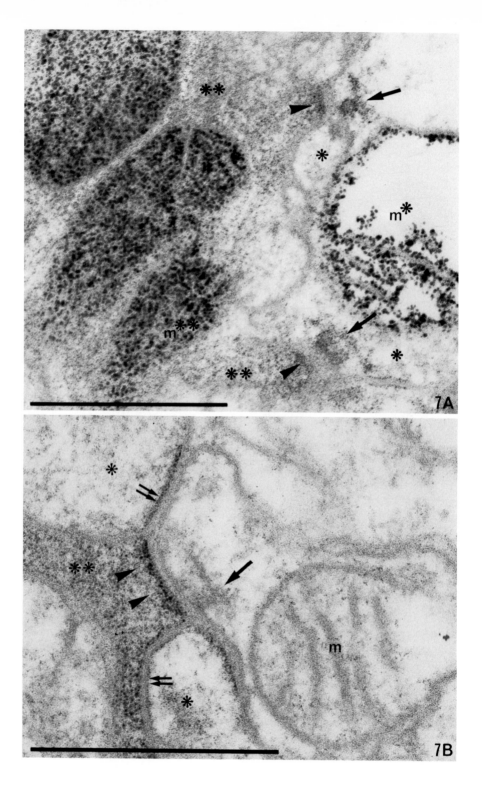

7A

7B

gap-junctionlike structures (Fig. 6C, D, F), are distinctly amplified by cobalt-silver precipitate grains. In cases where large-grained deposits predominate, the big grains may hide fine-structural details of synaptic connections (Fig. 3C).

c) "Transsynaptically" Filled Cells

Neurons filled by cobalt ions migrating specifically into them from a primarily filled nerve cell are termed "transsynaptically" filled (Strausfeld and Obermayer 1976). At high resolution their profiles are seen to contain significantly fewer electron-dense deposits than do profiles of primarily filled cells (after filling periods given on p. 20). Cytoplasmic precipitate grains are extremely small and, therefore, sections showing unspecific section contamination should not be used for analysis. However, in clean fills, electron dense grains are sometimes seen aligned along microtubules, and precipitate is always clearly visible in mitochondria (Figs. 7A; 8C−F), and at pre- and postsynaptic structures (Figs. 6F; 7A). Precipitates are also aligned at gap-junctionlike close membrane appositions (Figs. 6C, D, F; 7B) (Bassemir and Strausfeld 1983).

d) Extracellular Cobalt

Leakage of cobalt ions into the extracellular space ("overfilling," Tyrer et al. 1980) may be caused by poisoning and killing the filled neuron. Although not always identifiable in the light microscope, extracellular cobalt is easily detectable at the ultrastructural level. Cobalt-silver precipitate is no longer restricted to filled profiles: silver deposits are distributed throughout the extracellular space, often aligned between adjacent cell membranes (Fig. 6A, B).

Fig. 7. A Synaptic connections between a primarily filled (*double asterisk*) neuron (Giant Descending Neuron) and a "transsynaptically" labelled element (*single asterisk* Col A cell). Note the differences in mitochondrial label both in the primarily (*m***) and the transsynaptically filled (*m**) element. *Arrows* presynaptic ribbons outlined by silver precipitate; *arrowheads* postsynaptic precipitate accumulation (method H-intensified tissue, uranyl acetate stain). Bar = 0.5 μm. **B** Synaptic connection between a primarily filled (*double asterisk*) cell (Giant Descending Neuron) and another unfilled, unidentified presynaptic profile. Note that there is neither enhanced contrast of the presynaptic ribbon (*large arrow*) nor specific precipitate within the mitochondrion (*m*) of the unlabelled cell. In comparison "transsynaptically" filled elements (*single asterisk* Col A cells) show precipitate accumulations at gap-junctionlike close membrane appositions (*small double arrows*). *Arrowheads* postsynaptic cobalt-silver deposits. (method H-intensified tissue, uranyl acetate stain). Bar = 0.5 μm

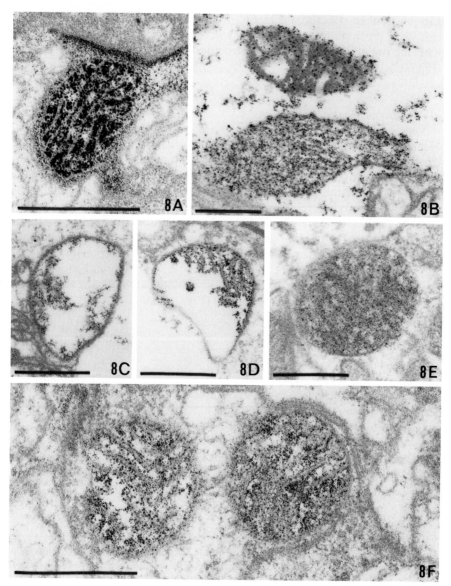

Fig. 8 A−F. Examples of cobalt-silver precipitate accumulations in mitochondria of primarily filled (**A** Giant Descending Neuron; **B** descending neuron of the V-cell system) and transsynaptically filled profiles (**C−F** Col A cells). Compare too with Fig. 7A. (method H-intensified tissue, uranyl acetate stain). Bars = 0.5 μm

Identification of Size and Nature of Precipitate

In both stained and unstained sections, small intensely electron-dense precipitates are round. Other shapes of precipitates are due to superimposition of grains as aggregates within the section. If there is any doubt about the nature of the precipitate, comparison between stained and unstained (Figs. 3A, B; 5A) sections is helpful (cf. Fig. 5A, B); and although uranyl acetate staining normally does not give rise to grainy deposits, the tissue can sometimes become covered by a fuzzy contamination due to clumsy section handling during the staining procedure.

In addition to cobalt sulphide grains, tissue may contain intrinsic sites that precipitate silver during the intensification procedure. This will certainly lead to misinterpretations. To eliminate this error it is necessary to examine silver-intensified control tissue (that does not contain cobalt) and compare it with cobalt-filled silver-intensified tissue. There should be no specific silver accumulation in the controls.

An elegant method to confirm the nature of cobalt, cobalt-silver or silver precipitate at specific cellular sites is electron microprobe analysis (see also Chap. 5). This needs, however, special facilities such as an X-ray detector system adapted to a high-resolution electron microscope. When tissue sections are exposed to an electron beam, X-rays are produced from the exposed region, that are characteristic of particular elements located in that area. The analysis can be carried out on stained and unstained ultrathin sections of about 80−90 nm thickness. Examples of analysed cellular sites in primarily filled cells are shown in Fig. 9A−E. Comparison of the energy spectra shows cobalt and silver peaks. The detectability of an element depends, amongst other factors, on the sensitivity of the instrument and on the local concentration of the element. The latter, however, is low in ultrathin sections and increasing section thickness results in a prominent loss of electron-optical resolution. For these reasons, in "transsynaptically" filled cells, where the cobalt concentration is low, cobalt often escapes detection. This being the case, indirect detection by silver precipitation onto cobalt sulphide is the only method available.

Discussion

Granular precipitates in cobalt-filled silver-intensified neurons have been described by Tyrer et al. (1980) from block intensification of whole mounts of locust thoracic ganglia. In their method intensification is carried out after osmication and the dehydration to 70% alcohol. Although these steps may be critical concerning loss of cobalt sulphide and increased background staining, one can expect that the local silver distribution is similar to that in our block intensification although Tyrer et al. (1980) do not comment on silver-cobalt specificity. Phillips (1980) showed an accumulation of coarse cobalt-

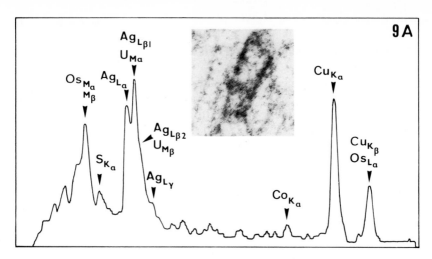

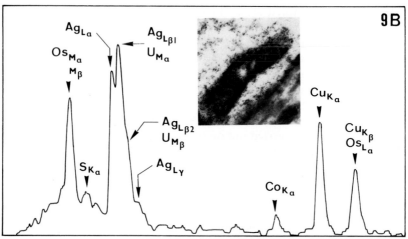

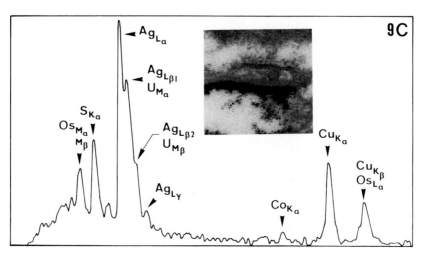

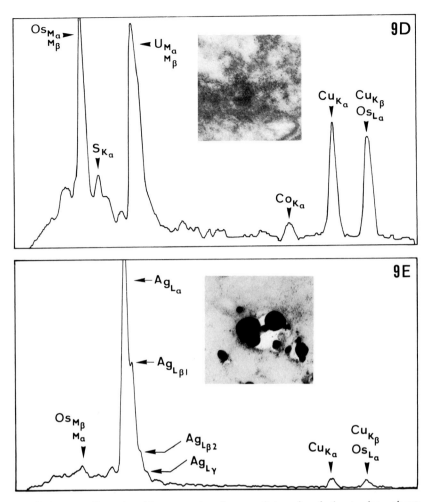

Fig. 9 A–E. Energy dispersive X-ray spectra from point-analysed (spot size: about 100 nm) selected cellular sites within cobalt-filled silver-intensified and in cobalt-filled unintensified neurons. *Small half-tone insets* represent analysed areas. *Arrowheads* identify the presence of cobalt (Co K_α), silver (Ag L_α, L_{β_1}, L_{β_2}, L_γ) and sulphur (S K_α). The presence of osmium (Os M_α, M_β, L_α) and uranium (U M_α, M_β) is due to osmication and uranyl acetate staining. Copper peaks (Cu K_α, K_β) are derived from the copper grids which were used to support the sections. A 75 kV electron beam, operated at a current of 10 μA, was used for the analyses. Recording time was 200 s. **A** X-ray spectrum from a cytoplasmic region, containing electron-dense background and microtubules. Cobalt and silver are detectable in the primarily cobalt-filled and silver-intensified neuron (uranyl acetate stain). **B** X-ray spectrum from a mitochondrion in the same cell as **A**. Cobalt and silver are detectable (uranyl acetate stain). **C** X-ray spectrum from a postsynaptic precipitate accumulation in the same cell as in **A** and **B**. Cobalt and silver are detectable (uranyl acetate stain). **D** X-ray spectrum from a presynaptic membrane specialization in a primarily filled profile. The tissue was not silver-intensified, therefore, a silver peak is lacking. Cobalt, however, is detectable (uranyl acetate stain). **E** X-ray spectrum of a big silver grain (large category, method P) in the cytoplasm of a primarily filled cell without additional uranyl acetate staining of the section. There is a clear silver peak. Cobalt is below the detection limits

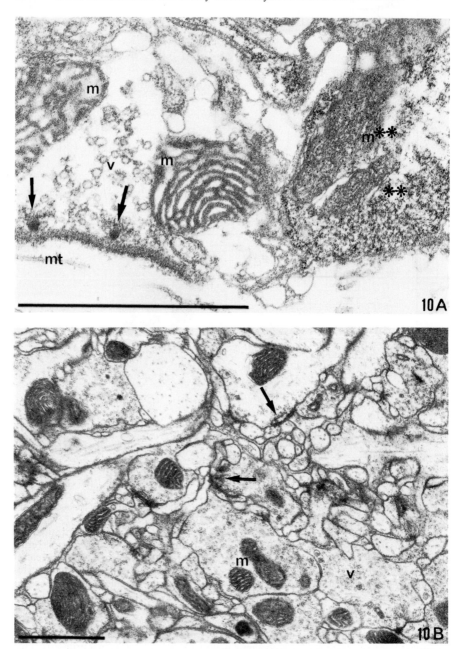

Fig. 10. A Uranyl acetate (20 min)/lead citrate (5 min)-stained section. *Left* profile has presynaptic T-bars (*arrows*), synaptic vesicles (*v*) and mitochondria (*m*). Its *right* neighbouring profile contains Co-Ag precipitate in the cytoplasm (*double asterisk*) and in the mitochondrion (*m***) Microtubuli (*mt*) are also well-defined. Due to the overall high contrast endowed by lead citrate staining, all ultrastructural details are well-delineated. But

silver deposits at synaptic densities of locust neurons as revealed by silver intensification of 2.5-μm plastic sections. However, in our experience, staining methods applied to thick sections are less sensitive for detecting cobalt than those used on whole-mount preparations.

The main disadvantage of mild block intensification, suitable for electron microscopy, is the relatively low contrast of intensified profiles for light microscopy. The 20–30 μm thick sections have a rather dark tissue background due to osmium treatment. Unfortunately this problem cannot be overcome entirely because osmication is indispensible for preserving ultrastructural details. Increasing the contrast of intensified profiles occurs at the expense of obtaining fine precipitate grains, thus diminishing the chances of seeing synaptic sites. There are, however, some areas for improving light microscopic identification of filled profiles. Embedding the preparations in a more translucent epoxy resin, such as Spurr's (1969) medium, reduces background colouration. Because mitochondria of cobalt-filled intensified cells often contain relatively high amounts of silver grains, they can serve as intensely staining guides, indicating the presence of a cobalt-silver-containing profile. Viewing with interference phase contrast allows better visualization of the palest profiles. Even taking these difficulties into account, there remain advantages over previous intensification methods. The obvious one is that entire neurons can be visualized before further tissue processing for electron microscopy. Tissue preservation is good both in unfilled and marked neurons. This enables unambiguous identification of pre- and postsynaptic sites even when the cytoplasmic label is weak. However, as demonstrated in Fig. 10A, B, optimal detection of cobalt-silver grains cannot be done after staining with lead citrate.

Using uranyl acetate staining only, mitochondria and synaptic sites remain enhanced, outlined by the cobalt-silver grains. It is interesting to compare this with neuronal markers that depend on the identification of osmiophilic reaction products, such as the heme proteins (Chaps. 3 and 6). At low marker concentrations, filled cells may be hardly distinguishable from unmarked neurons, and the synaptic sites do not specifically accumulate the marker. However, each method has its own particular advantages.

After filling a neuron, cobalt is precipitated to many thousands of cobalt-sulphide grains, each of which can act as a core that catalyses silver reduction. The loss of these cores during processing is probably much reduced by block intensification prior to osmication. Ideally, this would mean that each silver grain seen in the electron microscope should at some stage of intensification reflect the original cobalt distribution at the moment of sul-

---

exactly because of this an unambiguous definition of either the presence or absence of Co-Ag at the synaptic ribbons is impossible. Bar = 1 μm. **B** Uranyl acetate/lead citrate-stained section of brain tissue fixed and processed for conventional electron microscopy without the preceding cobalt-silver treatment. There is good preservation of ultrastructural details. *Arrows* presynaptic T-ribbons; *m* mitochondria; *v* synaptic vesicles. Bar = 1 μm

phide application. Also, high levels of unspecific silver background which may arise during intensification of ultrathin sections do not occur using the described block intensifications.

Applications

Electron microscopy of block-intensified tissue has resolved some outstanding questions concerning the synaptology of pathways leading from the brain to other ganglia. Unlike most Golgi methods, intensification allows both cell identification and the resolution of synapses. In particular, cell chains shown up by "transsynaptic" cobalt diffusion (Strausfeld and Obermayer 1976) have provided basic information about the multimodal sensory convergence onto descending neurons between the visual system of the compound eye and sensory cells from the antenna (Strausfeld and Bassemir 1983). The method has also shown us that certain large cells of the lobula plate, such as the vertical cells (Hausen 1976a, 1981), converge with certain ocellar neurons onto descending neurons that lead from the brain to the pro- and mesothoracic ganglia (Strausfeld et al. 1983). Analysis of these systems by conventional electron microscopy would be time-consuming since a cell could be identified only after reconstruction of many hundreds of serial 800-Å sections.

Another important aspect of this method is that it has enabled us to determine specific cytological aspects of "transsynaptic" cobalt diffusion. We now know that presynaptic structures of transsynaptically filled neurons characteristically contain cobalt deposits. Various filling times have enabled us to estimate the probable sequence of events during retrograde cobalt uptake by a presynaptic cell from an initially filled neuron. Cobalt affinity is first detectable at the presynaptic ribbon and at gap-junctionlike structures. Cobalt is next detectable at mitochondria and at distant pre- and postsynaptic sites and, finally, cobalt affinity is last seen at cytoskeletal structures such as neurotubuli (Bassemir and Strausfeld 1983).

The documentation of the transsynaptic phenomenon and the ease with which it can be achieved in certain cell systems of Diptera (e.g. giant descending neurons to Col A cells; the terminals of giant descending neurons of *Drosophila* to peripherally synapsing interneurons) suggests that cell contiguities shown by this stain are related to function (Strausfeld and Bassemir 1983). At least those systems that we have examined under the electron microscope are synaptically contiguous. Neuron-to-neuron staining is also extremely useful for the analysis of developmental mutants. Atavistic mutants or mutants that change the sensory pattern of the cuticle can be screened by cobalt diffusion (Palka et al. 1979; Palka and Schubiger 1980; Strausfeld and Singh 1980). Often cells are seen to have specific contiguities that differ significantly from the wild type, suggesting that an atypical receptor input into normal tissue may result in a cascade of developmental adjustment to ensure proper functional connectivity. The present electron microscopic data show that such contiguities are probably not artefactual.

Although the phenomenon of transsynaptic cobalt uptake is best documented from Diptera, low concentrations of cobalt from small-bore electrodes implanted into the brain also show up specific cell−cell contiguities, or assemblies, in Orthoptera (Bacon and Strausfeld 1980) and in bees (Mobbs, pers. commun.). This indicates that the phenomenon is not restricted to a single taxonomic order. Different insects react differently to the same method, and we are confident that changing the filling parameters will widen the applicability of the present technique. Cobalt has a specific affinity to synaptic membranes and it is a useful stain for synapses, in its own right, showing that postsynaptic membrane areas in neurons are much more extensive than were supposed. The present method (H) is especially good for detecting synaptic areas in large areas of neuropil at low magnification (e.g. 800 ×).

As described in Chap. 6, block intensification methods can precede a second (stochastic) impregnation by the Golgi method. Degeneration also provides an additional marker, and all three methods can be visualized in light microscopic sections. Multiple markers are essential for analysing complex neuropil where obvious structural reference points are not available.

Acknowledgements. We are grateful to Prof. Dr. H. G. Fromme and Dr. M. Grote, Institut für Medizinische Physik, Universität Münster, for use of the STEM and helpful advice concerning X-ray microprobe analysis of biological specimens.

Chapter 3

Horseradish Peroxidase
and Other Heme Proteins as Neuronal Markers

Dick R. Nässel

European Molecular Biology Laboratory
Heidelberg, F.R.G.

Introduction

Until 1965 neuroanatomists relied mainly on anterograde and retrograde degeneration methods introduced by Nauta and Gygax (1951) to demonstrate axonal projections in the nervous system. These methods are still valuable tracing techniques, especially for electron microscopy. An advance was made when Taylor and Weiss (1965) showed that introduction of ^{3}H-leucine into the retina resolved central projections of the retinal ganglion cells in autoradiograms. This technique relies on the axonal transport or axonal flow of the marker isotope (Weiss and Hiscoe 1948) and can be used for electron microscopy (Hendrickson 1969). Some years later it was shown that macromolecules, such as albumin labelled with Evans blue or tritium, and the enzyme horseradish peroxidase (HRP), are taken up at axon terminals and transported to perikarya (Kristensson 1970; Kristensson and Olsson 1971). The usefulness of HRP as a marker to study neuronal projections was also recognized by LaVail and LaVail (1972), and after their study on chick retinal projections, this technique caught on among neuroanatomists. The versatility of HRP can be seen in the prolific literature on the subject published in the last 10 years (reviewed by e.g. Cowan and Cuenod 1975; Winer 1977; Kristensson 1978; Spencer et al. 1978; LaVail 1978; Eckert and Boschek 1980; Alheid et al. 1981; Hanker et al. 1981; Kitai and Bishop 1981; Mesulam 1981; Warr et al. 1981).

What makes HRP and other heme proteins so advantageous as neuronal markers? I hope to answer this question by briefly describing the diverse problems that can be solved with HRP marking alone and in combination with other neuroanatomical and cytochemical techniques. The rest of this chapter comprises a summary of the chemistry of peroxidase-active proteins and the chemical reactions used to demonstrate them cytologically. This is followed by the method of peroxidase labelling, the subsequent histochemistry and the resultant pattern of labelling. Mechanisms suggested for uptake and transport of macromolecules such as heme proteins are discussed.

The Versatility of Exogenous Heme Proteins as Neuronal Markers

In mammals and other vertebrates, protein markers — especially HRP — have been used for tracing axonal projections, marking intracellularly recorded neurons, studying neuronal morphology, synaptology, transmitter vesicle turnover, and axonal transport (see reviews cited). Originally they were used to delineate extracellular space and to study diffusion barriers and endocytosis in a number of tissues. Another important use of HRP is to tag other molecules such as peptides, lectins, and antibodies so that they can be localized after uptake or binding.

Heme proteins have so far been rarely used to study invertebrate nervous system [leech segmental ganglia: Muller and McMahan (1976), Muller (1979), Orchard and Webb (1980), Macagno et al. (1981); mollusc neurons: Crow et al. (1979), Monsell (1980); lobster neuromuscular junctions: Holtzmann et al. (1971); crayfish peripheral nerve: Kristensson et al. (1972); and crayfish glia: Shivers (1976)]. In insects HRP was first used to trace contralaterally projecting optic lobe neurons in cockroaches (Roth and Sokolove 1976). HRP has since been used to label axonal projections from appendages of normal and homeotic mutants of *Drosophila* (Ghysen 1978, 1980; Green 1981), and to resolve motor neurons to flight muscles in *Drosophila* (Coggshall 1978), descending and optic lobe neurons in the brain of larger dipterous flies (Eckert and Boschek 1980; Nässel 1981, 1982; Nässel and Strausfeld 1982), and motor neurons in locust thoracic ganglia (Watson and Burrows 1981). One reason for the limited use of HRP in insects so far might be that the small size of these animals and their neurons allows many problems to be solved with other markers such as cobalt, Lucifer yellow, and the Golgi method. However, several features of heme protein markers make them also attractive for use in small animals: the simple methodology, their application to both light and electron microscopy, and their compatibility with other methods.

The simplest combination with peroxidase labelling is counterstaining using cresyl violet, neutral red or reduced silver (Adams 1977). Other markers can be introduced simultaneously for studying convergent neuronal pathways: cobalt ions, radiolabelled metabolic intermediates, ^{3}H-HRP, fluorescent markers, and iron dextran (Colwell 1975; Geisert 1976; Olsson and Kristensson 1978; Macagno et al. 1981; Yezierski and Bowker 1981; Stewart 1981). For this purpose the classical Golgi method or degeneration techniques can also be combined with HRP labelling (Somogyi et al. 1979). Other cytochemical or immunological methods can be used on peroxidase-labelled tissue to localize enzymes or putative transmitters (Ljungdahl et al. 1975; Berger et al. 1978; Blessing et al. 1978; Broadwell and Brightman 1979; Lewis and Henderson 1980).

HRP is also useful for the study of nervous system development. Since HRP injected into neural precursor cells in the early embryo is retained by the progeny many cell divisions later, cell lineages can be analyzed (Weisblatt et al. 1978; Hirose and Jacobson 1979; Jacobson and Hirose 1981).

Chemistry of Peroxidase-Active Proteins and Reactions for their Demonstration

The peroxidase-active proteins discussed here all contain a heme group. Heme is a molecule found in all organisms except anaerobic clostridia and lactic acid bacteria. In heme proteins it is derived from protoporphyrin IX (with a Fe III centre) formed by decarboxylation of two of the carboxyethyl side chains of uroporphyrin III to vinyl groups (Metzler 1977). Figure 1 shows the characteristic heme structure which is attached to amino acid residues: microperoxidase has 11 (Feder 1970); cytochrome c, 104 (Metzler 1977) and horseradish peroxidase, about 160 (see Eckert and Boschek 1980).

Two kinds of heme proteins are distinguished here according to their enzymatic activity. Those preferring H_2O_2 as a substrate are catalases and those preferring alkyl peroxides are peroxidases. However, both types of enzymes can use either substrate.

Peroxidases are found in plant tissues, at especially high concentrations in the roots of the horseradish and the sap of fig trees (Saunders et al. 1964). In cells they are often located in so-called peroxisomes. Peroxidases have also been demonstrated cytochemically in animal tissues: in leucocytes, liver Kupffer cells, colon, uterus, thyroid gland, salivary gland, lacrimal gland — in all cases in the nuclear envelope, endoplasmic reticulum, Golgi saccules, and secretory granules (see Essner 1974). Catalases are found in the liver and kidney, associated with peroxisomes. Some heme proteins show peroxidase activity although their main function is in oxygen transport (hemoglobin and myoglobin) or terminal electron transport (cytochrome c).

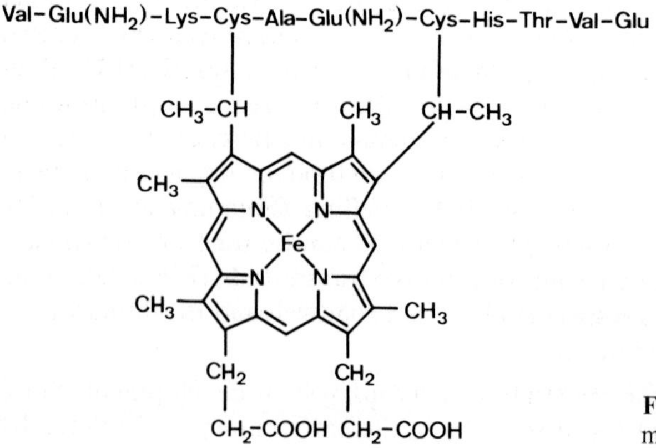

Fig. 1. The structure of microperoxidase (Feder 1970)

Peroxidases catalyze the following reaction (see Walsh 1979):

$$AH_2 + ROOH \rightarrow ROH + H_2O + A \tag{1}$$

where AH_2 = phenols, acryl and alkyl amines, hydroquinones, ascorbate, cytochrome c or glutathione; and where ROOH can be H_2O_2, giving

$$AH_2 + H_2O_2 \rightarrow 2\,H_2O + A \tag{2}$$

For catalases reaction (1) preferentially uses $AH_2 = H_2O_2$ and ROOH=H_2O_2 giving

$$2\,H_2O_2 \rightarrow 2\,H_2O + O_2 \tag{3}$$

The action of catalases and peroxidases on the substrates is not completely understood (see Williams et al. 1977, Walsh 1979).

What do these enzymes do in the living cell? Catalases are thought to detoxify other reactive products of oxygen metabolism. This occurs for example in peroxisomes which also contain oxidases producing H_2O_2 and which are thought to be involved in gluconeogenesis, purine catabolism, and non-phosphorylating respiration (Essner 1974, Walsh 1979). Thus, catalases prevent harmful accumulation of H_2O_2 which could, for example, oxidize hemoglobin to methemoglobin (Metzler 1977).

In plants, peroxidases probably regulate hormone activity (plant auxins are indole derivatives) and play a role in polyaromatic biosynthesis (Williams et al. 1977; Metzler 1977; Walsh 1979). In this biosynthesis, during co-oxidation of phenols and amines (as the peroxide is reduced), phenolic and aromatic amine radicals can couple and/or polymerize to polyphenolic products (Walsh 1979). Some peroxidases can even oxidize simple anions of halogen to halogen radicals and halogen cations, aggressive agents that can destroy bacteria (Williams et al. 1977). In animal tissues peroxidases probably also have a protective function, e.g., peroxidases in leucocytes also utilize H_2O_2 and a halide ion, as above, to attack bacteria (Metzler 1977). The thyroid peroxidases may catalyze the liberation of iodine from inorganic iodides by H_2O_2 (Pearse 1972).

Using reactions (1) and (2), peroxidases and catalases can be demonstrated cytochemically. The first cytochemical demonstration of oxidases (done before their recognition as peroxidases) was by Schulze (1909), using naphthol and dimethyl-p-phenylenediamine (the NADI reaction, Fig. 2), on fixed frozen sections yielding indophenol blue as a reaction product. Schulze's studies of peroxidases using the NADI reaction provided the basis for recent methods which rely on the oxidation of diamines to insoluble colored products even without naphthols. Fischel (1910), Graham (1918), and Strauss (1964) used benzidine in their cytochemical reaction (see Fig. 3). This compound and its derivatives (with extensive improvements in the histochemical procedures) are still used.

Five types of compounds, called chromogens because they produce colored reaction products, can be used to demonstrate peroxidase-active

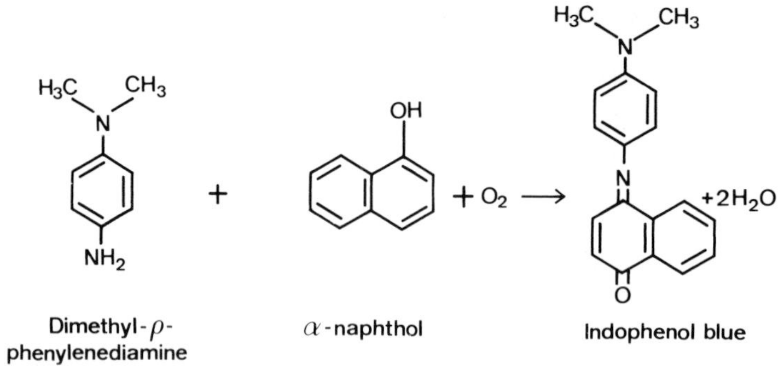

Dimethyl-p-phenylenediamine α-naphthol   Indophenol blue

Fig. 2. The NADI reaction used by Schulze (1909) given by Pearse (1972)

proteins. They are: (1) benzidine and its derivatives, (2) phenols and naphthols, (3) leucodyes, (4) indoles, and (5) aminocarbazoles (Pearse 1972). The more commonly used chromogens are listed in Table 1.

The benzidine derivatives are most commonly used. The benzidine reaction suggested by Lison (1936) is shown in Fig. 3. The peroxide-peroxidase system oxidizes benzidine to a blue or brown reaction product. The most frequently employed benzidine derivative is 3,3′-diaminobenzidine tetrahydrochloride (DAB), introduced by Graham and Karnovsky (1966), which is oxidized in the presence of H_2O_2 in a reaction catalyzed by a peroxidase [see reaction (1)]. The DAB molecule undergoes oxidative polymerization and cyclization as shown in Fig. 4 (Hanker et al. 1967; Seligman et al. 1968). The reaction product, the phenazine polymer, is osmophilic and after an osmium black reaction forms a precipitate that is brown to black in the light microscope and is electron-dense.

Different peroxidase-active proteins react with DAB and H_2O_2 under different conditions (Lewis 1977):

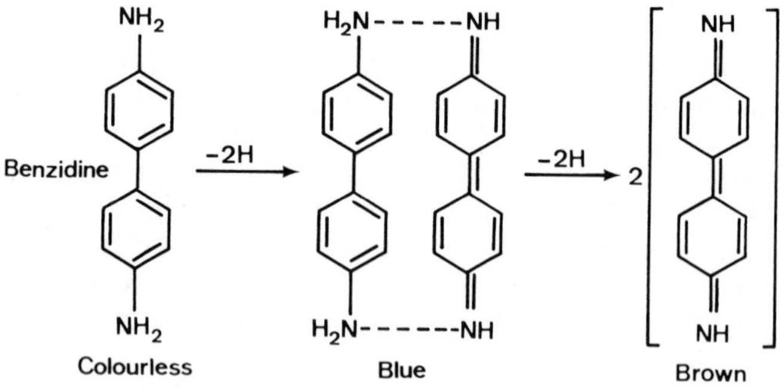

Benzidine   Colourless   Blue   Brown

Fig. 3. The benzidine reaction after Lison (1936). Oxidation of benzidine by the peroxide-peroxidase system gives a blue or a brown product

Table 1. Different chromogens used in peroxidase cytochemistry

Chromogen	Abbreviation	References
Benzidine		Fischel (1910), Graham (1918), Graham and Karnovsky (1966)
2,7-Diaminobenzidine ($+\alpha$- naphthol)		Ornstein (1968)
3,3-Diaminobenzidine	DAB	Graham and Karnovsky (1966)
2,4-Diaminofluorene ($+\alpha$-naphthol)		Ornstein (1968)
o-Dianisidine		Graham and Karnovsky (1966), Seligman et al. (1970)
Tetramethyl benzidine	TMB	Mesulam (1976)
o-Tolidine		Ornstein (1968)
N,N'-Bis(4-aminophenyl)N,N'-dimethylene-diamine	BED	
N,N'-Bis(4-aminophenyl)-1,3-xylylene-diamine	BAXD	Plapinger et al. (1968), Seligman et al. (1970)
p-Phenylenediamine (pyrocatechol)	PPD-PC	Hanker et al. (1977)
α-Naphthol (dimethyl-p-phenylenediamine)	NADI	Schulze (1909)
Orthophenylenediamine		Pearse (1972)
3-Amino-9-ethyl carbazol		Graham et al. (1965)
Acid fuchsin		Lison (1936)
Thioindoxyls		Pearse (1972)
5,6-Dihydroxyindole		Pearse (1972)

1. Peroxidases react with DAB in vivo and after glutaraldehyde fixation at neutral pH if thoroughly washed with buffer (aldehydes reversibly inactivate the enzyme).
2. Catalases react with DAB only after aldehyde fixation or at alkaline pH. It is difficult to inactivate catalases with aldehyde fixation.
3. Hemoglobin and myoglobin react with DAB at neutral pH in fixed and unfixed tissue, but require higher concentrations of H_2O_2 and longer incubation time.
4. Cytochrome c reacts with DAB at acid pH in fixed and unfixed tissue, requiring longer incubation and a higher concentration of H_2O_2.
5. Other endogenous compounds in cells react with DAB. Cytochrome oxidase reacts in vivo accounting for mitochondrial staining. After glutaraldehyde fixation, granules of adrenalin-containing cells of the adrenal medulla, peroxisomes of liver and nervous tissue, and lysosomelike bodies in some cells can also react with DAB.

Tetramethyl benzidine (Mesulam 1976) is another often used chromogen for sensitive light microscopic detection of peroxidases. Of the non-ben-

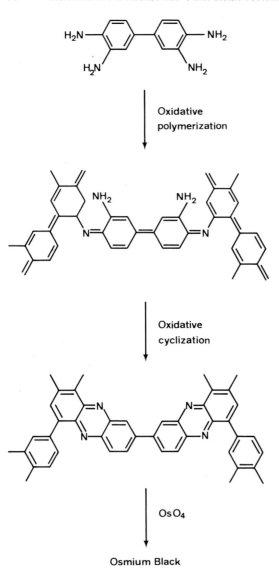

Fig. 4. The oxidative polymerization of DAB to an indamine polymer, followed by further quinoid addition to the primary amine resulting in oxidative cyclization to a phenazine polymer (Seligman et al. 1968)

Oxidative polymerization

Oxidative cyclization

OsO_4

Osmium Black

zidines, p-phenylenediamine combined with pyrocatechol (Hanker et al. 1977) is the most reliable chromogen for light and electron microscopy. Figure 5 shows the structure of six commonly used chromogens. Although some of the chromogens in Table 1 were originally used to demonstrate endogenous peroxidases and others to demonstrate exogenous ones, all chromogens react with either type. Because of this, one has to control for possible endogenous activity when using peroxidases as exogenous markers.

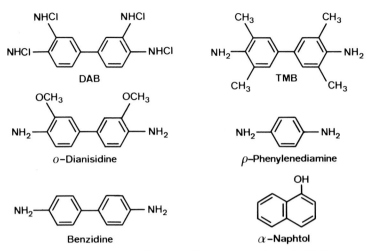

Fig. 5. Six commonly used chromogens used in peroxidase cytochemistry (*DAB* diamino-benzidine; *TMB* tetramethyl benzidine)

Methods

This section is a detailed treatment of the steps in peroxidase labelling of insect neurons. Variations of the conventional method are included when especially appropriate. A schedule of the complete routine is given in Appendix 1.

Application of the Proteins

The Protein Solutions

Neuronal uptake and transport have been tested on HRP (the most commonly used marker), microperoxidase, cytochrome c, myoglobin, hemoglobin, lactoperoxidase, and catalase, as well as various isoenzymes of HRP and turnip peroxidase (see Malmgren and Brink 1975; Malmgren et al. 1978; Bunt and Haschke 1978; Nässel et al. 1981; Nässel 1982, 1983; Chan and Haschke 1981; Giorgi and Zahnd 1978). Of the tested proteins only HRP, myoglobin, and cytochrome c are useful for sensitive and detailed neuronal marking in conventional uptake experiments. Although catalase, lactoperoxidase, hemoglobin, and microperoxidase are taken up and transported by injured nerve cells, enzyme uptake (and reactivity) is too low for detailed marking (Nässel 1983). These proteins may, however, be good markers for studies of diffusion barriers, extracellular space, and endocytosis (see Reese et al. 1971; Bennet 1973a, b; Graham and Karnovsky 1966; Karnovsky and Rice 1969; Graham and Kellermayer 1968; Zacks and Saito 1969; Heuser

Table 2. Some properties of heme proteins used as neuronal markers

Heme protein	M.W.	pI	Source	References
Microperoxidase	1 900	5.4	Horse heart cytochrome c	Feder (1970)
Cytochrome c	12 000	10.5	Horse or beef heart	Karnovsky and Rice (1969)
Myoglobin	18 000	7.0	Horse or whale skeletal muscle	Anderson (1972)
Peroxidase (Sigma type VI)	40 000	8.2	Horseradish root	Graham and Karnovsky (1966)
Hemoglobin	68 000	7.0	Bovine blood	Goldfischer et al. (1970)
Lactoperoxidase	82 000	8.0; 9.2	Bovine milk	Graham and Kellermayer (1968)
Catalase	240 000	5.7	Beef liver	Venkatachalam and Fahimi (1969)

and Reese 1973; Shivers 1976). Proteins with low molecular weight and small molecular radius such as microperoxidase and cytochrome c might be useful as intracellularly injected markers (Labhart and Nässel, in prep.).

The literature describes a large range in the concentration of protein solutions used, from 0.5% to pure protein lyophylate or solid enzyme pellets. In many vertebrate studies 10—30% HRP has been used. In flies, much lower concentrations are used (2—4%), since results obtained are similar to those using more concentrated protein solutions (10—20%).

A 2% solution of protein is normally made up in 0.1 M KCl. 0.1 ml enzyme solution is more than enough for 20 insects. The solution is made simply by mixing 2 mg enzyme "powder" with 0.1 ml KCl in a 1-ml syringe and then stored. It is, however, vital that the enzyme solutions are freshly made from frozen protein lyophylates. The different proteins used as markers are listed in Tables 2 and 3. Of all HRP preparations Sigma Type VI is the most commonly used. Other solvents for the enzyme can be used. Insect salines according to Case (1957) or O'Shea and Adams (1981) are good alternatives,

Table 3. Examples of different peroxidase preparations used

Peroxidase	References
HRP, Boehringer, Mannheim grade I	Gilbert and Wiesel (1979)
grade II	Giorgi and Zahnd (1978)
HRP, Serva Biochemical ⎫	
HRP, Miles Laboratories ⎬	Keefer (1978)
HRP, Sigma Type II ⎭	
HRP, Sigma Type VI (pI = 8.2) ⎫	
Type VII (pI = 3.5) ⎪	Malmgren et al. (1978),
Type VIII (pI = 4.4) ⎬	Bunt and Haschke (1978)
Type IX (pI = 9.5) ⎭	
Turnip peroxidase isoenzmye P7	Welinder and Mazza (1977), Bunt and Haschke (1978)

especially when applying the solution to intact CNS or incubating in vitro. Distilled water or Tris-HCl can be used on injured nerve cells. Several compounds which can be added to improve uptake are listed below. For all proteins addition of 3% lysolecithin is recommended to enhance uptake, but it should be used with caution, as described later.

Griffin et al. (1979) recommended using solid HRP pellets for extensive uptake from a localized application zone. HRP and cytochrome c pellets have been tested in the fly CNS, made from a concentrated solution of enzyme in e.g. 0.1 M KCl. A drop of solution is put in a small silicon or wax mould and air-dried in a covered Petri dish. The pellet is then broken up with a needle into pieces which are inserted into lesions in the nervous system or in peripheral organs. Pellets can be stored in covered containers in a freezer.

Introduction of the Enzyme

How the enzyme is applied depends on the experimental question. There are two main types of uptake, by intact or by injured neurons. For uptake by an

Table 4. Uptake time of HRP in different experiments using flies

Application of HRP	Minimum uptake time		
	Musca	*Calliphora*	*Drosophila*
Lesion retina	2 – 3 h	3 – 4 h	0.5 h
Lesion antenna	3 – 4 h	4 – 6 h	
Lesion ocelli	2 h	3 h	
Lesion leg	3 – 4 h	4 – 6 h	
Lesion thoracic ganglia for ascending/descending neurons	2 – 3 h	4 – 6 h	0.5 – 1 h
Lesion thoracic ganglia for "transneuronal uptake"	3 – 6 h 16 – 24 h (4 °C)	6 – 8 h 16 – 24 h (4 °C)	1 – 1.5 h
Lesion in brain to demonstrate thoracic projections	2 – 3 h	3 – 4 h[a]	
Lesion in brain or ganglia for local projections	1 – 4 h	2 – 4 h	
Intramuscular injection (thorax)	2 – 3 h[a]	3 – 4 h	16 – 20 h[b]
Intact brain or ganglia	3 – 6 h	3 – 6 h	
Sensilla or bristles			5 – 12 h[c]
Extracellular microinjection		10 – 30 min[d]	
Intracellular microinjection		10 – 30 min[d]	

[a] Uptake times estimated since experiments are not yet performed.
[b] Coggshall (1978).
[c] Ghysen (1980).
[d] After pressure or iontophoretic ejection of HRP (see text).

intact cell the enzyme is applied near it without damaging it, and the neuron actively incorporates the enzyme by e.g. endocytosis. Injured cells take up protein directly by diffusion into damaged axons or dendrites, followed by transport to the soma. The enzyme solution is usually injected into the CNS from a Hamilton syringe. Nerve cells transected by the needle rapidly incorporate the protein. Protein can also be applied to crushed or transected nerves or tracts or directly injected into cells.

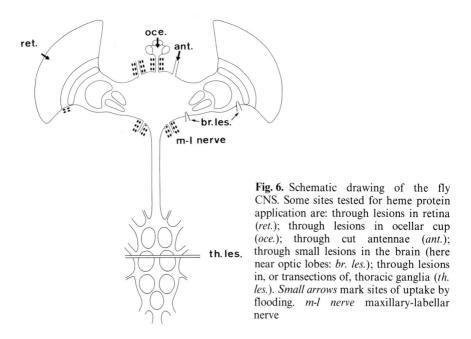

Fig. 6. Schematic drawing of the fly CNS. Some sites tested for heme protein application are: through lesions in retina (*ret.*); through lesions in ocellar cup (*oce.*); through cut antennae (*ant.*); through small lesions in the brain (here near optic lobes: *br. les.*); through lesions in, or transections of, thoracic ganglia (*th. les.*). *Small arrows* mark sites of uptake by flooding. *m-l nerve* maxillary-labellar nerve

These types of experiment will now be described for use on insect nervous tissue. In all cases insects are "anaesthetized" on ice or in the refrigerator and then immobilized with wax on a slide. Enzyme uptake times for different experiments are for Diptera (see Table 4), but can be extrapolated to other sizes of insects. Figure 6 schematically shows the fly CNS and some sites of enzyme application discussed in this chapter.

1. Filling Nerves, Cut Appendages and Sensory Cells. Labelling central projections from the periphery involves isolating a particular nerve, appendage (or part of it), or external sensory structure and applying enzyme to the cut surface. Uptake time depends on the length of the nerve projection. The following approximate values are appropriate for filling peripheral neurons or receptors with HRP: 3–6 h in *Calliphora*, 2–4 h in *Musca*, 0.5–2 h in *Drosophila*. For the other proteins the uptake time should be doubled. The experiments are performed as follows.

Cut peripheral nerves attached to the CNS are isolated in a vaseline trough (see Altman and Tyrer 1980), filled with protein solution. Appendages (e.g. legs, antennae, wings, halteres) can be cut off at the desired length and the stump inserted in a glass microcapillary filled with protein solution (Fig. 7). The tip of the microcapillary should be drawn to fit the stump of the appendage. The simplest way to hold the micropipette in position is with a plasticine block (see Bacon and Strausfeld 1980). Since HRP (but no other peroxidase) is also taken up by intact sensory cells in the appendage, one cannot be certain that the labelled axons are only from cells distal to the cut. Unambiguous labelling of lesioned axons is achieved using cytochrome c or myoglobin. The results of filling cut antennal axons are shown in Figs. 8 and 9.

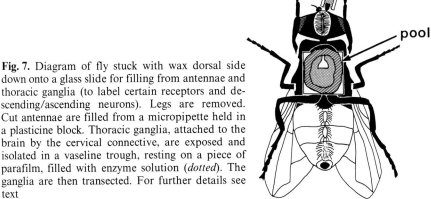

Fig. 7. Diagram of fly stuck with wax dorsal side down onto a glass slide for filling from antennae and thoracic ganglia (to label certain receptors and descending/ascending neurons). Legs are removed. Cut antennae are filled from a micropipette held in a plasticine block. Thoracic ganglia, attached to the brain by the cervical connective, are exposed and isolated in a vaseline trough, resting on a piece of parafilm, filled with enzyme solution (*dotted*). The ganglia are then transected. For further details see text

A similar strategy can be employed to fill axons from bristles and sensilla. The bristle is broken off and its stump inserted into a protein-filled microcapillary. Alternatively the bristle or bristle group can be isolated in a wax or vaseline trough filled with marker solution before being broken (see Ghysen 1980).

Visual receptors can be filled simply by cutting the cornea and injuring receptors with a razor blade, applying a drop of protein solution to the cut, and sealing it with vaseline (see also Fig. 23). The lesion size roughly determines the number of labelled receptors.

2. Lesions in the CNS. Lesions can be made in two ways: (1) small lesions localized in neuropil or cell body layers or (2) large transections of tracts, connectives, or commissures after cutting a window in the cuticle to expose part of the brain. Small lesions can be made in the desired area with a sharpened tungsten (or stainless steel) needle, a fragment of razor blade, or a

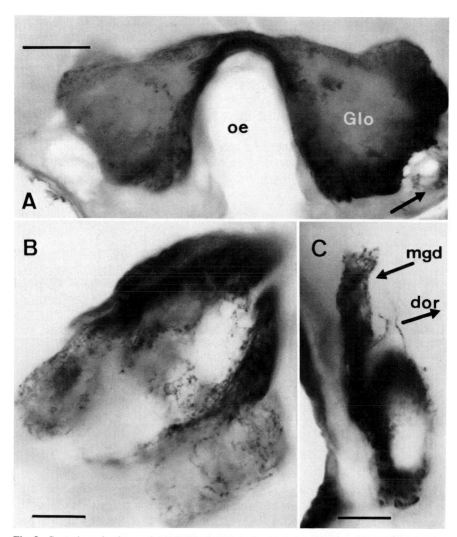

Fig. 8. Central projections of HRP-filled antennal afferents. Routine biochemistry followed by osmium treatment (as in Appendix 1). *Calliphora* **A** The two antennal lobes filled
from a single antennal nerve (*arrowed*). Note contralateral projection over the oesophagus
(*eo*). *glo* primarily filled antennal glomeruli. Bar = 50 μm. **B** Details of labelled antennal
glomeruli. Bar = 25 μm. **C** Antennal projection into the posterior deutocerebrum. (Micrograph rotated relative to **A, B**: *large arrow* points dorsally.) *mgd* axonal projection to ventral
dendrite of giant descending neuron (see Fig. 18D). Bar = 25 μm

Fig. 9. Axon branches in *PRO*- and *MESO*thoracic ganglia from the large campaniform sensillum of the second antennal segment. Routine histochemistry, with osmium fixation. *Calliphora.* Bar = 100 μm

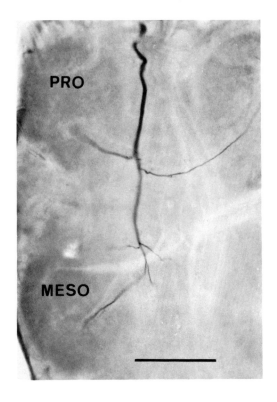

drawn glass micropipette. The protein can then either be applied as a drop over the lesion, by diffusion from a micropipette, or by enzyme pellets. The area should be covered with vaseline to prevent desiccation. Uptake time again varies but for more local projections 1−2 h in *Musca* and 1−3 h in *Calliphora* are sufficient. Labelling visual interneurons by small lesions in the brain is shown in Fig. 10. In this type of experiment one must be aware that HRP can be taken up by intact cells distant from the lesion site.

More or less complete projections between thoracic and cerebral ganglia (or between segmental ganglia) (see Figs. 11, 17−20) are displayed by the following procedure (Fig. 7), based on Strausfeld and Obermayer's (1976) method for cobalt backfilling through large lesions in the thoracic ganglia.

Mount the fly upside down with wax on a piece of microscope slide. Cut off legs, remove cuticle on the ventral thorax, and expose the fused thoracic ganglia. Peripheral thoracic and abdominal nerves are cut leaving the ganglia attached only to the brain by the cervical connective. After lifting the ganglia towards the head, vaseline is put into the body cavity and a piece of parafilm put onto the vaseline. The ganglia are then lowered onto the parafilm and a trough of vaseline applied from a 1-ml syringe with a fine hy-

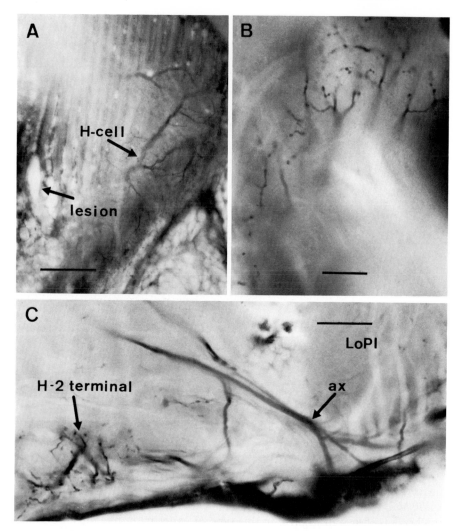

Fig. 10. HRP labelling after uptake through brain lesions. Routine histochemistry, follow-ed by osmium treatment. **A** Lesion sited in *Calliphora* lobula plate. Note tissue degener-ation and unspecific labelling of a large volume of tissue in addition to dendrites of a damaged H-cell. Bar = 50 μm. **B** Lobula plate cell processes labelled by HRP uptake in the contralateral lobula plate (not shown). *Musca.* Bar = 25 μm. **C** Projections of lobula plate neurons of *Musca* after HRP uptake from a lesion in the contralateral lobula plate (not shown). A terminal from the contralateral *H-2* cell (Hausen 1981) is seen ending on un-filled ipsilateral H-cell terminals. Some other contralaterally derived lobula plate neurons (*ax*) are also labelled. *LoPl* lobula plate. Bar = 50 μm

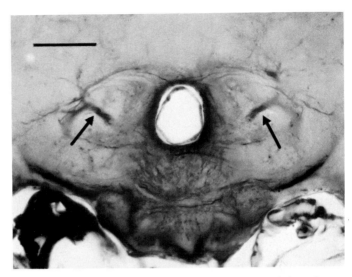

Fig. 11. A pair of "giant descending neurons" (*arrows*) in a *Drosophila* brain, backfilled from the thoracic ganglia with HRP. Several other descending and ascending neurons are labelled. Routine histochemistry, with osmium treatment. Bar = 50 μm

podermic needle is built around them. Ganglia are then transected at the desired site (e.g. between pro- and mesothoracic ganglia) and the vaseline trough filled with enzyme solution and sealed with vaseline. Uptake time for larger Diptera varies between 3–24 h for HRP and should be 24 h for other proteins. When using longer uptake times keep the flies at 4 °C (see Table 4). *Drosophila* being smaller can be filled in a different way. The thorax of a fly mounted in plasticine or O.C.T. medium for frozen specimens is simply cut at the level of the pro-mesothoracic ganglia and a drop of enzyme solution applied to the cut which is then covered with vaseline. This rather crude procedure nevertheless labels neurons in a manner compatible with results obtained from bigger flies with the more elaborate procedure (see Fig. 11).

The reverse experiment, showing thoracic parts of descending and ascending neurons, is as follows. The fly is mounted legs down on wax. The head is bent as far forward as possible without rupturing the neck, and immobilized with wax. A small opening is made in the cuticle on the back of the head just above the cervical connective. The lower part of the brain (in flies the suboesophageal ganglion) is transected, a drop of protein solution placed inside the head capsule, and the opening sealed with vaseline. Uptake time for bigger flies should be 2–6 h for HRP; twice that for other proteins. Thoracic ganglia terminations of some descending neurons filled with HRP are shown in Fig. 12.

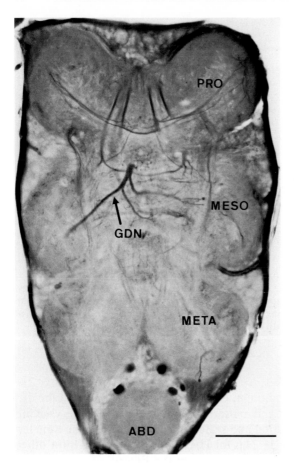

Fig. 12. Thoracic components of descending neurons in *Calliphora* filled from the brain with HRP. Note the termination of one of the two "giant descending neurons" (*GDN*). This preparation shows selective labelling, with only few descending neurons containing reaction product. *PRO* pro-; *MESO* meso-; *META* metathoracic neuromere. Note cell bodies anterior to the abdominal ganglia (*ABD*). Routine histochemistry with osmium. Bar = 100 μm

3. Injections into CNS with Microcapillaries

a) Extracellular injection. Most insects are too small for injecting protein solution into the brain with a syringe and require glass micropipettes instead. The pipette is drawn to a tip and implanted into a selected region using a micromanipulator or by holding the pipette in a plasticine block. The tip of the pipette is first broken to a diameter of 10−50 μm, depending on the amount of labelling desired.

The protein solution can simply be left to diffuse out of the pipette tip but this is not very efficient even with added detergent or dimethyl sulfoxide. Ejection of the protein by current or pressure gives better fills. A pulse of pressure can be applied by attaching the micropipette to an infusion pump or a syringe (see Wässle and Hausen 1981). Uptake time depends on the diameter of the pipette tip. Commonly a pulse of pressure is given only for 10−30 s, and the protein is left to diffuse for 10−30 min. Diffusion alone requires 2−4 h in brains of *Calliphora* and *Musca*.

b) Intracellular injection. This is combined with intracellular recordings using a glass microelectrode. Muller and McMahan (1976) describe the following technique to inject leech neurons using 2% HRP in 0.1 M KCl filtered with a 0.22 µm pore filter. Electrodes had resistances of about 100 MOhm. For ejection a pressure pulse of 0.3−1 atm (30−100 kPa) was applied to the open end of the barrel for 5−30 s during recording and repeated after 5 min. The preparation was left for 30−90 min so the enzyme could spread through the injected neuron (for injection apparatus see Kater et al. 1973). Eckert and Boschek (1980) injected HRP intracellularly into fly neurons by applying a pressure of 2 atm for 10−30 s.

Several authors have used intracellular iontophoresis of HRP (see reviews by Spencer et al. 1978; Tweedle 1978). Basically, an electrode with a tip of 0.5−2 µm is inserted into the cell and a current of about 20−500 nA is applied. With 50−150 nA a few minutes of ejection is enough (Spencer et al. 1978). This is followed by 10−30 min for the enzyme to spread through the cell. HRP concentrations ranging from 0.5% (Spencer et al. 1978) to 25% (Cullheim and Kellerth 1976) have been used, but many authors recommend lower concentrations and a microcapillary tip bigger than 1 µm. To my knowledge only one other peroxidase-active protein has been used for intracellular marking: cytochrome c (Labhart and Nässel, in prep.). For more details on intracellular HRP injection see Graybiel and Devor (1974), Jankowska et al. (1976), and Snow et al. (1976).

4. HRP Flooding of Intact Ganglia. Since intact nerve cells incorporate HRP especially in peripheral nerves, simple flooding of ganglia and/or intact peripheral nerves will label selected neurons (Fig. 13). In the fly brain and thoracic ganglia this can be done by applying one or two drops of HRP to the desired region in the living animal. Since the enzyme is also carried in the hemolymph, it labels other peripherally projecting neuron populations elsewhere. In this simple fashion the method displays several such projections. The exposed tissue is covered with vaseline and allowed 3−6 h for uptake. Addition of dimethyl sulfate (Keefer 1978) does not seem to improve this labelling in flies. Lysolecithin is not recommended since it appears to vesiculate and disrupt membranes.

5. Intramuscular Injection. The first demonstration of HRP uptake and transport was after intramuscular injection of the enzyme (Kristensson and Olsson 1971). This method relies on HRP being taken up by intact neurons at the neuromuscular junctions by recycled transmitter vesicle membrane (endocytosis) and active transport through the axon (Zacks and Saito 1969; Kristensson and Olsson 1971; Heuser and Reese 1973). Some labelling might also be due to axons damaged by the injection. In larger animals (and big muscles) the enzyme solution (0.5−1 µl) can be injected from a Hamilton syringe. In smaller muscles, such as those of insects, a glass micropipette with a tip broken to 5−20 µm can be used for pressure injection, as

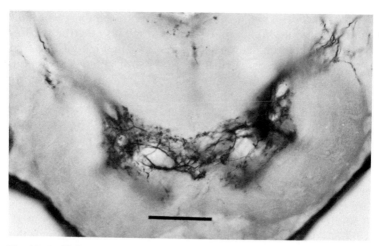

Fig. 13. Projections of maxillary-labellar sensory afferents into suboesophageal ganglion labelled after flooding an intact *Musca* brain with HRP.
Routine histochemistry with osmium. Bar = 50 μm

described above. Coggshall (1978) pressure injected HRP into muscles of
Drosophila and let the enzyme spread through the axons for 16−20 h. A
much simpler but cruder method for intramuscular uptake is to penetrate
the desired muscle fiber(s) with a fine needle and apply a drop of protein
solution on the muscle. After sealing with vaseline 2−4 h is sufficient for extensive uptake (Fig. 14).

The enzyme can spread to neurons in adjacent muscle fibers or en passant into axons in the same nerve. These artefacts can be minimized with
shorter uptake times or by using cytochrome c or myoglobin instead of HRP
since these proteins are only taken up by injured neurons.

6. The Use of Enzyme Pellets. Enzyme pellets (Griffin et al. 1979) can be inserted into small lesions. In flies pellets have been used in thoracic ganglion
lesions to demonstrate descending and ascending connections with the brain
and in peripheral lesions such as in retina, ocelli, legs, and muscles. An advantage of pellets is a high localized concentration of protein giving extensive localized uptake. Uptake times are the same as for solutions.

7. Labelling of Larval Neurons. Axons in the larval nervous system of Diptera
can be filled with HRP by removing the CNS and immersing it in a small
well of wax or vaseline filled with a 2% HRP solution in fly saline (Case
1957; O'Shea and Adams 1981). Although many peripheral nerves are transected, many imaginal ones remain intact, connected to their discs. One use
of this method is to study the ingrowth of imaginal retinal photoreceptors.

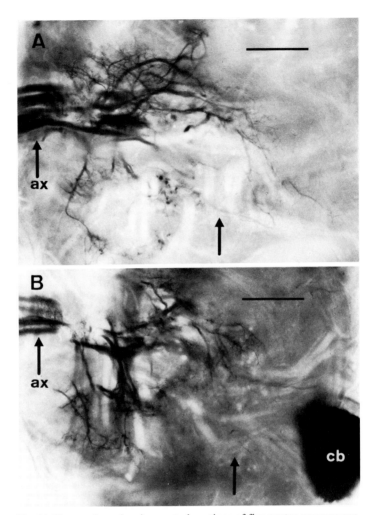

Fig. 14. Two sections showing central portions of five motor neurons supplying the dorsal longitudinal flight muscles after intramuscular injection of HRP (*Calliphora*). Labelling reveals motor neuron axons (*ax*), large cell bodies [one contralateral cell body (*cb*) is shown] as well as profusely branching processes in the mesothoracic neuromere. HRP with 5% Nonidet-P4O. Routine histochemistry with osmium. *Arrow* points anteriorly. Bar = 50 μm

The optic stalk, connecting the eye imaginal disc to the brain, is cut (the disc is still attached to the brain elsewhere). After 1.5 h uptake at 4 °C labelling of the transected photoreceptors occurs anterograde to the CNS and retrograde to the optic disc (Fig. 15). Presumably, other neural projections can be studied in a similar way.

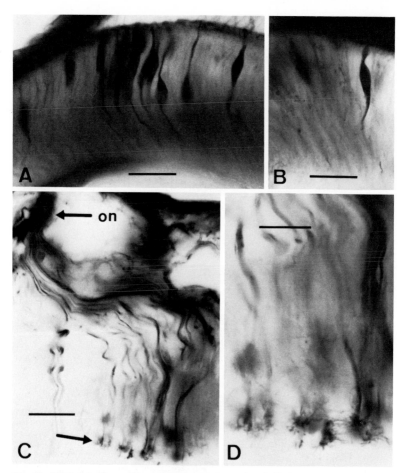

Fig. 15. HRP labelling of larval photoreceptors in *Calliphora*. Optic stalks of third instar larvae were transected between eye-imaginal disc and brain and the cut ends immersed in HRP solution. **A** and **B** selective labelling of backfilled photoreceptor cells in the eye disc (retinal anlage). **C** forwardfilled photoreceptor axons entering the brain through the optic stalk "nerve" (*on*) before radiating out into the lamina anlage (first optic neuropil) where they terminate (*arrowed*). Numerous filopodia of endings are shown enlarged in **D**. Bar **A–C** = 25 μm; **D** = 10 μm

Admixtures to the Enzyme Solutions to Improve Labelling

Most of the compounds reported to improve HRP labelling have been tested on the fly nervous system, but only a few significantly enhance labelling. They can work in one of three ways: (1) to facilitate tissue penetration or entry into cut axons (DMSO, lysolecithin and NP-40), (2) to induce endocytosis (protamine and adenosine), and (3) to promote protein transport in the axon (lectins, nerve growth factor, and liposomes).

In the fly primary labelling of intact and damaged neurons is "cytoplasmic" or "diffuse" and shows great structural detail. Additives of type (1) and (3) merely shorten the uptake time needed for labelling.

The most common additive is DMSO (dimethyl sulfoxide), a chemical often used to increase tissue penetration of other compounds. A 2% DMSO solvent for HRP was recommended by Keefer (1978) to increase the number of labelled cells and increase the amount of reaction product. In experiments on flies neither DMSO nor the detergent Nonidet P-40 (Lipp and Schwegler 1980) in a 5% solution made much difference. On the other hand a 3% solution of lysolecithin (1-lauroylpropanediol-3-phosphorylcholine, Serva) significantly improved the rate of enzyme uptake by injured and intact neurons and increased transneuronal labelling by HRP (see also Frank et al. 1980; Nässel 1981, 1982). Lysolecithin is strongly recommended for other heme proteins to shorten uptake times necessary for detailed resolution of nerve cells. Since lysolecithin is known to vesiculate membranes (see Lucy 1970; Poole et al. 1970), it should be used to study long projections (e.g. between appendages and CNS or between brain and thoracic ganglia) to a site distant from that of protein application.

Jirmanova et al. (1977) found 0.1 M protamine sulfate essential to induce uptake of HRP by endocytosis in mouse skeletal muscle. In flies, protamine added to the HRP only marginally improved uptake. Spencer et al. (1978) found that addition of 0.05% adenosine enhanced HRP labelling, but this has not yet been tested on flies. Preliminary results show that 25 mM $La(NO_3)_3$ in the HRP solution enhances uptake in flies especially by photoreceptor cells. Heuser and Miledi (1971) showed that a much lower concentration of La^{3+} causes depletion of neurotransmitter vesicles at the motor endplate. Thus an increased membrane turnover caused by lanthanum ions could result in more HRP taken up by endocytosis.

Improvement of axonal transport relies on binding HRP to other molecules, such as lectins or nerve growth factor, or trapping the protein inside liposomes. Except for preliminary experiments using liposomes, these molecules have not yet been tested with HRP on insects. For further information see Schwaab (1977), Gonatas et al. (1979), Brushart and Mesulam (1980), Staines et al. (1980) and Anderson et al. (1981).

Histochemical Procedures

Histochemical reactions, performed after enzyme uptake and tissue fixation, reveal the distribution of exogenous enzyme. A vast literature on these procedures describes varieties and modifications of all parameters, such as choice of fixative and buffers, concentration and choice of chromogen and substrate, timing, pH, and reaction temperature. The schedules here are standard for use on insects, and the procedures can be applied to intact ganglia or tissue slices. Subsequent histological treatment can aim for light microscopy, electron microscopy or whole-mount observation.

1. Fixation of Tissue. The fixation procedure is the same for all the techniques. After buffer rinses to remove excess extrinsic protein, the tissue is fixed in aldehyde. Since paraformaldehyde seems to irreversibly impair enzyme reactivity (Adams 1977; Malmgren and Olsson 1978), it has been more or less abandoned. Some workers still include it in low concentration (0.5−1.0%) with glutaraldehyde. Glutaraldehyde alone (1.5−4% in buffer) is recommended since it does not lower enzyme reactivity if tissue is thoroughly washed in buffer after fixation (Adams 1977; Malmgren and Olsson 1978). Fixation for 3−4 h in 2.5% glutaraldehyde in buffer gives excellent preservation of fly nervous tissue for electron microscopy without lowering enzyme reactivity. Buffered formaldehyde alone or in combination with glutaraldehyde gave inferior results.

The buffer, pH, and osmolarity should be adjusted to the particular tissue and species used. Within limits these variables probably make little or no difference to the subsequent enzyme reaction. Three different buffers with 2.5% glutaraldehyde (and 6% sucrose to adjust osmolarity) have been successfully used on Diptera: (1) 0.1 M PIPES buffer (Salema and Brandao 1973; Bauer and Stacey 1977), (2) 0.1 M phosphate buffer, (3) 0.16 M sodium cacodylate buffer; all at pH 7.0−7.4. I recommend cacodylate buffer since it can be used in the subsequent histochemical procedures. Before the enzyme reaction can be performed the tissue must be thoroughly washed in buffer. Malmgren and Olsson (1978) recommend routinely leaving specimens overnight in buffer at 4 °C and washing once more before tissue processing. All excess aldehydes are removed from the tissue by washing it for 30 min in 0.1 M $NaBH_4$ in buffer. This step, however, can be skipped if tissue is in buffer overnight. The methods described below were performed on entire ganglia (or brains) dissected out and with external trachea and air sacs removed. The methods for frozen sections or tissue slices are slightly different and are described later.

2. Enzyme Reaction in Whole Tissues Using DAB. This enzyme reaction is designed for HRP but is usable for the other peroxidase-active proteins. The best chromogen is DAB if the preparations are to be used for electron microscopy. Tissue is preincubated in a solution of 0.04% DAB in either 0.12 M Tris-HCl (pH 7.6), 0.16 M sodium cacodylate buffer (pH 7.0) or 0.1 M phosphate buffer (pH 7.2) in the dark at 4 °C for 1−3 h to ensure complete DAB permeation. DAB penetrates tissue slowly; its rate in lacrimal gland is estimated at 20 μm in 2 h (Herzog and Miller 1972; see Essner 1974).

Preincubation is followed by a 1- to 2-h incubation in fresh DAB solution (as above) with 2 drops of 3% H_2O_2 per 5 ml at room temperature and in the dark. Sodium cacodylate buffer is the optimal carrier, and the pH should be 5.5−7.0. Malmgren and Olsson (1978) recommend pH 5.5, having tested a wide pH range, but in the fly no significant difference was seen in this range. The reaction is stopped by immersing the tissue in several changes of buffer.

Following Lundquist and Josefsson (1971) H_2O_2 can be substituted by 0.2 mg ml^{-1} glucose oxidase (Sigma, Type V) and 0.2 mg ml^{-1} D-glucose with the above DAB concentration for 3 h at 4 °C. The solution should be oxygenized to ensure rich production of H_2O_2 in the reaction catalyzed by glucose oxidase.

3. Enzyme Reactions in Whole Tissues Using Chromogens Other Than DAB.
Two further techniques for resolving extrinsically applied peroxidases have been tested on insects. These use chromogens that are more sensitive than DAB, but have drawbacks in insect preparations intended for electron microscopy.

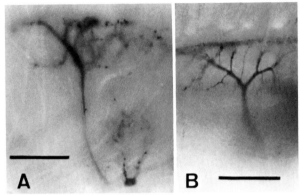

Fig. 16. Comparison between PPD-PC (**A**) and DAB histochemistry (**B**) of a transneuronally filled Col A neuron in the lobula. *Musca* (see text). Bar = 50 μm

a) Hanker et al. (1967) and Carson et al. (1980) introduced p-phenylene-diamine mixed with pyrocatechol (PPD-PC) as a chromogen. Although the method is very sensitive and turns HRP-containing neurons black, it did not give consistent results on fly ganglia or brains. Neurons were labelled intensely even after short incubation times. However, the reaction product often appeared to be partly washed out of the labelled neurons into the extracellular space (Fig. 16). This might be due to excessive enzyme reaction on the large amounts of heme protein in the cell.

The procedure eventually adopted is as follows. Incubation is in 0.08% PPD-PC (Sigma) in 0.12 M Tris-HCl (with 6% sucrose) at pH 7.6 with 2 drops of 3% H_2O_2 per 5 ml for 1 h. Preincubation in PPD-PC without peroxide was unnecessary. The reaction product is osmophilic, like the one achieved with DAB.

b) Mesulam (1976, 1978) described a method for histochemical demonstration of HRP using tetramethyl benzidine (TMB) as a chromogen. This method has been used only in preliminary experiments for electron microscopy (Carson and Mesulam 1981) but was extremely sensitive when applied to frozen sections of vertebrate nervous tissue (Mesulam and Rosene 1979). TMB has been tried in preliminary experiments on fly nervous system (whole brains) using the following procedure (after Mesulam 1978).

1. Three washes (5 min each) in distilled water with 8% sucrose.
2. Incubation for 1.5 h at room temperature in solutions A + B. Solution A contains 92.5 ml distilled water, 100 mg sodium nitroferricyanide and 5 ml 0.2 M acetate buffer pH 3.3. Just before use add solution B which contains 5 mg TMB (Sigma) in 2.5 ml absolute ethanol. (To dissolve TMB in ethanol heat solution to 40 °C.)
3. Run the enzymatic reaction for 1 h at room temperature in the above solution (with freshly added TMB) with 3 ml 0.3% H_2O_2 per 100 ml solution.
4. Stabilize for 1 h at 0−4 °C in a mixture of 45 ml distilled water, 50 ml absolute ethanol, 9 g sodium nitroferricyanide, and 5 ml acetate buffer (pH 3.3).

If a blue precipitate forms in the incubation step with TMB and H_2O_2 replace the solution with a fresh one. To avoid dissolving reaction product in an ethanol series a "one-step" dehydration (Muller and Jacks 1975) can be used: Immerse the tissue in three washes (7 min each) of HCl-acidified 2,2-dimethoxypropane (Merck), and then go via propylene oxide into araldite or

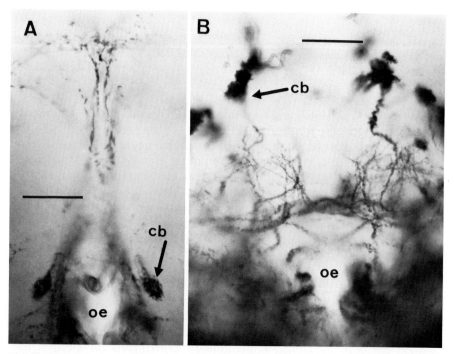

Fig. 17. Tetramethyl benzidine (TMB) histochemistry, on neurons originating in the brain, backfilled with HRP from thoracic ganglia. Micrographs are from whole mounts cleared in methyl salicylate and embedded in Canada balsam after dehydration in 2,2-dimethoxy propane. *Calliphora*. Note grainy labelling of cell bodies (*cb*) and neural processes, especially in axons of the medium bundle (*A*). *oe* oesophagus. Bar = 50 μm

via methyl salicylate to whole mounts in Canada balsam (Fig. 17). The reaction product using TMB-nitroferricyanide as the chromogen is not brown-to-black but bluish. Counterstaining with a 1% aqueous solution of neutral red has been combined with TMB histochemistry.

The procedure is modified for electron microscopy (Carson and Mesulam 1981). Dissolve TMB in distilled water instead of ethanol and stabilize (step 4) in non-alcoholic solution. After stabilizing and washing, osmication can be performed. In this step the temperature and pH is critical. Carson and Mesulam (1981) found that the osmium tetroxide should have a pH of 6.0 and that the osmication should be at 45 °C. The reaction product is crystalline, electron-dense, and insoluble.

Other chromogens in Table 1 have not been tested on insects, since they are not sensitive enough for detailed resolution of nerve cells.

4. Subsequent Processing for Whole Mounts and Enhancement of Reaction Product Contrast. When DAB is the chromogen and OsO_4 fixation is not intended (for whole mounts or frozen sections) the contrast of the reaction product can be enhanced (Fig. 18) by immersing the tissue or sections in a dilute solution of cobalt chloride (0.5% cobalt chloride in 0.12 M Tris-HCl) for 30 min before preincubation in DAB (for sections 10 min is enough) (Adams 1977). The preincubation in DAB and the enzyme reaction can thereafter be performed as usual (tissue should be rinsed in buffer between cobalt infiltration and DAB).

For whole mounts the brains or ganglia should be stripped of as much of the tracheae and air sacs as possible. After dehydration the tissue can be cleared in methyl salicylate and/or xylol and embedded in cavity slides using Permount or Canada balsam (Fig. 17).

5. Subsequent Processing of Sections for Light and Electron Microscopy. After the enzyme reaction the tissue is thoroughly washed in buffer before osmication. Osmium treatment is recommended for two reasons. First, OsO_4 reacts with the osmophilic oxidized DAB polymers (osmium black reaction shown in Fig. 4) enhancing the labelling contrast (Seligman et al. 1968; Hanker et al. 1967) as shown in Fig. 18. Second, if tissue is not fixed with osmium it shrinks substantially during dehydration. This shrinkage distorts the shapes of the labelled neurons (axons and other processes became wiggly as shown in Fig. 18 A). With osmium fixation the neuronal profiles resemble those found after Golgi impregnation or labelling with cobalt or fluorescent dyes (Fig. 18).

After three 10-min washes tissue is fixed in 1% OsO_4, for 2 h at 4 °C, carried in phosphate or cacodylate buffer. The osmium fixation is followed by three 10-min washes in buffer, dehydration, and embedding. Dehydration can either be performed by the "one-step" method, mentioned in connection with TMB histochemistry, or by using a graded ethanol series ending in 2,4-propylene oxide. Spurr's resin (Spurr 1969) or araldite were found the most

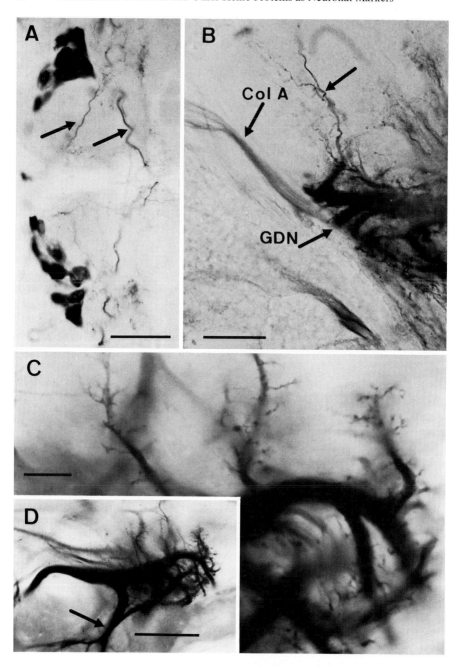

useful and versatile for light and electron microscopy sections. The components were mixed to give a soft-to-firm resin that could be cut both thick (25−40 μm) and thin (silver to grey). Complete tissue penetration and polymerization is critical if the soft polymer is to be cut at both thicknesses without tissue fragmentation.

A sliding microtome (e.g. Reichert-Jung, Nussloch F.R.G) with steel knives can be used for thick sections. I recommend using an infrared lamp to heat the resin block while cutting thick sections. This lowers the risk of cracks forming in the tissue. The sections are then soaked briefly in xylol and mounted in Permount under cover slips. Thick sections can also be mounted in resin between acetate sheets (see Chap. 1, this vol.), but with some loss of optical quality.

Thick sections that have been analyzed and documented can be "re-embedded" for electron microscopy as follows:

1. Remove the cover slip (or part of it).
2. Soak the section briefly with xylol.
3. Lift it off with fine forceps and place it for some minutes in xylol.
4. Soak the section (2×30 min) in unpolymerized resin.
5. Place the section in unpolymerized resin between two polyvinyl acetate sheets (e.g. sheets used for overhead projection) and press the sheets together to flatten the section in a thin film of resin.
6. Polymerize the resin.
7. Remove the upper acetate sheet and put a resin-filled Beem capsule (Balzers Union) flat end down on the section.
8. Again polymerize the resin.
9. Excess resin is trimmed off around the Beem capsule and the acetate sheet is stripped off. The thick section is now firmly mounted on the large end of the resin block and, after appropriate trimming, can be cut for electron microscopy.

Routine sectioning is followed by staining for 30 min in 2% uranyl acetate in 50% ethanol. Sections are washed in distilled water and contrasted for 20 min in 0.3% lead citrate in distilled water (with NaOH) after Venable and Coggeshall (1965).

Fig. 18. A−D Improvements of contrast and resolution of HRP-labelled *Musca* descending neurons, backfilled from thoracic ganglia. **A** Histochemistry omitting osmium treatment. Note the wiggly axons (*arrows*) and low contrast. Only cell bodies and larger axons are clearly shown. Bar = 50 μm. Dorsal is here to the left. **B** Part of the giant descending neuron (GDN) primarily labelled with HRP and contiguous with transneuronally labelled Col A axons originating in the lobula. Osmium omitted. HRP reaction product intensified with CoCl$_2$ increases contrast and resolution but still the wiggliness remains (*arrow*). Bar = 50 μm. **C, D** Part of a giant descending neuron as it appears after "routine histochemistry" (as in Appendix 1) with osmium treatment. Wiggliness is almost gone. Contrast is greatly increased, enabling resolution of very fine dendritic spines (enlargement, **C**). *Arrow* indicates ventral dendrite (see Fig. 8 E). Bar **C** = 50 μm, **D** = 10 μm

6. Histochemistry on Frozen Sections and Tissue Slices. Most studies on vertebrate CNS employ sections, since reagents do not penetrate deeply into large pieces of nervous tissue. Although these techniques are usually unnecessary for small insect brains, a brief description of available techniques is included (see also LaVail and LaVail 1974; Adams 1977; Hanker et al. 1977; Malmgren and Olsson 1978; Mesulam 1978; and Mesulam and Rosene 1979).

Fixatives and buffers are the same as for reactions in blocks. Frozen sections are cut on a cryostat or tissue slices are cut with a vibratome or tissue chopper from fixed, washed tissue.

a) Frozen sections. Tissue is mounted in O. C. T. mounting medium on a specimen holder and rapidly frozen in isopentane contained in liquid nitrogen. To avoid damage from ice crystal formation tissue can be first infiltrated through a graded sucrose series (6%, 10%, 20%, 30%) at 4 °C for 24 h (see Adams 1977). Frozen tissue is sectioned at e.g. 25 μm on a freezing microtome or cryostat. Sections are mounted on albumin-coated slides and air-dried. They can then be stored in buffer at 4 °C in Coplin jars.

b) Tissue can also be cut with a vibratome or a tissue chopper at the desired thickness (about 100 μm). Sections can be stored in buffer. They are treated histochemically floating in a container.

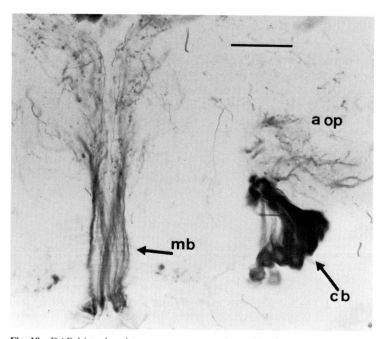

Fig. 19. DAB histochemistry on cryostat section of brain showing median bundle neurons (*mb*), cell bodies (*cb*) and dendrites of the anterior optic tubercle (*a op*). HRP filling from *Calliphora* thoracic ganglia, contrast enhancement by CoCl$_2$. Bar = 50 μm

1. Preincubation (optional) in 0.04% DAB in buffer (Tris-HCl, phosphate or cacodylate) for about 10 min.
2. Incubation in a solution of 10 ml 0.04% DAB (in buffer) and 4 drops of 3% H_2O_2 for 20 min. The sections can be observed and the reaction stopped when appropriate staining has been achieved.
3. Wash in buffer, dehydrate, and mount with Permount under cover slip.

For sections I recommend incubation in 0.5% $CoCl_2$ in Tris-HCl for about 5 min before preincubation in DAB. This substantially increases labelling contrast. Weakly labelled profiles appear grey or bluish grey; more densely labelled structures are brown to black (Fig. 19).

7. Counterstaining Peroxidase-Labelled Preparations. Counterstaining labelled neurons is best done on unembedded sections. In general any stain compatible with glutaraldehyde-fixed frozen sections can be used. More elaborate techniques, such as Bodian's reduced silver stain (Adams 1977), have also proved useful. Toluidine or methylene blue can be used as for routine staining of semithin araldite sections (see Osborne 1980). Thick plastic sections (e.g. 25 μm) have to be flattened and stuck to the microscope slide on a hot plate (60 °C). The dye should be left on the sections for only a few seconds to avoid overstaining. The intrinsic contrast of osmium-fixed tissue is usually sufficient to resolve larger unlabelled profiles and general cytoarchitecture of neuropils.

Safety Precautions

Benzidine and its derivatives are potential carcinogens and should be handled carefully. Utensils and working space exposed to DAB or other chromogens should be washed with 5% sodium hypochlorite (laundry bleach), which denatures the benzidine (Adams 1977). Sodium cacodylate, the recommended buffer, should be handled with care. It is a potent poison. Other chemicals such as osmium tetroxide and components of resins constitute health hazards that should not be overlooked.

Methods Combinable with Heme-Protein Labelling

This section describes the principal neuroanatomical and histochemical methods that can be combined with HRP and other heme-protein tracers (see also Fig. 26). More details are found in the original papers listed as examples.

1. Golgi-HRP. Combined HRP and Golgi impregnation joins two powerful anatomical techniques, one labelling selected neurons, the other stochasti-

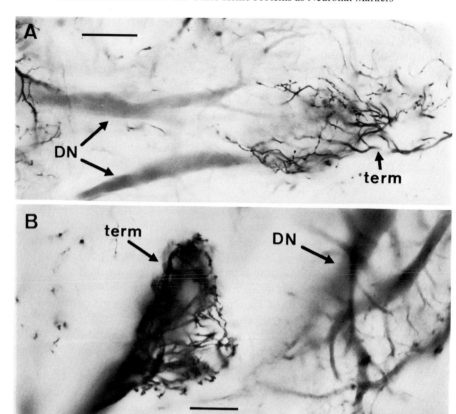

Fig. 20 A, B. Double-labelling with HRP and Golgi method 2 (see text). HRP backfilled from *Calliphora* thoracic ganglia labelled descending neurons (*DN*) originating in the brain. Histochemistry was followed by Golgi impregnation. A convergence between terminals (*term*) of Golgi-impregnated lobula neurons and HRP-labelled dendrites (see Fig. 21); in **B** the HRP and Golgi-labelled cells never overlap. Note the obvious difference of density of the two markers: Golgi labelling is dark brown or black, HRP is pale to medium brown. Bar **A** = 50 µm; **B** = 25 µm

cally impregnating nerve cells (Fig. 20 A, B). With a bit of patience and luck one might find Golgi-impregnated neurons contacting HRP-filled cells for electron microscopical analysis (Somogyi et al. 1979).

Three variants of the Golgi method have been used with HRP (see also Chap. 6), all after peroxidase histochemistry. The methods listed below use tissue washed in buffer after the DAB-H$_2$O$_2$ reaction.

Method 1

a) Fixation in 1% OsO$_4$ in 0.16 M sodium cacodylate for 2 h at 4 °C.
b) Wash in buffer.

c) Wash in 2.5% K$_2$Cr$_2$O$_7$ 3×10 min.
d) Incubate in 25% glutaraldehyde and 2.5% K$_2$Cr$_2$O$_7$ (1:4) for 2 days.
e) Incubate in 0.75% AgNO$_3$ for 2 days.

Method 2

a) Wash in 2.5% K$_2$Cr$_2$O$_7$.
b) Fixation in 0.5% OsO$_4$ in 2.5% dichromate for 3 h at 4 °C.
c) Wash in 2.5% dichromate.
d) and (e) as in Method 1

Method 3

a) Wash in 2.5% K$_2$Cr$_2$O$_7$.
b) Incubate in 2% OsO$_4$ and 2.5% dichromate (1:9) for 2 days.
c) Incubate in 0.75% AgNO$_3$ for 2 days.

The incubation steps can be repeated to stain more neurons. The HRP reaction product and Golgi precipitate can easily be distinguished in both the light and the electron microscope (Figs. 20A, B and 21). In the light microscope HRP-labelled neurons appear light-to-dark brown; Golgi-impregnated cells, black. In the electron microscope the HRP reaction product is granular and confined to cell membranes and organelles (Figs. 21, 22), whereas the Golgi precipitate is much more electron-dense and normally fills the neuron.

2. Cobalt-HRP (see Chap. 6). These two markers are applied in a controlled fashion by filling cut axons. HRP requires longer uptake times than cobalt

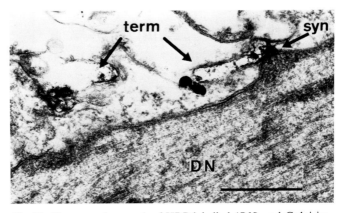

Fig. 21. Electron micrograph of HRP-labelled (*DN*) and Golgi-impregnated neurons (*term*). The original light microscopy preparation was similar to that for Fig. 20A. Note the difference between granular HRP reaction product and black Golgi precipitate, some of which has been heat-displaced due to electron beam exposure. *Arrow* indicates probable synaptic site (*syn*). Bar = 0.5 μm

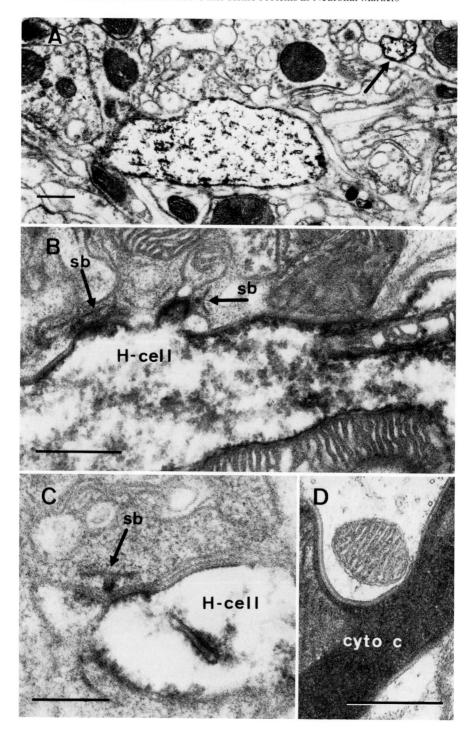

and should be introduced first. After brief $(NH_4)_2S$ exposure to precipitate the CoS, tissue is fixed and processed for demonstrating peroxidase activity. Silver intensification of CoS precipitate is performed afterwards (see Chap. 6). A drawback of this technique is that silver intensification can occur at HRP-filled profiles as well.

3. HRP Combined with Fluorescent, Radiolabelled and Other Tracers. Most fluorescent dyes used for intracellular labelling and axon filling (see Chap. 7) can be combined with HRP; e.g. Lucifer yellow and HRP used to separately fill identified neurons in leech ganglia (Macagno et al. 1981; DeRiemer and Macagno 1981). A number of other compounds used to fill cut axons might be useful for "double marking." Examples are: bisbenzimide, Evans blue, 4,6-diamino-2 phenylindole 2 HCl (DAPI), granular blue, nuclear yellow, propidium iodide, and "true blue" (Kuypers et al. 1979; Björklund and Skagerberg 1979; DeOlmos and Heimer 1980; Yezierski and Bowker 1981). Fluorescein- or tetramethylrhodamine-tagged HRP have also been used as a neuronal marker (Hanker et al. 1976). Sections are analyzed by fluorescence microscopy.

Other double marking on vertebrate CNS has employed iron dextran (Olsson and Kristensson 1978) or radioactive tracers (e.g. enzymatically active or inactive ^3H-HRP, tritiated amino acids or ^3H-N-acetyl-D-galactosamine: Geisert 1976; Hayes and Rustiani 1979; Colwell 1975; Thompson and Bailey 1979), but are of doubtful use and resolubility on insect CNS.

4. HRP Labelling and Neuronal Degeneration. This combination has been used only on vertebrates, using electron microscopy (Beckstead 1976; Somogyi et al. 1979) or light microscopy employing Fink-Heimer silver staining (Blomquist and Westman 1975). Degenerated axons can be identified in insect CNS (Griffiths and Boschek 1976; Schürmann 1980) and there should be no drawback to applying this approach to its analysis.

5. HRP Labelling Combined with Cytochemical and Immunological Methods. In vertebrates, HRP labelling has been combined with histochemical or immunological methods that display sites of enzymes or putative neurotransmitters. These methods include reactions for acetylcholine esterase (AChE)

Fig. 22. Electron microscopy of HRP (**A**−**C**) and cytochrome c (**D**) labelled neurons in *Musca.* Note association of reaction product with the inner phase of the plasmalemma, microtubules (**A**), postsynaptic sites and outer mitochondrial membranes (**B**). A Medulla tangential neurons transneuronally labelled after thoracic ganglia backfilling. Note the resolution of even small profiles (*arrow*). Bar = 0.5 μm. **B** Lobula plate H-cell dendrite transneuronally labelled. *Small arrows* indicate table-shaped presynaptic bar *sb.* Bar = 0.5 μm. **C** Higher magnification of H-cell dendrite postsynaptic to unlabelled cell. Bar = 0.2 μm. **D** Even distribution of reaction product in dendritic spine of mesothoracic leg motor neuron after backfilling. Bar = 0.5 μm

(Lewis and Henderson 1980), cytochrome oxidase (Livingstone and Hubel 1983), and catecholamine fluorescence techniques (Blessing et al. 1978; Berger et al. 1978). For AChE localization HRP histochemistry is followed by reaction for AChE, postfixing in osmium tetroxide, and electron microscopy (Lewis and Henderson 1980).

Catecholamine fluorescence employs either of two approaches.

a) HRP histochemistry is followed by induced fluorescence on vibratome sections by glyoxylic acid followed by formaldehyde vapor (Berger et al. 1978).

b) Fluorescence is induced by tissue perfusion with formaldehyde and glutaraldehyde (the FAGLU technique). Analysis of catecholinergic neurons precedes HRP histochemistry (Blessing et al. 1978). HRP labelling of nerve cells, combined with immunological demonstration of tyrosine hydroxylase as well as serotonin, was used by Ljungdahl et al. (1975) and Priestly et al. (1981). All these methods should be usable on insect CNS.

6. HRP and Autoradiographic Techniques. In principle, it should be possible to combine other autoradiographic methods with HRP histochemistry: binding of toxins, uptake of transmitters and their analogs, as well as ^{3}H-thymidine labelling. The latter has been used to determine the birth dates of neurons later identified with HRP filling (Nowakowski et al. 1975).

Cytology of Peroxidase-Labelled Neurons

First I will describe the histological features of peroxidase-labelled neurons and how to identify reaction products of extrinsic heme proteins. This is followed by the results of different filling techniques, including the phenomenon of transneuronal labelling. Finally, problems of interpretation and artefacts are discussed.

In the fly nervous system both intact and damaged neurons accumulate enzyme. After histochemistry and osmium staining reaction products (rp) are seen distributed diffusely in cytoplasm, along microtubules, and at the inner phase of the plasmalemma (Fig. 22). Rarely is rp detected inside cellular organelles. In insects, labelling was cytologically identical for the seven heme proteins tested. Different uptake times are required to produce the same rp density. This is probably due to the different peroxidase activities of the proteins (Nässel, 1983).

Depending on the uptake time, the rp density varies. The darker the neuron appears in the light microscope, the more diffuse the electron microscopical appearance of rp in the cytoplasm. Typically, labelled neurons are brown to black, resembling Golgi-impregnated cells but usually with darker background. This Golgi-like labelling shows fine detail of processes, neurites, and cell bodies, making possible the resolution of entire neurons. One drawback is that occasionally rp is only seen in processes distal to the uptake site. This may be due to an actual distal shift of the enzyme or to insufficient penetration of reagents (DAB and H_2O_2), or both.

Only labelled neurons that are clearly resolved in the light microscope can be easily seen at low magnification in the electron microscope (Fig. 22) and identified with certainty as being labelled. HRP and other peroxidases are good markers for synaptological studies since pre- and postsynaptic structures of the labelled neurons are not obscured by rp as can be the case with e.g. Golgi-impregnated neurons (but see Chap. 1, this Vol.).

In flies no parts of neurons were found to react with DAB and H_2O_2 in such a way that they could be mistaken for labelled neurons without being exposed to exogenous peroxidases. Intrinsic activity could, however, sometimes be detected in fine grains throughout some synaptic neuropils. These are possibly related to mitochondria and cannot be confused with labelling achieved by application of extrinsic enzymes. Controls should, however, be made to screen for intrinsic activity. These simply involve reacting fixed brains, unexposed to extrinsic enzyme, with DAB and H_2O_2.

1. Labelling After Filling Peripheral Nerves, Sensory Fields and Appendages

HRP has proved valuable in studying wild type and homeotic mutants of *Drosophila* (Ghysen 1978; Green 1981), demonstrating central projections of sensory neurons and leg motor neurons. In larger flies fairly complete primary projections can be demonstrated with HRP and other heme proteins, such as the central projections from antennae shown in Figs. 8 and 9.

Neurons other than the sensory or motor neurons of the severed appendage (e.g. leg) often take up HRP that has leaked into the thoracic cavity. Also, intact axons proximal to the cut can incorporate HRP. Severing the ocellar cups fills receptors and ocellar interneurons with HRP, the latter being resolved in great detail in the brain. Retinal fills with HRP can also label intact relay neurons to optic lobe neuropils (Fig. 23). To exclude these artefacts cytochrome c or myoglobin should be employed to label only lesioned neurons (Nässel et al. 1981; Nässel 1983).

2. Labelling After Uptake Through Lesions in the CNS

Applying peroxidases to lesions in the thoracic ganglia fills descending and ascending neurons connecting these regions to cerebral ganglia (Nässel 1981; Nässel et al. 1981) and allows quantitative analysis of their cell bodies and their morphology in the brain (Figs. 11, 17−20). When using HRP this type of filling also transneuronally labels certain optic lobe interneurons. Applying HRP to lesions in the brain fills neurons leading to the thoracic ganglia (Fig. 12). This experiment gives less consistent results because it is more difficult to sever all the ascending and descending neurons in the brain. Also unspecific HRP leakage occurs more readily from the brain, into the head capsule, and then to the thoracic cavity than vice versa.

Applying HRP to small lesions in the brain is appropriate for studying contralaterally projecting nerve cells (Fig. 10) with long axons. Enzyme tends to diffuse from the lesion site. Dense reaction product obscures neural components close to it.

3. Labelling After Heme Protein Injection

Extracellular injection of heme proteins labels relay neurons in fly brains (Eckert and Boschek 1980) if current or pressure is used to eject the enzyme from the micropipette. Passive diffusion from the micropipette results in poor labelling, most probably because HRP will not readily diffuse from a fine-tipped pipette. A fairly recent use of HRP is to mark a cell after intracellular recording. Response properties and the light microscopic anatomy of the cell can be followed by electron microscopic analysis of it and its neighbours (Muller and MacMahan 1976; Eckert and Boschek 1980).

4. Labelling After HRP Flooding of Intact Ganglia and Nerves

Flooding a ganglion or brain with HRP easily labels certain populations of neurons outside the blood-brain barrier. Exposing intact fly brains to HRP labels neurons of the maxillary-labellar nerve (Fig. 13), antennal sensory axons, ocellar neurons, and retinal photoreceptor axons. Intact sensory and motor neurons of leg nerves take up HRP after flooding thoracic ganglia.

5. Intramuscular Injections of Marker

This "classical" technique also works well on insects, as first shown by Coggshall (1978). The protein-uptake mechanism by the neuron at the neuromuscular junction is not known but could, in part, be accounted for by endocytosis. In the small muscles of Drosophila and other Diptera the motor neurons may be damaged during injection, allowing uptake into the neuron by direct diffusion. In any case all motor neurons from the injected muscle readily take up the HRP backfilling their cell bodies and the central arborization. HRP can also fill sensory axons accompanying motor neuron axons en passant. When HRP is injected into dorsal longitudinal flight muscles, peripherally synapsing interneurons (see Coggshall 1978; King and Wyman 1980) take it up (see Fig. 14). This does not occur with cytochrome c or myoglobin.

6. Transneuronal Labelling with HRP

Filling lesioned axons of descending neurons from the thoracic ganglia reveals their processes in the brain in great detail. In addition, arrays of certain neurons intrinsic to the brain take up HRP although they have not been mechanically damaged. This "transneuronal" labelling reproducibly displays nerve cells closely apposed (sometimes presynaptic) to primarily filled neurons (Nässel 1981, 1982). Although we do not know the mechanism of this phenomenon, it is a convenient means of labelling some central neurons without damaging them (Figs. 23–25). This labelling indicates how many neurons there are of a certain kind if the relevant cells all take up the marker (Figs. 23 and 24). HRP filling, like cobalt backfilling, (see Strausfeld and Obermayer 1976) can indicate connections between descending neurons and their optic lobe inputs (see Fig. 18 B), especially when resolved by electron microscopy (Fig. 22). Possible transneuronal uptake of HRP, particularly in brains that have been less studied, can cause problems of interpretation. Initially, shorter uptake times or other markers (such as cytochrome c) which do not fill neurons secondarily should be used.

7. Artefacts and Problems in Interpretation

Some most commonly encountered artefacts and interpretational difficulties are summarized below.

a) *Filling.* Some pathways may be particularly difficult to demonstrate possibly because their neurons reseal and fail to incorporate the marker enzyme. Another problem is that pathways can flood with excess enzyme especially when filling severed appendages or sensory organs. Addition of 3% lysolecithin can help to overcome poor filling and in systems that easily flood the uptake time should be reduced or another heme protein should be used. The latter is also relevant in pathways where transneuronal passage of HRP occurs.

b) *Histochemistry.* Difficulties in histochemical processing are often related to the condition of the reagents, especially DAB since it auto-oxidizes at room temperature (accelerated by light) and its specificity and chromogenic potency deteriorates. A deteriorated solution of DAB shows a black precipitate and general darkening. Partly oxidized DAB gives less intensely colored HRP-containing profiles. Also, the tissue is darker and grainy. The grains might represent mitochondria containing peroxidase- or oxidase-active proteins reacting with the now less specific oxidized DAB (see also LaVail 1978).

In larger blocks of tissue rp is found only in superficial neurons. Impregnation can be improved by increasing incubation times or by decreasing the block size. Reactions on frozen sections or tissue slices largely eliminate

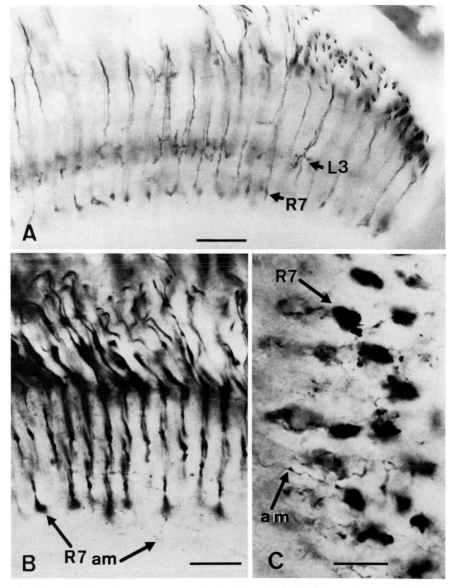

Fig. 23. Transneuronal labelling from HRP-filled retinal photoreceptors. Routine histochemistry, with osmium. **A** Medulla terminals of long photoreceptor axons (R 7), primarily labelled with HRP and 3% lysolecithin, and secondarily labelled L 3 lamina monopolar
cells. Bar = 25 μm. **B** R 7 endings and transneuronally labelled amacrine cells (*am*) in the
medulla. Bar = 10 μm. **C** Cross section of R 7 and amacrines. Bar = 10 μm

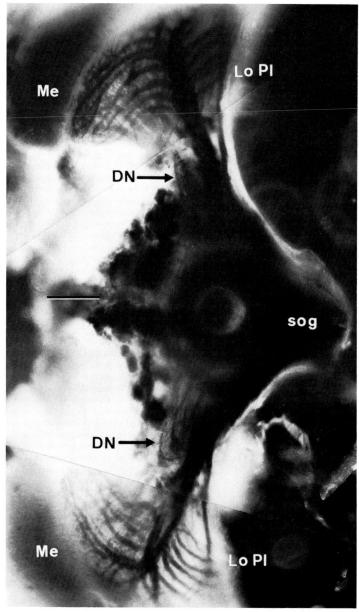

Fig. 24. Dark-field micrograph of *Musca* brain whole mount. Lobula plate H- and V-cells are transneuronally HRP-labelled after backfilling descending neurons from thoracic ganglia. The whole mount omitted osmium or CoCl₂ intensification and was cleared in xylol and embedded in araldite. *DN* descending neurons; *LoPl* lobula plate; *Me* medulla; *sog* suboesophageal ganglion. Bar = 100 μm

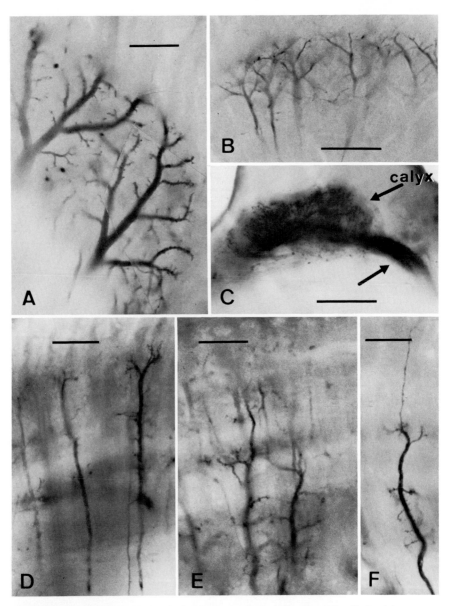

Fig. 25. Transneuronally HRP-labelled neurons. **A** Lobula plate H-cell dendrites trans-neuronally labelled after thoracic ganglia backfilling. *Calliphora*. Bar = 25 μm. **B** Dendrites of similarly transneuronally labelled Col A cells in *Calliphora's* lobula. Bar = 50 μm. **C** Terminals in the calyx of interneurons projecting via the antennoglomerular tract (*arrowed*) from the antennal lobes. Primary label was into the antennal nerve afferents into the antennal lobe. **D–F** "Tertiary" labelling of medulla columnar neurons after thoracic backfilling with HRP and 3% lysolecithin. Bar = 25 μm

partial staining. Longer incubations should be performed at 4 °C, in the dark, renewing reagents every hour.

Postfixational diffusion of heme protein or enzyme rp occurred using PPD-PC as a chromogen, but not DAB. It appears as a brown halo around a filled profile, but did not "label" other neurons. Possibly the enzyme or its reaction product shifts inside the filled profiles.

Without osmium fixation substantial tissue shrinkage is seen after embedding and sectioning. Curiously, HRP-labelled profiles appear to shrink less than surrounding tissue. This alters their shape. Axons, dendrites, and other processes appear wavy where they are straight in fresh tissue (see Fig. 18). Osmium fixation eliminates this artefact. In frozen sections and whole mounts, the amount of shrinkage depends on how tissue is processed.

c) *Lysolecithin.* Long uptake times using HRP and lysolecithin can result in lysolecithin diffusing centrally and disrupting membranes to give unspecific labelling.

Mechanisms of Neuronal Uptake and Transport of Heme Proteins

Relatively little is known about the mechanisms of uptake and transport of heme proteins and other macromolecular markers by neurons. Data collected over the last 10 years or so sheds some light on these processes. Phenomena underlying the final labelling of a neuron can be divided into three components: (1) the entry of the marker into the neuron, (2) the passage of the marker from the entry site to other parts of the neuron, (3) the final fate of the marker inside the cell. Mechanisms underlying each of these apparently depend on experimental conditions and possibly on the cell system studied. The mechanism of marker entry and transport probably also depends on what part of the neuron is exposed to it.

The nervous system has barriers against diffusion of large molecules such as heme proteins. These are the perineurium and tight junctions between cells. However, once inside these barriers, markers can spread extensively in the intercellular spaces (Kristensson and Olsson 1973; Dunker et al. 1976). Some parts of the nervous system have no such diffusion barriers: intramuscular (Kristensson and Olsson 1971), intraventricular, intraperitoneal, and intravenous injections of HRP result in uptake by nerve cells (see Broadwell and Brightman 1976, 1979; Yu 1980).

Possible Mechanisms for Uptake by Intact Neurons

HRP molecules are commonly believed to enter a neuron by means of endocytosis through its membrane. Although endocytosis can occur anywhere along the membrane, it is most vigorous (and best studied) at the axon terminal where it has been related to membrane recycling from synaptic vesicles (Holtzmann et al. 1971; Heuser and Reese 1973; Turner and Harris 1973; Zimmermann 1979). Uptake thus simply represents liquid phase endocytosis of HRP due to membrane retrieval resulting in vesicle-enclosed marker. Membrane recycling and uptake of HRP can be increased by stimulation of transmitter release as shown by the above cited authors and Teichberg et al. (1975) and Schacher et al. (1974). Depending on the rate of transmitter release, some vesicles may be directly re-used as transmitter vesicles while others might move elsewhere in the neuron (Teichberg et al. 1975; Zimmermann 1979) accounting, in part, for retrograde HRP transport (Teichberg et al. 1975; Holtzmann 1977). This is supported by increase of retrograde HRP transport after stimulating transmitter release (Litchy 1973; Singer et al. 1977) in the presence of high extracellular HRP concentrations.

There are some indications that HRP uptake is not linked only to synaptic activity. One is that HRP can be taken up by other parts of the neuron (Holtzmann and Peterson 1969; Olsson and Hossman 1970; Turner and Harris 1974; Broadwell and Brightman 1979; Kristensson and Olsson 1978). Also, cells without synaptic activity, such as ependymal cells or cultured neuroblastoma cells, take up HRP by endocytosis (Mensah et al. 1980; Chan et al. 1980). Uptake is not entirely unspecific. Different heme proteins are incorporated differently into neurons. Of the heme proteins, catalase, lactoperoxidase, hemoglobin, HRP, myoglobin, cytochrome c, and microperoxidase, only HRP is taken up in detectable amounts by intact neurons (Malmgren et al. 1978; Nässel et al. 1981; Nässel 1982, 1983). Chan et al. (1980) reported some endocytosis of lactoperoxidase by neuroblastoma cells. Furthermore, the basic isoenzyme C of HRP is taken up much more readily by intact nerve cells than its acidic isoenzyme A (Bunt et al. 1976; Bunt and Haschke 1978; Giorgi and Zahnd 1978). The two differ in charge and carbohydrate moiety; one or both of these factors could account for differences in uptake (Bunt et al. 1976; Malmgren et al. 1978).

Uptake by Injured Nerve Cells

When cut or crushed axons are exposed to HRP, it first diffuses through the cut or leaky membrane. Cytochemistry shows a solid mass of HRP in the axon near the uptake site, whereas a few millimeters distant, HRP reaction product is incorporated in organelles as when taken up by an undamaged neuron (Kristensson and Olsson 1976). Initial influx seems to be non-specific since all the other tested heme proteins enter and label injured neurons in the fly CNS (Nässel 1983). Usually, "diffuse" distribution of rp as opposed to "granular" membrane-enclosed product is sufficient indication that the labelled neuron was injured (see Keefer 1978; Vanegas et al. 1978). However, "diffuse" labelling can also occur without detectable mechanical damage (see Broadwell and Brightman 1979; Nässel 1981, 1982).

Transport of Marker by Intact Neurons

HRP incorporated at axon terminals is transported retrogradely to the cell body. HRP incorporated into dendrites or the cell body is transported to the axon terminal. Retrograde HRP transport occurs at a rate of $2-3$ mm h^{-1} or $48-72$ mm day^{-1} in rabbit sciatic nerve (Kristensson 1975) and $72-84$ mm day^{-1} in the chick visual system (LaVail 1975). As a comparison, retrograde transport of acetylcholine esterase occurs at a rate of $5.6-10$ mm h^{-1} (Lubinska 1975). The anterograde transport velocity is estimated to be twice that of the retrograde. For example in the mammalian nervous system the anterograde transport rate is $150-200$ mm day^{-1} (Kristensson 1978) and is thought to involve agranular endoplasmic reticulum (LaVail and LaVail 1974).

In vertebrates, HRP is probably carried retrogradely in vesicles (diameter $100-125$ nm), multivesicular bodies, cup-shaped organelles, and small tubes of agranular endoplasmic reticulum (see LaVail 1975; Chan et al. 1980). Mechanisms of transport are not clear but are dependent on metabolic energy since uncoupling of oxidative phosphorylation by 2,4-dinitrophenol blocks transport. Also, the integrity of neurotubules and neurofilaments seems important since transport is impaired by application of vinblastin and colchicine (Kristensson and Olsson 1973). In flies, however, 2,4-dinitrophenol, vinblastin, and colchicine did not completely block HRP passage inside neurons (Nässel 1983). Transport of HRP inside the cell is the same for enzyme taken up by liquid phase endocytosis and adsorptive endocytosis (Goude et al. 1981).

Transport in Damaged Neurons

As noted earlier, there is an initial large unspecific influx of protein marker into the cytoplasm, but further along the axon specific retrograde transport occurs inside organelles as it does in intact cells (Kristensson and Olsson 1976). Since the unspecific influx can extend a few millimeters into the axon, neurons with short axons may become labelled entirely in this manner when they are injured. This seems to be the case in insects since no neurons could be detected that were labelled in a granular fashion (Nässel 1981, 1983; Nässel et al. 1981). However, it is not clear whether this represents passive diffusion or active transport of protein.

Fate of HRP in Nerve Cells

In mammals HRP taken up by nerve cells cannot be detected cytochemically after $3-4$ days. The marker is thought to end up at the perikaryon in multivesicular bodies forming secondary lysosomes in which HRP is enzymatically degraded (Kristensson and Olsson 1973; Kristensson 1978; Broadwell and Brightman 1979). The fate of HRP in insect neurons has not been studied.

Transneuronal Passage of Marker

HRP transneuronally labels the same neurons as cobalt ions (Strausfeld and Obermayer 1976; Hausen and Strausfeld 1980). Although cobalt ions and the fluorescent dye Lucifer yellow (Stewart 1978) may pass directly between nerve cells coupled by gap junctions (see Politoff et al. 1974; Bennet 1973b; Stewart 1981; Strausfeld and Bassemir 1983), HRP is probably too large (MW 40,000) to pass through an intact gap junction. Certain insect gap junctions allow passage of molecules of up to MW 1800 (Flagg-Newton et al. 1979). Two possible routes for HRP transfer between neurons are:

1. Uptake of HRP by injured neurons is accompanied by its leakage or release from dendrites into a local extracellular pool from which adjacent intact axon terminals incorporate HRP. This model does not require transneuronally labelled neurons to be in synaptic contact with primarily filled nerve cells although they may be. It does require a selectivity of secondary uptake since very few types of neuron are transneuronally labelled.

2. Extracellular diffusion of HRP along certain diffusion pathways is followed by selective secondary uptake by intact neurons in contact with these pathways. This model requires: (a) compartmentalization of the extracellular space of the central nervous system and (b) eventual cellular incorporation of most of the HRP since little extracellular HRP reaction product can be resolved with the electron microscope. In this model HRP could label synaptically coupled cells if these share extracellular compartments.

Concluding Remarks

As pointed out in the introduction, HRP and other heme proteins have proved extremely useful and versatile for vertebrate neuroanatomy and are gaining popularity in invertebrate neurobiology. Heme proteins are suitable for detailed analysis using light and electron microscopy. The compatibility of heme protein labelling with a large number of other anatomical, cytochemical and immunological methods (Fig. 26) suggests that in the years to come these substances will figure prominently in neuroanatomical research.

Acknowledgements. I thank Harjit S. Seyan and John A. Berriman for skillful technical assistance and helpful suggestions. The experimental work was done in the laboratory of N. Strausfeld and he as well as my other colleagues are gratefully acknowledged for fruitful discussions. Hilary Anderson kindly read and commented on the manuscript. I am grateful for the photographic assistance of K. Christensen, J. Stanger and A. Wechsler.

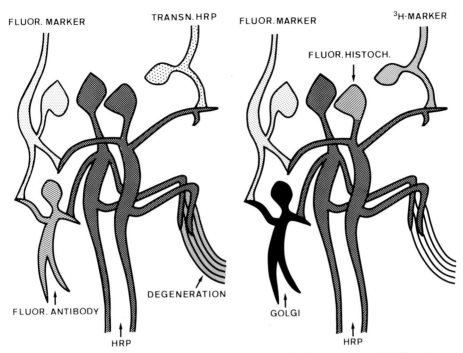

Fig. 26. Summary of techniques that can be combined with heme protein labelling (see text)

Addendum

Since this chapter was submitted further publications have appeared that contain detailed descriptions of the HRP labelling technique (mainly as applied to vertebrates) and improvements in the histochemical detection of the marker (Mesulam 1982; Liposits et al. 1982; Gallyas et al. 1982).

Liposits et al. (1982) and Gallyas et al. (1982) report a method for intensifying DAB-HRP reaction products by deposits of metallic silver (for light microscopy) and substituting this with metallic gold for electron microscopy. Possibly this technique would be useful for intensifying weakly labelled neurons in insects after transneuronal HRP uptake.

An exhaustive description of labelling techniques combinable with HRP and other heme proteins is by Steward (1981). One new double labelling technique, useful for both light and electron microscopy, employs separate injection of HRP and Lucifer yellow to demonstrate convergence of neurons (see also Figs. 8, 9, Chap. 6). If intensely illuminated by blue light Lucifer yellow will photo-oxidize DAB to a reaction product (Maranto 1982). This can be distinguished from that of HRP in the electron microscope.

A novel application of HRP labelling in insects has been performed on the ant *Cataglyphis bicolor* in order to demonstrate the projection of special regions of the retina into the optic lobes (Meyer 1983). A tiny drop of 2% HRP in 0.1 M KCl was forced out of the fine tip of a glass microcapillary mounted on a hypodermic syringe and left to air dry. The solid pellet, now at the tip of the capillary, was advanced into the medulla neuropil by means of a micromanipulator. After 2 h long visual fibres from the retina were filled by retrograde HRP passage from the vicinity of the pellet.

There are several new reports of transneuronal HRP uptake (see Gomez-Ramoz and Rodriguez-Echandia 1981; Hongo et al. 1981; Lechan et al. 1981; Triller and Korn 1981; Gerfen et al. 1982; Itaya and VanHoesen 1982; Ruda and Coulter 1982; Wilcynski and Zakon 1982). These occur between neurons. However, transfer of HRP from neurons to associated glia cells occurs in the gastropod *Aplysia* (Goldstein et al. 1982).

From some of the cited papers it appears that mechanisms of transcellular HRP passage is by means of localized exo- and endocytosis. This transfer seems to require substantial membrane contacts between primarily and secondarily labelled cells. It remains to be demonstrated, however, that transneuronal HRP passage occurs only at specific synaptic junctions, as seems to be the case for cobalt and Lucifer yellow passage between certain neurons.

Appendix 1: Method

Routine Schedule for HRP Uptake and Histochemistry

The following was designed for HRP but is also used for other peroxidase-active proteins.

1. Make a fresh 2% solution of HRP (e.g. Sigma, Type VI) in 0.1 M KCl. Add 3% lysolecithin (Serva) to optimize HRP uptake (optional).
2. Apply HRP as described. (Uptake of other peroxidase active proteins is twice as long).
3. Fix nervous tissue for 4 h in 2.5% glutaraldehyde and 6% sucrose in 0.16 M sodium cacodylate buffer. Then 4 washes in 6% sucrose in cacodylate buffer 3×10 min and leave in buffer at $4\,°C$ overnight.
5. Final dissection of tissue. Brain or ganglia should be well exposed.
6. Preincubation in DAB (BIOHAZARD). Use 0.04% DAB in cacodylate buffer (pH 7.0) with 6% sucrose. Incubate 1.5 h on a rotator at room temperature and in the dark. For larger pieces of tissue leave in DAB for 3 h at $4\,°C$.
7. Enzyme reaction. Use 10 ml of the above DAB solution (fresh) with 4 drops of 3% H_2O_2. Incubate for 1.5 h on rotator at $20\,°C$ in the dark. For larger tissue incubate $3-4$ h at $4\,°C$ and then bring tissue to $20\,°C$ and run the reaction for a further 1 h.
8. Wash in cacodylate buffer (3×10 min).
9. Fix in 1% OsO_4 in cacodylate buffer for 2 h at $4\,°C$.
10. Wash in buffer (3×10 min).
11. Dehydrate and embed in araldite or Spurr's medium.

For whole mounts, tissue should be dehydrated after step 8. A 30 min immersion of tissue in 0.5% $CoCl_2$ (in 0.12 M Tris-HCl) before step 6 enhances contrast of the DAB reaction product. This is followed by washes (3×10 min) in Tris-HCl.

Variations in steps 1–9 of the routine schedule

1. HRP can be dissolved in distilled water, insect saline, 0.2 M potassium citrate, or 0.12 M Tris-HCl (and many other buffers or salines).
3. Instead of cacodylate buffer, 0.1 M PIPES or 0.12 M phosphate buffer can be used. Fixation can be performed at 0–4 °C.
6. Instead of cacodylate 0.12 M Tris-HCl or phosphate buffer (pH 7.6) can be used.
7. Tris or phosphate buffer can be used. The concentration of H_2O_2 can be varied to achieve optimal results. Dimethyl sulfoxide (2%) added to the incubation solution may improve DAB permeation. H_2O_2 can be substituted for glucose oxidase (Sigma, Type V), 0.2 mg ml^{-1} and D-glucose, 0.2 mg ml^{-1} in the DAB solution; incubation time 3 h at 4 °C.

Appendix 2: Solutions

Buffer Solutions and Other Chemicals

1). 0.16 M sodium cacodylate buffer (poison)

Add 34.24 g sodium cacodylate to 1 l distilled water. Adjust pH with 1 M HCl.

2). 0.12 M Tris-HCl

Stock solution: Add 14.53 g Tris salt to 1 l distilled water. Take 250 ml of the stock solution. Adjust to pH 7.6 with 1 M HCl. Bring to 1 l with distilled water. Tris salt = Tris(hydroxymethyl amino methane).

3). 0.1 M PIPES

Add 30.23 g PIPES to 1 l distilled water. Adjust pH with 1 M NaOH. Initially use NaOH pellets to rapidly bring pH close to 7.

4). 0.2 M Sodium acetate

Mix 20 ml 1 M sodium acetate and 19 ml 1 M HCl. Add distilled water to 100 ml. Adjust pH to e.g. 3.3 with 1 M HCl.

5). 0.1 M Phosphate buffer (Sorenson)

Solution A: 27.8 g NaH_2PO_4 in 1 l distilled water.
Solution B: 53.6 g $Na_2HPO_4 \cdot 7 H_2O$ in 1 l distilled water.
pH 7.2: Add 28 ml A to 72 ml B. Add water to 200 ml.
pH 7.6: Add 13 ml A to 87 ml B. Add water to 200 ml.

Araldite mixture for combined light and electron microscopy

I use Durcupan-ACM-Fluka mixed as follows. Component A/M: 10.8 g; component B: 8.9 g; component C: 0.5 g; component D: 2.0 g. Resin was polymerized for 24 h at 60 °C.

Chapter 4

Intracellular Staining with Nickel Chloride

Fred Delcomyn

Department of Entomology
University of Illinois, Urbana, Illinois U.S.A.

Introduction

The development of methods for the staining of individual neurons (especially intracellular staining) has undoubtedly been the most important technical advance in invertebrate neurobiology since the introduction of the glass microelectrode. Publication of the original methods for staining with Procion yellow (Stretton and Kravitz et al. 1968), cobalt chloride (Pitman et al. 1972) and Lucifer yellow (Stewart 1978) was followed not only by many papers describing applications of these techniques to particular preparations, but also by papers describing ways to increase the effectiveness or versatility of the techniques (e.g. Kater and Nicholson 1973; Strausfeld and Miller 1980). The widespread use of these staining techniques has undoubtedly been aided by the fact that different specific methods (using metallic ions, fluorescent dyes, or other markers) are available, since this means that in circumstances where one method is difficult to apply, another may sometimes be substituted. In this article another variation on established methods is described, the use of nickel instead of cobalt ions for injection.

Among insect neurobiologists, the substances used most commonly for intracellular staining have been cobalt and Lucifer yellow. Although both can give excellent definition of neuronal branching patterns, they both also have certain drawbacks, drawbacks that I encountered in a study of cockroach interganglionic interneurons. The difficulty with cobalt-filled electrodes is that their well-known tendency to block (i.e. to fail to allow current to pass) is for unknown reasons even greater in cockroaches than in many other insects. The difficulty with Lucifer yellow is that extremely long fill times are required in order to get good definition of branches of interneurons in adjacent ganglia as far apart as are cockroach thoracic ganglia (about 3–4 mm). Therefore, when a colleague suggested $NiCl_2$ as a possible alternative substance for injection (described for molluscs by Quicke and Brace 1979), I tried it. In the cockroach, at least, this turned out to be a happy choice. Tests revealed that nickel-filled electrodes not only have lower resistances (comparable to electrodes filled with 1 M potassium acetate), but also pass current more readily than electrodes filled with $CoCl_2$, making them easier and more reliable to use (Delcomyn 1981).

Method

Nickel chloride is used as an electrolyte in electrodes just as cobalt chloride is. Solutions of 250 mM $NiCl_2$ (about 3%) prepared in distilled water yield good electrodes. Other concentrations would presumably also give good results. An electrode containing a glass filament along its length can have its tip filled by dipping the open end of its shaft in the solution of $NiCl_2$ for a minute or so. Nickel can be ejected from the electrode by passing pulsed depolarizing (positive) current, at a rate of about 1 pulse every 500 ms, each pulse lasting about 250 ms. Electrodes tolerate the passage of greater amounts of current under this schedule than they do if the current is passed continuously, or if it is pulsed with durations over 500 ms. Good electrodes with resistances as high as 30 MOhm usually allow the passage of 10–14 nA of current without difficulty. Higher currents are tolerated by some electrodes, but more commonly, attempts to pass more than 15–16 nA per pulse result in irreversible block of the electrode within a few minutes.

Electrode block is the most severe problem associated with the use of cobalt-filled electrodes. Nickel-filled electrodes suffer from the same problem, but not nearly to the same degree. Many electrodes in the 20–30 MOhm range can pass current for as long as 30 min with only moderate increases in resistance. Electrodes with lower initial resistances perform even better (meaning that even fewer show any block). In fact, my experience with the cockroach *Periplaneta americana* is that sometimes the animal itself seems to be a more important variable than the electrode, since in some individual insects no electrode seems able to pass current easily.

Although nickel-filled electrodes block less readily in cockroaches than cobalt-filled ones, once they do block it is nearly impossible to continue to use them. Cobalt-filled electrodes can often be unblocked at least temporarily, by passing brief hyperpolarizing pulses through them, so that small amounts of cobalt can slowly be injected even from an electrode that does not readily allow current to pass. This procedure rarely works with nickel-filled electrodes. An effective strategy to avoid wasting time on what may turn out to be a bad electrode is briefly to test the current-carrying capacity of an electrode as soon as it has penetrated the nerve sheath. Few electrodes that pass this test ever block subsequently.

Nickel spreads in neurons at rates comparable to those for cobalt (about 3–5 mm h^{-1}). Ten to 15 min is usually sufficient for good staining of all neurites of a neuron in the ganglion in which the neuron is impaled. One and a half to 2 h are required for adequate staining of neurites in adjacent ganglia. Unfortunately, preparations in which nickel is left to spread for times longer than 4−6 h rarely yield results as good as those in which fixation is begun before that time, and often yield no results at all. I interpret this observation to mean that neurons slowly lose nickel ions after the ions are injected, since by other criteria such long-term fills should have been excellent.

Nickel may be precipitated in a neuron by addition of either ammonium sulfide or rubeanic acid (Quicke and Brace 1979). Rubeanic acid ($C_2S_2N_2H_4$, a derivative of oxalic acid and also known as dithiooxamide), is nearly insoluble in water, but a small amount can be dissolved in 100% ethanol. Placing a few drops of saturated ethanolic solution on the exposed nerve cord in situ usually results in a deep blue-black precipitate becoming visible within a minute. Leaving the preparation for 5−10 min more ensures that most of the available nickel is precipitated (see also Chap. 5).

After being dissected out of the animal, the nerve cord can be treated with any of several different fixatives. I have found alcoholic Bouin's solution (1 h), Carnoy's fixative (1 h), glutaraldehyde (1 h) or paraformaldehyde (0.5 h) equally effective. After fixation, the tissue can be dehydrated and cleared for whole-mount viewing as usual. The precipitate formed by rubeanic acid is slightly soluble in alcohol (Altman and Tyrer 1980), so the tissue should not be stored in ethanol.

Whole mounts of nickel-filled neurons can also be Timm's intensified in the standard way (Bacon and Altman 1977). Precipitation with rubeanic acid offers a special advantage: it is possible to eliminate fixation and start intensification immediately after precipitating the nickel. (This may work because of the fixation provided by the ethanol in which the rubeanic acid is dissolved). Tissue to be intensified without specific fixation should be placed in distilled water for about 5 min after being dissected out of the animal, then transferred to water at 60 °C, and carried through intensification as usual. No differences in the quality of the fill are apparent in fixed as compared to unfixed tissue.

Inadvertently overintensified ganglia can conveniently be destained by dipping them briefly (10−30 s) in the destaining solution of Pitman (1979). A good method for destaining overintensified spots on the surface of a ganglion is to apply the destaining solution locally with a broken microelectrode, then rinse, and repeat application as necessary. An example of the fine detail of neuronal structure that it is possible to obtain with these techniques is shown in Fig. 1.

The impact on invertebrate neurobiology of the techniques for intracellular staining of individual neurons has been enormous. Part of the reason for this has been the continued elaboration and refinement of the methods to make them more versatile or easier to use. The cobalt technique is used extensively in spite of the well-known difficulties it presents. However, as shown by Delcomyn (1981) and discussed in this article, nickel can be as effective for intracellular staining as cobalt, while suffering only little from some of the difficulties presented by cobalt. Nickel-filled electrodes have re-

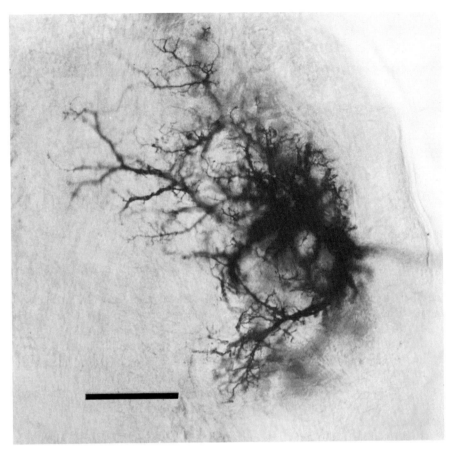

Fig. 1. Dorsal view in one plane of focus of a nickel-filled, silver-intensified motoneuron in the metathoracic ganglion of the cockroach *Periplaneta americana*. The cell soma is visible as a circular shadow at the anterior margin of the ganglion. The ganglion has been slightly destained (Pitman 1979) to enhance contrast. Bar = 100 μm

sistances similar to electrodes filled with potassium acetate, pass depolarizing current readily, and block infrequently in most preparations. Nickel-filled electrodes will certainly not eliminate all problems associated with staining individual neurons. However, the ease with which most such electrodes can be used should encourage investigators to try them in a variety of preparations.

Acknowledgements. The technique described in this chapter was developed while I was on sabbatical leave at the Department of Physiology, University of Alberta. I thank Dr. K. G. Pearson, in whose laboratory I worked, Dr. E. Saunders, who provided microscope facilities, and Dr. M. Schachter, head of the department, for their hospitality. Supported by NIH grant NS 15 632 to FD and an MRC grant to KGP.

Chapter 5

Rubeanic Acid and X-Ray Microanalysis for Demonstrating Metal Ions in Filled Neurons

Carole M. Hackney and Jennifer S. Altman

Institute of Science and Technology
University of Manchester
Manchester, England

Introduction

The staining of neurons with cobalt chloride, either by intracellular injection or infusion through the cut ends of axons, and subsequent precipitation of the black insoluble sulphide within the cells has become a well-established method for determining neuronal morphology (Strausfeld and Obermayer 1976; Altman and Tyrer 1980). Various other metal chlorides can also be used to stain neurons and the filling of different cells with different chlorides in the same preparation would enable the relationships between groups of neurons to be examined if some means of distinguishing between them could be found. Unfortunately the sulphides of the most commonly used metal ions, cobalt and nickel, are similar in appearance, and cannot always be distinguished.

Two possible techniques are currently available for differentiating between introduced metal ions in neurons. For light microscopy, the organic reagent, rubeanic acid (RA) gives complexes which differ in colour depending on the metal chloride used (Quicke and Brace 1979). For electron microscopy, X-ray microanalysis permits metals with atomic numbers above that of sodium to be located and identified. Both these methods have only recently been introduced, and their applications in insect neurobiology are still at a developmental stage. In this chapter, we describe our current work on both techniques and suggest their uses for extending the potential of the cobalt marking method, especially for the analysis of contacts between two identified neurons or sets of neurons.

Use of Different Metal Ions

Choice of Metals

For infusion through the cut ends of nerves in the locust (*Schistocerca gregaria*, Forskål), cobalt chloride ($CoCl_2 \cdot 6\,H_2O$) gives the best results when used at relatively low concentrations (1.5−5.0%), overnight (18−20 h) at low temperature (4 °C), or for shorter periods (2−3 h) at physiological temperatures (27 °C). Under these conditions, a large proportion of preparations fill successfully and artefacts, such as parallel filling (Altman and Tyrer 1980; Hackney and Altman 1982), are avoided. Using these parameters, only a faint precipitate is visible in the filled neurons after ammonium sulphide development, but subsequent silver intensification reveals a wealth of fine branching in the neuropil (Bacon and Altman 1977; Tyrer et al. 1980).

Several other metal chlorides have been tested using the same conditions. Nickel chloride ($NiCl_2 \cdot 6 H_2O$) produces fills which are similar to cobalt chloride in both extent and density (Chap. 4), whereas cupric chloride often fails to fill the whole neuron. Successful fills have not been obtained with ferric chloride. We have therefore selected nickel chloride as our second metal ion for staining. It can be used for peripheral infusion and intracellular injection at similar concentrations to cobalt chloride. It gives a blacker precipitate with ammonium sulphide than cobalt chloride and can be intensified with silver in exactly the same way.

Differentiation Between Metals with Rubeanic Acid (RA)

The following colours are obtained when metal ions are precipitated with RA.

Fe^{2+} Yellow
Co^{2+} Yellow
Ni^{2+} Blue
Cu^{2+} Olive   (Pearse 1960)

Mixtures of cobalt chloride and nickel chloride, depending on the proportions of the two metal ions, produce a range of colours from orange-pink (Co^{2+}, 95:Ni^{2+}, 5) → red (50:50) → purple (25:75).

In locust nerve preparations the yellow colour obtained with cobalt chloride alone is difficult to distinguish, as contrast with the yellowish tissue is poor. We prefer to use nickel chloride, which gives a deep blue precipitate, and a 50:50 mixture of cobalt and nickel chlorides, which gives a vivid red precipitate (Fig. 1). In other species, cobalt chloride alone or mixed in different proportions with nickel chloride [such as in the cricket *Gryllus bimaculatus* (Sakai and Yamaguchi 1983)] may be more appropriate depending on the background colour of the tissue after fixation.

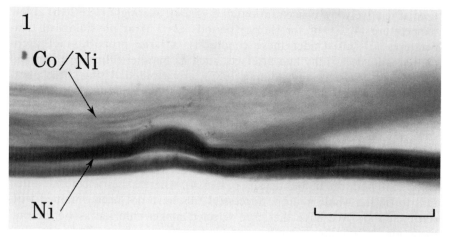

Fig. 1. A locust peripheral nerve preparation with one branch filled with 5% nickel chloride and the other with a 5% mixture of cobalt and nickel chlorides, developed with rubeanic acid. The nickel-containing axons are dark blue and those containing the cobalt/nickel mixture are red. No intermingling of ions or extra-axonal staining can be seen. Bar = 100 μm

Rubeanic Acid Development

The Reaction

The coloured precipitate produced when certain metals react with rubeanic acid (dithio-oxamide) is thought to be due to the formation of an inner complex salt of the di-imido form of rubeanic acid e.g. with copper

$$H_2N-\underset{\underset{S}{\|}}{C}-\underset{\underset{S}{\|}}{C}-NH_2 \rightarrow HN=\underset{\underset{S}{|}}{C}-\underset{\underset{\underset{\underset{Cu}{\diagdown\diagup}}{S}}{|}}{C}=NH$$

(Feigl 1956)

In alkaline solution, only Cu, Co and Ni salts give a positive reaction. If the medium contains both ethanol and acetate ions, Co and Ni rubeanates are soluble.

Method for Locust Nervous Tissue

Peripheral nerves are filled using standard methods (Altman and Tyrer 1980). Solutions of nickel chloride and a 50:50 mixture of cobalt and nickel chlorides at a final concentration of 5–10% (0.21–0.42 M) are used to produce deep colours with good contrast. Care must be taken, however, since artefacts such as parallel filling may occur at these concentrations (see p. 101). For light microscope whole-mount preparations, the sterna bearing the ganglia with stained nerves are cut out and placed in a watch glass containing insect saline. A few drops of a saturated ethanolic solution of rubeanic acid are then added. Development is controlled by observation under a binocular microscope and is usually complete within 10 min. Any flecks of orange rubeanic acid that have come out of solution are removed by washing in several changes of saline. The tissue is transferred to alcoholic Bouin's fixative for 2 h. [Bouin's fixative gives the tissue a yellowish colour which

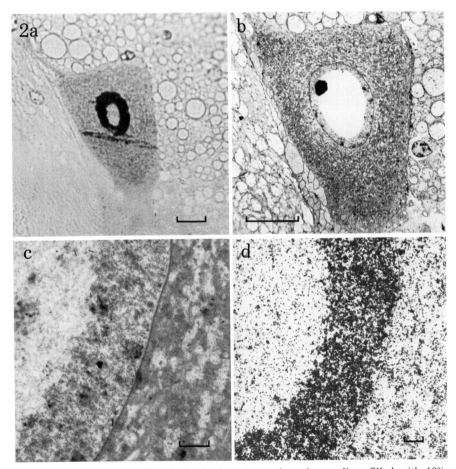

Fig. 2. A motor neuron cell body in the locust metathoracic ganglion, filled with 10% NiCl$_2$, is seen in a 3-μm intensified resin section (**a**), which was subsequently resectioned (**b–d**). In the ultrathin sections, the nickel deposit is not clearly visible (**b, c**), but when the section is reintensified, the silver precipitate indicates its location (**d**). Bars **a, b** = 10 μm; **c, d** = 1 μm

obscures the yellow complex formed by cobalt. If necessary, 70% ethanol can be used as a fixative or other colourless fixatives could be tried – see Altman and Tyrer (1980) for advantages and disadvantages of various fixatives]. The ganglia are dissected out in 70% ethanol, dehydrated in graded alcohols, cleared in methyl benzoate or salicylate and mounted in Canada balsam. Quicke and Brace (1979) note that Co- and Ni-rubeanic complexes may be unstable in acidic media and recommend buffering the saline to pH 8.0–8.5. We have never found this necessary, but it should be borne in mind if other fixatives are used. They also suggest that preparations should not be stored in alcohol for long periods.

For electron microscopy, the development procedure is the same except that 0.05 M sodium phosphate buffer containing 0.4 M sucrose (550 mOsm, pH 7.2) is used instead of saline. After development, the tissue is fixed in 2.5% glutaraldehyde in 0.05 M sodium phosphate buffer containing 0.2 M sucrose (pH 7.2) for 2 h at room temperature. The ganglia are dissected out in the last few minutes of glutaraldehyde fixation and postfixed in 1% OsO_4 in 0.05 M sodium phosphate containing 0.2 M sucrose (pH 7.2) for 1 h, dehydrated in an ethanol series and embedded in Spurr's resin (Spurr 1969).

Both whole-mount ganglia and semithin ($2-3 \mu m$) or ultrathin (70 nm) resin sections can be intensified by usual methods (see Quicke et al. 1980; Tyrer et al. 1980) (Fig. 2). This results in a brown or black precipitate of silver over the filled neuron that obscures any colours which were previously visible. It does, however, allow one to check that a neuron has been completely visualised.

Applications of Rubeanic Acid Development

Light Microscope Analysis

The most important advantage of the method is in the search for contacts between identified neurons, which is facilitated if two markers can be used. Neurons thought to be in contact can be filled with cobalt chloride, nickel chloride or a mixture of the two, through two intracellular electrodes, two peripheral nerves or a combination of electrodes and peripheral nerves. After RA development the two neurons will appear differently coloured, allowing any juxtapositions to be identified more easily.

A second application suggested by Quicke and Brace (1979) is the rapid identification of neurons with axons in more than one nerve root. If one root is filled with cobalt chloride and the other with nickel chloride, then neurons that have axons in both, such as the common inhibitors and dorsal unpaired median (DUM) cells, should appear red or purple as the ions mingle in the cell body, neurite and arborisation.

Filling different neurons with nickel or cobalt/nickel mixtures also provides a method of distinguishing between leakage and true transneuronal staining where one neuron can be filled directly with cobalt and filled trans-neuronally with nickel from another fibre. The double-filled neuron should appear pink or red, rather than yellow (Strausfeld, pers. comm.)

Precautions

Preparations made with metal chloride concentrations of less than 5% and developed with RA result in fills with poor contrast in whole-mounts, particularly when using cobalt chloride in locust preparations. If concentrations of 5–10% are used, preparations of similar contrast to unintensified, sulphide-developed preparations are obtained. It is possible to resolve branches down to 0.25 μm but with eye strain. As the concentration of the metal chloride is increased, the colours obtained with RA development are more intense, but the risk of contamination and artefacts caused by ions leaking from filled axons is greater. The main manifestations of this problem are a background haze of colour, artefactual staining of unintended neurons and neuron branches appearing swollen, often with huge ovoid "blebs." A compromise is therefore necessary: the distribution of filled neurons and areas of contact may be mapped using higher concentrations (10–15%), but a series using low concentrations (1.5%) followed by silver intensification must also be run as controls. Features common to both types of preparation may be considered valid.

Experimental Uses of Rubeanic Acid Development

As the location of different metal ions can easily be determined by observing the distribution of their coloured precipitates, RA development provides a tool for analysing the mechanisms of metal ion transport within neurons. The conditions causing leakage of ions from filled neurons can be examined and the course of ion uptake and transport through axons investigated.

1. Leakage and Artefactual Filling. Fredman and Jahan-Parwar (1980) and Slade et al. (1981) have described artefacts in molluscan preparations in which neurons became filled with cobalt even though they did not have cut ends in the cobalt solution, thus casting doubt on the accuracy of the method for defining neuronal pathways. Altman and Tyrer (1980) suggest that this parallel filling artefact is caused by using extreme preparation conditions, in particular high concentrations of cobalt ions and the application of current across the nerve to be filled; and it also results from damage caused by cutting nerves (Hackney and Altman 1982). To check whether this artefact occurs under conditions used routinely for peripheral filling in our laboratory (5% cobalt or nickel chloride for 18 h at 4 °C), we filled a sensory branch of the metathoracic wing nerve of the locust with 5% nickel chloride and a motor branch of the same nerve with a 5% mixture of cobalt and nickel chlorides.

In all preparations the motor neurons, which stained red, were clearly distinct from the sensory neurons, which appeared blue. There was no evidence for intermingling of ions either in the main nerve 1 (Fig. 1) or in the ganglion, indicating that ions were not leaking from axons and parallel filling was not occurring (Hackney and Altman 1982).

2. Sequential Filling. Little is yet known about the mechanism by which metal ions are taken up and transported along axons. One way to tackle this problem is to infuse with the metal chloride solution for a short time and observe its course through the neuron. Movement as a front would produce a sharp boundary, whereas diffusion would result in a concentration gradient. To do this experiment, a replacement for the test solution is required, preferably one with similar characteristics. Here cobalt and nickel are ideal as they have close atomic weights and structures and their sesquihydrate chlorides have the same molecular weight. If a nerve branch is placed in a solution of one chloride for a fixed period followed by another period in the other chloride, the distribution of the two ions can be monitored by the colours resulting from development with RA.

Sequential filling of a branch of the metathoracic wing nerve in the locust has produced results which indicate that both nickel and cobalt ions are transported along a gradient. Using nickel chloride first, followed by cobalt chloride, the motor neuron cell body and neurite, and the sensory projection in the ganglion were stained blue (nickel), while the peripheral nerve was stained purple (nickel > cobalt mixture) at the edge of the ganglion grading into red (nickel = cobalt mixture) and then orange-yellow (cobalt > nickel mixture) towards its cut end. Further experiments of this type are in progress.

Sectioning RA Material for Light and Electron Microscopy

Ganglia containing RA-developed, metal-filled neurons can be embedded in wax or resin and sectioned for light microscopy (LM) or transmission electron microscopy (TEM).

1. Wax Sections. Unintensified cobalt sulphide precipitates in 10-μm wax sections are very difficult to see in LM except in large profiles where the concentration is high (Tyrer and Bell 1974). We therefore assume that the same will be true of RA-developed material, although it is possible that the bright colours may make them more visible. The enhancement technique given below for resin sections may be useful.

2. Semithin (3–5 μm) Sections. Sections of resin-embedded material can be cut on a dry glass knife. Where high concentrations of metal ions are present,

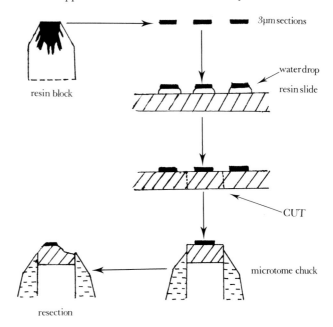

Fig. 3. The resin slide method for resectioning semithin resin sections, modified from Sigee (1976). See text

for example in a peripheral nerve, the colours can be distinguished with the LM. They can be enhanced by treating the sections with a few drops of saturated ethanolic rubeanic acid solution in 5 ml distilled water for 10–15 min. This is useful for identifying the profiles containing particular ions, as a guide for subsequent EM examination (see p. 104). The colours are faint, however, and we have not yet been able to distinguish them in neuron profiles within the CNS.

In order to identify profiles within a ganglion in semithin sections, it is necessary to intensify the sections with silver. The standard procedure described by Tyrer et al. (1980) is used (Fig. 2a). Once sections have been intensified, the distinction between the ions is lost.

3. Ultrathin Resin Sections. For TEM, either ultrathin sections can be cut direct from the block (Altman and Tyrer 1980) or semithin sections resectioned after LM examination (Phillips 1980). With both methods, it is necessary to intensify at least one thin section to locate the metal ions in the TEM, following the method described by Tyrer et al. (1980) (Fig. 2b–d).

4. Resectioning Technique. Of the variety of resectioning techniques in use we find a modification of Sigee's (1976) resin slide method the most reliable. As the process is a little involved, we give it in detail here (Fig. 3).

a) Make resin slides by pouring Spurr's resin into latex moulds cast from thick glass slides. Harden overnight at 60 °C.

b) Cut serial sections $2-3\,\mu m$ thick with a dry glass knife. Place on individual drops of water on a resin slide.
c) Place the slide on filter paper on a warm plate and cover with a Petri dish lid, to produce a chamber which has a high initial humidity. The sections expand and flatten slowly before drying down and sticking to the side.
d) Silver-intensify (Tyrer et al. 1980) and identify areas to be resectioned with the light microscope.
e) Saw the piece of resin bearing the selected section out of the slide to make a piece suitable for clamping in an ultramicrotome chuck.
f) Trim block face under a binocular microscope and resection using an ultramicrotome, carefully realigning the block face and the knife.

The silver deposited on the surface of the original semithin section does not penetrate into the block face and is scraped off during resectioning (see p. 110). The ultrathin sections may be silver-intensified again if held on polythene rings (Tyrer et al. 1980) to locate the position of the metal-containing profile in TEM (Fig. 2d). We have not observed any differences between nickel- or cobalt-containing profiles in the appearance or distribution of the silver precipitate in the TEM.

Advantages of Rubeanic Acid

1. Different metal ions give different-coloured complexes with RA. By using a variety of metal chlorides or their mixtures for filling, the relative positions and regions of interactions of neurons can easily be examined.
2. There are experimental advantages of obtaining a range of colours with different proportions of cobalt and nickel ions, since the movements of the different ions can be observed. This can be used to check for leakage artefacts or in experiments to determine the mechanism of metal ion transport.
3. Neurons filled with cobalt or nickel chloride and developed with rubeanic acid can be silver-intensified in whole-mount, semithin or ultrathin sections as readily as sulphide-developed preparations, in order to check morphological details.
4. Rubeanic acid is far more pleasant to handle than toxic, noxious-smelling ammonium sulphide as a method for obtaining an insoluble precipitate with cobalt or nickel ions. It can be used in the open laboratory and the progress of development can be monitored more easily and safely.
5. Rubeanic acid development does not appear to have a greater detrimental effect on the ultrastructure of locust nervous tissue than ammonium sulphide treatment, and so can be used generally as a substitute for ammonium sulphide.
6. Hazards. RA is not known to be toxic (unless taken orally) or carcinogenic. It does, however, stain skin indelibly black and should therefore be handled with gloves.

X-Ray Microanalysis for Detection of Metal Ions

X-ray microanalysis may be used to detect the position of cobalt ions in nervous tissue (Kirkham et al. 1975). By using X-ray microanalysis combined with standard TEM procedures for locating the ions, the advantages of using two different metals for analytical neuroanatomy can be extended to the ultrastructural level. X-ray microanalysis also provides a method for checking on procedures such as silver intensification and for experimental analysis of the mechanisms of metal ion transport through neurons (see also Chap. 2). These studies are in their infancy, and below we outline the advantages, limitations and potential uses of the technique.

Theory

X-ray microanalysis depends on the fact that every element has a unique X-ray energy spectrum by which it may be classified. Each atom has a set of electrons in orbits at well-defined energy levels. An incident electron may cause the ejection of an orbital electron, the gap created being filled immediately by an electron transferred from a shell of higher energy. An electron ejected from the innermost shell would be replaced by one from the next shell out. The X-ray photon given off by this transition has a specific energy and it is the configuration of energy levels in the atomic structure of each element that gives the unique X-ray energy spectrum.

When a specimen is bombarded by an electron beam in an EM, X-rays are given off. In a static probe transmission X-ray microanalytical microscope, such as the AEI Corinth microanalytical TEM (CORA), the electron beam is focussed by a third condenser lens to give a fine electron probe which ensures that only a small area of the specimen is generating X-rays at any time. Alternatively, a normal transmission image can be obtained so that a precise correlation between the fine structure of cells and tissues and the location of elements within them can be determined by a non-destructive method.

In CORA, the X-rays from all the elements present above sodium are received by a solid state energy dispersive detector and converted into electrical signals. A Link computer, which includes a multichannel analyser, plots the incoming data as histograms showing the X-ray spectrum from all the elements present. The X-axis in electron volts (eV) represents the elements present and the Y-axis the number of counts obtained for each. Lighter elements are not detected as efficiently as the heavier ones because the smaller size of the atoms means they get hit less frequently, and because lower energy X-rays are more often absorbed by the surrounding material of the specimen and do not pass so readily through the window of the detector.

Besides X-rays characteristic of the various elements, background or white radiation is generated simultaneously. This results from the deceleration of electrons passing through the atoms but not striking orbiting electrons. It is a function of all the elements present in the specimen, and it may be taken as a measure of section thickness for a specimen of relatively constant density (Hall 1971). This background count allows a comparison of results from different sections by using the peak-to-background ratio to indicate the relative amounts of various elements present. A complex extension of this procedure may be used for quantitative purposes to calculate the mass fractions of elements present in a section (Hall 1971, 1972; Hall et al. 1973).

Two factors limit the detection of any element. First, the element of interest may be swamped by larger quantities of another element with a similar eV. Second, a large amount of any element may raise the background radiation level to such an extent that small peaks are obscured; the heavier the element, the worse the problem.

Other electron optical systems are available for X-ray microanalysis besides static probe transmission, including scanning (SEM) and scanning transmission (STEM) modes (for a detailed account, see Chandler 1978). These may be used for thicker specimens and can produce an X-ray scanning image of the specimen, but the overall resolution is at present poorer than in the TEM. They tend to be most useful for high concentrations of elements.

Specimen Preparation Methods

Biological material for X-ray microanalysis can be conventionally fixed and embedded as for TEM. Sections 80 nm thick are mounted on Formvar-carbon-coated grids and stained for observation. Although these preparation methods mean that the sections can also be examined by normal TEM, they have two disadvantages: first, soluble components may be removed from or relocated in the specimen during fixation and embedding; and second, heavy elements added during processing, such as osmium, lead and uranium, will raise the background level of radiation and possibly obscure peaks of interest. Copper grids may also cause this problem. Standard preparation procedures can be used for detecting insoluble precipitates of introduced elements such as cobalt or nickel, which are likely to be present in relatively high concentrations, but the peaks for these elements may not be very prominent, due to low peak-to-background ratios.

Several means of enhancing the peak-to-background ratio for the elements of interest may be employed, but all have their disadvantages and so should only be used if absolutely essential. With introduced metals, they are usually unnecessary.

1. Omit the osmium postfixation step. Poor specimen preservation, especially of synaptic apparatus, and lack of contrast may however result.
2. Omit section staining. This reduces contrast and may make it difficult to determine the structure of the tissue and locate areas of interest.

3. Use aluminium or nylon grids. These are expensive and difficult to handle.
4. Freeze the specimen in nitrogen slush and use an ultracryostat to section it. This is a difficult procedure, but essential if the in vivo state of naturally occurring elements in the specimen is being investigated. Again, poor contrast makes it difficult to distinguish features in the specimen.

Improved detection of cobalt, nickel and silver can most easily be obtained by omitting the lead citrate and uranyl acetate staining procedure and by using aluminium grids. If material has been embedded in resin, then it is only worth doing qualitative analysis, or at best semi-quantitative analysis by comparing peak-to-background ratios. Absolute quantities for different elements are probably meaningless after so much processing.

Applications

X-ray microanalysis can be used to solve several problems in conjunction with the introduction of different metal ions into the insect nervous system. These include experimental procedures to investigate the problems of leakage of introduced metal ions from one axon to another, and of the transport mechanism which results in the distribution of introduced metal ions through stained neurons; and determining the location of introduced metal ions in different sub-cellular components. Technical information about the silver intensification procedure can also be obtained, for instance, whether silver deposits only on the introduced metal ions themselves, and whether it penetrates through the resin of a semithin section, to enable labelled neurons to be identified after resectioning. Lastly, X-ray microanalysis can be used in analytical neuroanatomy for the identification of metal-ion-containing profiles at the ultrastructural level, in conjunction with TEM of silver-intensified material. We will discuss the possible applications of this method for determining the ultrastructural areas of contact between identified neurons.

1. Location of Introduced Metal Ions. It has been calculated that the minimum amount of naturally occurring nickel detected in resin-embedded dinoflagellates is 10^{-16} g using a spot size of 0.5 μm (Sigee and Kearns 1980). It is likely that at least this amount would be present in metal-filled neurons.

We have used X-ray microanalysis to determine whether there is any leakage of introduced ions from filled neurons in the locust (Hackney and Altman 1982). The motor branch of the locust metathoracic wing nerve was filled with 5% $CoCl_2/NiCl_2$ (50:50) and one of the sensory branches with 5% $NiCl_2$. The ganglion was developed with rubeanic acid, fixed and embedded for electron microscopy. The axons in the nerve were identified in a 3-μm transverse section of the nerve proximal to the junction of the branches examined with the LM (Fig. 4a), where the red and blue colours of the RA-

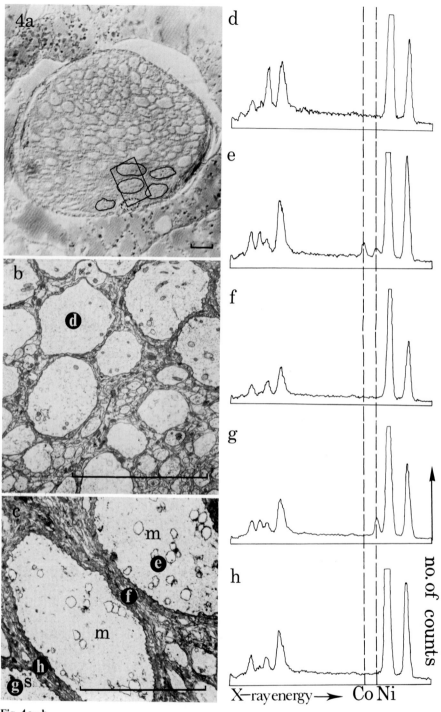

Fig. 4 a–h.

metal complexes could be discerned. Adjacent sections were examined in CORA. The distribution of cobalt and nickel in one such specimen is shown in Fig. 4b−h. Cobalt and nickel were found in the motor axons and nickel only in the sensory axon as expected. Neither cobalt nor nickel could be detected with the probe centre 0.25 μm or 1 μm outside the axonal membrane or over profiles originating in the unfilled nerve branch, which are adjacent to the filled profiles. We concluded that neither cobalt nor nickel ions had leaked from the filled axons under the preparation conditions used (Hackney and Altman 1982).

2. Distribution of Introduced Metal in Neurons. Evidence from silver-intensified ultrathin sections suggests that cobalt collects in mitochondria, in presynaptic densities and synaptic vesicles (Phillips 1980; Altman, Shaw, Tyrer and Hackney, unpubl.). Phillips (1980) suggests that it may replace calcium ions in these structures. Nickel and cobalt have also been seen to accumulate around the edge of the motor neuron nucleus (Hackney, unpubl.). X-ray microanalysis confirms that the small electron-dense deposits seen in unintensified axon profiles contain cobalt (or nickel) and clear areas of cytoplasm in between do not contain detectable amounts. The deposits are often associated with the mitochondria (see also Chap. 2).

3. X-Ray Microanalysis of Silver-Intensified Material. For detection of silver, unstained material is preferable since the uranium and silver lines overlap and are difficult to distinguish in the spectrum. A large peak of silver is obtained from the deposits on intensified sections (Fig. 5a). A silver peak may be detected on intensified sections even when silver grains are too small to be visible in the TEM. Examination of intensified sections indicates that when silver is present, cobalt or nickel ions can no longer be detected. This may be because they are displaced during the intensification procedure, or it may be that the quantity of heavy silver present raises the background level of radiation and obscures the cobalt and nickel peaks. In adjacent, unintensified sections, cobalt and nickel can be readily detected (Fig. 5b). Fur-

---

Fig. 4a−h. The location of cobalt and nickel ions in metathoracic nerve 1 of the locust, after filling the motor branch with a 5% mixture of $CoCl_2$ and $NiCl_2$ and the sensory branch with 5% $NiCl_2$, demonstrated by X-ray microanalysis. **a** 3-μm plastic section showing the positions of the axons examined. A *solid outline* indicates motor profiles (*m*), containing cobalt and nickel and a *dotted line* encloses the largest sensory axon (*s* stretch receptor) containing nickel. Axons in the upper half of the nerve come from another sensory branch which was not filled. A group of these is shown in **b**. An electron micrograph of three of the filled neurons is shown in **c**. Bars **a, b, c** = 10 μm. **d−h** Tracings of X-ray microanalytical spectra, indicating the occurrence of nickel and cobalt at the labelled points shown in **b** and **c**. The analyses were all carried out under the same conditions and the spectra are all to the same scale. The other peaks are, *from left to right*, osmium, sulphur, chlorine, uranium, copper and copper/osmium, and come predominantly from fixation materials and the copper grid

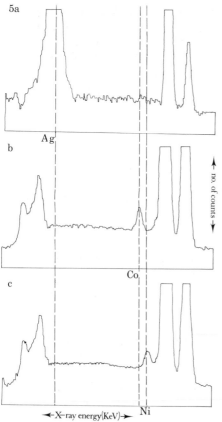

Fig. 5a–c. Tracings of X-ray microanalytical spectra comparing silver-intensified and unintensified sections. **a** A cobalt-filled motor axon silver-intensified on the ultrathin section shows a huge silver peak which swamps the osmium and uranium peaks. No cobalt is detectable although it is clearly present in the adjacent unintensified section **b**. This may be due to the raised background level resulting from the presence of silver, or because the silver has replaced the cobalt ions. **c** An ultrathin section cut from an intensified semithin section of a motor neuron containing nickel in which no silver was detectable, suggesting that silver does not penetrate into the resin during intensification

thermore, no silver could be detected in sections cut from an intensified semithin section, demonstrating that silver does not penetrate into the resin during intensification (Fig. 5c).

4. TEM Anatomy. Static probe X-ray microanalysis is used for selective analysis of small areas of tissue. Building up a picture of the distribution of nickel and cobalt throughout a section is therefore laborious, making this type of X-ray microanalysis less convenient for TEM of cobalt- and nickel-filled neurons than silver intensification. It is, however, a useful ancillary technique, mainly because it can locate the two ions in different profiles in the same section.

The most exciting potential application of the method is to label with different metals two neurons which are thought to synapse. The sites of contact between the two can first be located in silver-intensified sections and the profiles ascribed to their parent neurons by X-ray microanalysis of adjacent unintensified sections, in which contacts between the two can also be examined by standard TEM.

So far we have shown that this is possible in peripheral nerves where two sets of axons are filled with different metals. Here it is relatively simple to separate the motor neurons (filled with a mixture of cobalt and nickel) from the sensory neurons (filled with nickel only), both at the light microscope level using RA development and in the TEM with X-ray microanalysis (Hackney and Altman 1982). We are currently examining profiles in the CNS.

5. *Mechanism of Cobalt Uptake.* An extension of the experiment of sequential filling with nickel and cobalt already discussed for LM could be used at the ultrastructural level to map the course of ion movements through a neuron. In order to obtain accurate instantaneous information about ion movement, it would be necessary to freeze the tissue in nitrogen slush and to cut frozen ultrathin sections, as precipitating the ions with ammonium sulphide or RA may move them from the transporting structures.

Conclusion

Neither RA development nor X-ray microanalysis is seen as a total substitute for the classic processing methods for tissue containing metal-filled neurons. Both, however, offer useful extensions of existing methods with the exciting possibility of being able to identify different neurons in one preparation in the LM and TEM by the metals they contain, so that synaptic contacts between two or more known neurons can be examined.

The advantages of this over other multiple-marking techniques such as combining cobalt and HRP is that the same histochemical methods are used to demonstrate both markers (nickel and cobalt), which simplifies tissue processing and reduces the hazards of long and complicated preparation procedures.

With skilful manipulation, these techniques, combined with the methods already available, make detailed studies of neuronal connectivity, and investigations of the mechanism of the metal uptake and transport systems much more feasible than they have hitherto been.

Acknowledgements. We would like to thank Mr. D. Dickason, who made many of the initial experiments with various metal chlorides and rubeanic acid, and Mr. S. Butler for subsequent technical help. We are grateful to Dr. R. D. Butler and the staff of the E. M. Laboratory in the Department of Zoology, University of Manchester, for the use of CORA and to Mr. I. Miller for photographic assistance. The work was supported by a grant from the SERC to Drs. N. M. Tyrer and J. S. Altman.

Chapter 6

Double Marking for Light and Electron Microscopy

Harjit Singh Seyan, Ursula K. Bassemir
and Nicholas J. Strausfeld

European Molecular Biology Laboratory
Heidelberg, F.R.G.

Introduction

It is useful to see individual neurons against features of the neuropil in which they lie. Counterstaining Golgi-impregnated material with general stains, such as eosin and haematoxylin, or with cresyl violet helps identify structures, such as nuclei, in the surrounding tissue. Interference phase contrast helps identify other outstanding features, such as tracts and neuropils. Here we describe some methods of selective double staining and their applications to insect CNS.

There are two types of neuron-specific double staining. The first is light microscopic, showing cobalt-filled neurons superimposed on the neurofibrillar architecture of the brain. The second is combined light and electron microscopic, used to identify cobalt or heme protein-filled neurons in conjunction with Golgi-impregnated nerve cells. In this method, silver-intensified cobalt fills or DAB reaction products (on an HRP or cytochrome c substrate) are treated for electron microscopy and afterwards impregnated by the Golgi method. Double staining with HRP and cobalt has also been achieved, but we find it to be of dubious advantage. Intensified cobalt—Golgi combinations are reported here for the first time. Unintensified cobalt—Golgi was first used by Mobbs (1976) and HRP—Golgi was introduced on vertebrates by Somogyi et al. (1979). For its use on insects see Chap. 3.

Golgi impregnations and backfillings are used for light and electron microscopy. They rely on photon- and electron-dense deposits (of cobalt, heme protein or Golgi precipitates) being separately identifiable. The main drawback of the HRP method is that tissue preservation may suffer from extensive exposure to the reagents. In the cobalt—Golgi method tissue preservation is slightly inferior to that of conventionally treated tissue. This is due to the relatively late fixation (about 6 min after death; see Chap. 2). However, imperfect fixation, for example of mitochondria and membranes, is offset by the enormous advantage that double marking holds over conventionally treated material or singly marked preparations for electron microscopy. Suspected contacts between two differently stained neurons seen by light microscopy can be confirmed by electron microscopy.

Double Marking for Light Microscopy

The Cobalt-Silver–Cajal Block Silver Method

"Timming," or silver reduction at cobalt sulphide precipitates, relies on a developer carried with silver nitrate in a protective colloid (Timm 1958 a, b). In contrast, the early block-reduced silver techniques of Cajal and de Castro (1933) employ a developer stage without colloid and make use of only the silver cations carried over in the tissue. The present method exploits this simple technique to intensify cobalt-filled cells. After backfilling, tissue is sulphided, fixed in Cajal's ammoniacal alcohol, washed in distilled water and incubated in a silver nitrate solution. Some days later tissue is rinsed briefly and put in a pyrogallol reducing solution. After reduction it is embedded in araldite and serially sectioned. Hydroquinone, the usual reducer for Timming, impregnates only large fibres in the block Cajal method. Pyrogallol gives rise to a delicate impregnation of small to medium diameter fibres and deposits sufficient amounts of metallic silver on cobalt sulphide.

Method

1. Fill neurons with cobalt chloride, either through cut connectives or by neuropil diffusion (Altman and Tyrer 1980; Vol. 1) using 6% $CoCl_2$ in distilled water containing 0.03 g bovine serum albumin (Merck type 12 018 or Sigma type V).
2. Precipitation of cobalt: add 2 drops fresh ammonium sulphide $(NH_4)_2S$ to 10 ml 70% alcohol. Open the head and body cuticle. Immerse animal in sulphide solution for 3−5 min. Our method originally used absolute alcohol as the sulphide carrier, but a yellow precipitate develops after only 5 min; 70% ethanol-sulphide solution is good for about 10 min.
3. Wash twice in 70% alcohol.
4. Fix for 2−4 h in ammoniacal alcohol: 2 ml 25% ammonium is added to 98 ml absolute alcohol.
5. Wash in 70% ethanol.
6. Wash several times over 2 h in distilled water.
7. Incubate in 2.5% $AgNO_3$ in distilled water at 37 °C in the dark for 5 days.
8. Rinse in distilled water for 30−60 s with continuous agitation. This stage probably determines the amount of silver carried over to the developing solution.
9. Develop in 4% pyrogallol at 37 °C for 8−24 h in the dark.

10. Wash in distilled water.
11. Dehydrate through graded ethanols to propylene oxide (twice 10 min each).
12. Bring to araldite embedding mixture using propylene oxide: araldite mixtures of 1:1, 1:2 and 1:4, for 2 h each. Let stand in pure araldite overnight and then block out and polymerize for 24 h at 60 °C. A useful "soft" araldite resin mixture is

Fluka Durcupan ACM Component A 10.8 g
        Component B 8.9 g
        Component C 0.5 g
        Component D 2.0 g

Sections are cut at 12–25 μm. With a microsyringe, dab three parallel rows of water spots on a slide and place the sections in series on them. Place the slide on a hot-plate (50 °C). As the water warms and evaporates, the sections flatten and stick to the slide. Draw a strip of Permount liquid embedding medium on a coverslip. Wet this with a few drops of xylol and place the coverslip on the slide. Let cool.

Results. Cobalt-filled neurons are black or dark brown. Unfilled axons are reddish, pale brown or, very occasionally darker brown. Distinguishing then from cobalt-filled cells is described below.

Analysis

The degree of silver intensification of a cobalt-filled neuron depends, in part, on its content of cobalt sulphide. Errors of interpretation are made when neurons are poorly intensified and some background axons show strong colouration although they contain no cobalt sulphide at all (Figs. 1, 2). It is therefore essential to make control preparations using various concentrations of silver (ranging from 0.5% to 8%). A mixture of 2:1 (vol/vol) 4% pyrogallol : 2% hydroquinone will stain the whole range of small and large fibres. A further variable is the timing of step 8.

Where background staining is quite dark, cobalt-filled profiles can be distinguished from others by using a red filter to neutralize all colours except the black cobalt-silver deposits. This is essential for interpreting the sections (Fig. 2).

Gold Toning and Paraffin Sections

A direct comparison between Bodian – reduced silver preparations (see Gregory 1980) and cobalt-intensified cells can be made by gold toning. After reduction (step 9) the tissue is washed in distilled water. Subsequent steps are:

10. Clear for 2×20 min in benzol (CAUTION, BIOHAZARD: Use fume cupboard and protective gloves. Do not breathe fumes).

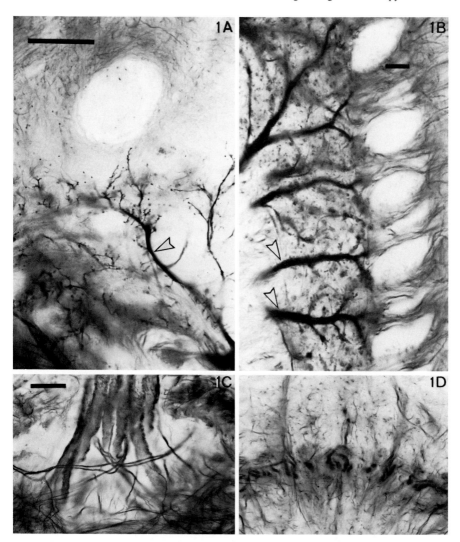

Fig. 1 A−D. Cobalt-silver Cajal block silver combination. **A** Protocerebral dendrites of the Giant Descending Neuron (*arrowhead*) backfilled with CoCl₂ and block-intensified by Cajal's method. Fibroarchitecture shows the typical concentric arrangement of the neuropil around the palely stained pedunculus of the mushroom body. **B** Transsynaptically filled dendrites of lobula plate horizontal cells stained by Cajal's method. Note the good resolution of other fibres (cobalt-filled profiles *arrowheads*) which enable an exact retinotopic map to be made of filled neurons against reduced silver-stained columnar cells entering from the medulla (to the *right*). **C** and **D** Controls with Cajal block silver alone. Silver reduction is stronger. Some fibres appear very dark brown or even black. **C** shows ocellar interneurons entering the midbrain just posterior to the protocerebral bridge. **D** shows cross-sections of large fibres in the flanges of the central body. All preparations are of *Musca domestica* (female). Bars **B, C** with **D** = 10 μm; **A** = 40 μm

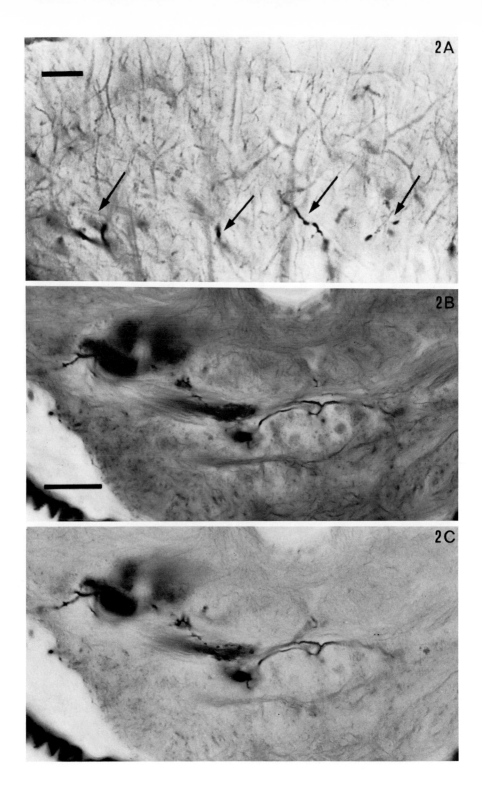

11. Infiltrate with Histowax or Paraplast (at least 3 h, preferably overnight).
12. Block out in Histowax (R. Jung, 6901 Nussloch F.R.G.) or Paraplast (Ted Pella Inc. Tustin CA 92680, USA).
13. Cut ribbons, sectioning at $8-10 \mu m$.

Ribbons are mounted on clean slides using glycerin-gelatin adhesive prepared as follows:
Dissolve 1 g gelatin in 100 ml distilled water, heating it gently. Then cool and add 15 ml glycerin. During use keep at 50 °C. Flood the slide with the adhesive and lay the paraffin ribbons on it. Stretch these by warming on a hot-plate. Draw off the excess adhesive with the edge of a cut filter paper. Blot the slides with fluff-free lens tissue or tissue paper. Dry slides overnight in a dust-free incubator at $42 ° -46 °C$.

14. Gold toning. Deparaffinized hydrated sections are placed in slightly acidified 0.5% gold chloride (3 drops glacial acetic acid/100 ml). Tone for 7 min.
15. Rinse in distilled water.
16. Place in 2% oxalic acid for 5 min. Sections should now be deep blue or purplish.
17. Rinse in distilled water and then fix in 5% sodium thiosulphate for 5 min.
18. Wash three times in distilled water, for 10 min, dehydrate, clear and mount in a suitable medium, e.g. Entellan (E. Merck, 61 Darmstadt, F.R.G.).

Axons and neuropil are stained purple, pink or dark blue. Cobalt-filled neurons are black or grey against this background. Shatter can occur if cobalt floods the tissue or if many fibres are stained.

Double Marking for Light and Electron Microscopy

Silver-intensified Cobalt−Golgi Methods

The basic method was introduced by Mobbs in 1976, using unintensified material. After sulphiding and fixation in glutaraldehyde, tissue was pro-

◄―――――――――――――――――――――――――――――――――――

Fig. 2 A−C. Cobalt-silver Cajal block silver combination. **A** Black profiles of very small diameter centrifugal fibres (*arrows*) filled with cobalt shown against the background of columnar and tangential neurons in the lobula. **B** and **C** Darkly stained neurofibrils using Cajal's method can be confused with cobalt-filled neurons (see controls in Fig. 1 C and D). In this preparation afferents of mechanosensory receptors from the left antenna are shown in the lateral deutocerebrum and suboesophageal ganglion. In **C** a red filter effectively suppresses reddish-brown background and helps to discriminate Cajal block silver-stained fibres and cobalt-silver-stained fibres. The preparations are from *Musca domestica* (female). Bars **A** = 5 μm; **B** (also for **C**) = 25 μm

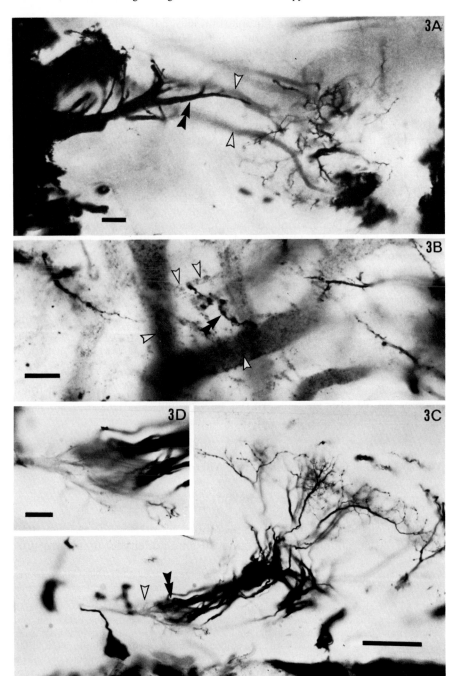

cessed by the Golgi−Colonnier technique. Strongly filled, unintensified profiles contain enough CoS to be identified with ease in ultrathin sections (see Chap. 2).

We employ a minor variant of the Bacon and Altman (1977) block intensification method or use a modification of the intensification method of Görcs et al. (1979). The former results in fine granular cobalt-silver precipitates within profiles (see Chap. 2), the latter results in sparsely distributed but clumped silver precipitates (Fig. 3B). The method is basically straightforward; first silver intensification, then osmication and finally Golgi staining.

Method

1. Backfill neurons through their cut axons or by extracellular application of Co^{2+} into neuropil.
2. Precipitate to cobalt sulphide using $2−3$ drops pale yellow (fresh) ammonium sulphide $[(NH_4)_2S]$ in 10 ml cacodylate buffer − 0.16 M sodium dimethylarsonate $[NaAsO_2(CH_3)_2 \cdot 3 H_2O]$. Dissolve 3.42 g sodium cacodylate (dimethylarsonate) in 100 ml distilled water. Adjust pH to $7.0−7.4$ with 1 N HCl then add 3 g sucrose (0.1 M). Use fume cupboard, wear protective gloves. CoS precipitation should not exceed 4 min.
3. Wash twice for 1 min in buffer.
4. Fix in 2% glutaraldehyde in buffer for 2 h at room temperature.
5. Wash in buffer, cool to 4 °C.
6. Dissect out nervous tissue.

Store overnight in buffer at 4 °C (optional step).

7. Wash out buffer with 0.1 M sucrose at 4 °C. This is a crucial step to remove all excess dialdehydes and buffer.
8. Incubate for 1 h at 40 °C in 0.1 M sucrose.
9. Place in developer made up as follows:

◄───

Fig. 3. A HRP−Golgi double marking. Terminals of Golgi-stained vertical giant neurons ("VS" 2 and 3) of the lobula plate (*double black arrowhead*) terminate on HRP-backfilled descending neurons (*open arrowheads*). Electron microscopy shows the vertical cells and descending neurons to be in contact. **B** Double marking with cobalt-silver intensification (Görcs et al. 1979) followed by the Golgi−Colonnier impregnation. *Black double arrowhead* shows ending of a T_4 neuron from the medulla on the transsynaptically filled dendrites of vertical cells (VS 7 and 8) (*open arrowheads*) of the lobula plate. Electron microscopy shows T_4 presynaptic to the vertical cells. **C and D** Double labelling with cobalt and Golgi. A silver-intensified (Chap. 2) descending neuron (*open arrowhead*) is shown with converging Golgi-impregnated axons of columnar cells from the lobula (*black double arrowhead*). The difference in contrast between the silver-intensified cobalt-filled cell and the Golgi−Colonnier impregnated neurons is shown in the inset (**D**). All preparations are from *Calliphora erythrocephala* (female). Bars **A** and **D** = 10 μm; **B** = 5 μm; **C** = 50 μm

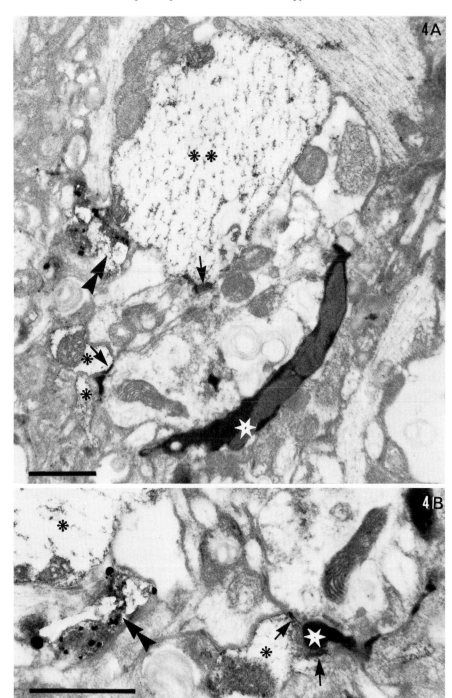

100 ml distilled water
2—5 g unpurified gum arabic
0.17 g hydroquinone
3 g sucrose

Dissolve 7.5 g anhydrous citric acid in 10 ml distilled water and add this drop by drop to the above solution to adjust it to between pH 2.8 and 3.2.

10. After 1 h place in 9 vol developer and 1 vol 0.1% $AgNO_3$. Incubate 1 h at 40 °—50 °C.
11. Place in fresh solution made of 9 vol developer and 1 vol 1% $AgNO_3$.
12. Replace with a fresh solution if the first becomes tinted. In any event it is advisable to change solutions after 0.5 h.
13. Check tissue occasionally under the binocular microscope using indirect illumination.
14. When the tissue turns pale brown, place it in warm 0.1 M sucrose. Let cool. Wash in fresh changes of sucrose, finally cooling to 4 °C.
15. Wash in cold cacodylate buffer with 0.1 M sucrose.
16. Postfix for 1 h at 4 °C in 0.1% osmium tetroxide in cacodylate, omitting sucrose from the buffer.
17. Wash in cold cacodylate, then bring to room temperature.
18. Wash several times in 2.5% potassium dichromate.
19. Place in 1 vol 25% glutaraldehyde in 4 vol 2.5% potassium dichromate. Leave for 4 days.
20. Drop tissue in 0.1% $AgNO_3$ and swirl until there is no more precipitate.
21. Place in 0.75% $AgNO_3$. Leave for 3 days.

Steps 9—14 and 19—22 are in the dark. Developer, silver and sucrose solutions are maintained at 50 °C throughout intensification. Mix developer and silver solutions immediately before use.

Step 14 can be followed by step 19, but substituting the Colonnier solution with a Golgi rapid solution. This comprises 1 vol 2% osmium tetroxide

---

Fig. 4A, B. Cobalt-silver Golgi double marking for electron microscopy (see Fig. 3c). **A** There are two degrees of cobalt-silver labelling: the densely labelled profile (*white asterisk*) can be resolved both in light and electron microscopy. The other less densely labelled profile (*black asterisk*) has sparsely distributed cobalt-silver grains. (Under the light microscope such profiles are poorly resolved against the brownish osmicated background of 20-μm sections.) A Golgi-impregnated profile (*double arrowhead*) contacts the large diameter cobalt-silver profile. **B** Higher magnification of the same preparation as in **A**, but several thin sections deeper. The Golgi profile (*double arrowhead*) no longer contacts the sparsely cobalt- silver-labelled fibre (*black asterisk, top left*). Note large silver chromate deposits within the amorphous electron-dense Golgi impregnation (*double arrowhead*). Typically, silver chromate precipitate is lost or translocated during ultramicrotomy and staining, leaving behind empty areas. *Arrows* indicate cobalt-silver accumulations at postsynaptic sites within the sparsely cobalt-silver-labelled neuron. *White asterisk* densely cobalt-silver-labelled profile. Ultrathin sections are stained with uranyl acetate. Bars **A**, **B** = 1 μm

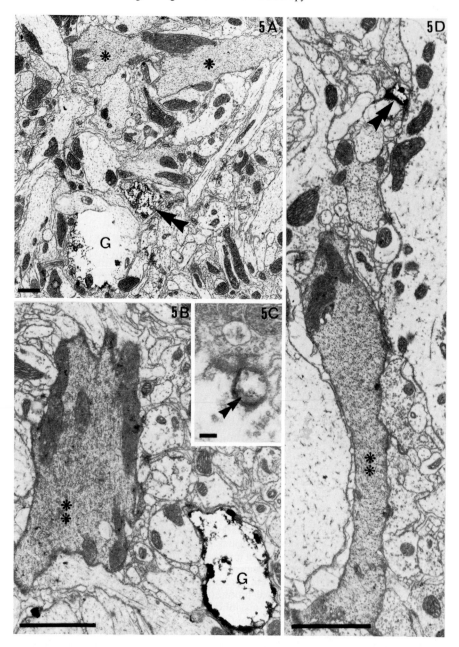

Fig. 5A–D. HRP–Golgi double marking for electron microscopy (see also Chap. 3).
A Low-power electron micrograph of HRP-labelled (*black asterisks*) and Golgi-impregnated neurons (*G* and *double arrowhead*). In profile *G* part of the impregnation was lost during
ultramicrotomy and section staining, leaving "empty areas." Note the relative low contrast
of HRP label. **B** In this electron micrograph the concentration of HRP-label (*black as-*

in distilled water with 20, 30 or 50 vol 2.5% potassium dichromate to achieve different degrees of impregnation. Incubation is for 3–8 days.

This completes the method for light microscopy. After dehydration and embedding in "soft" araldite, serial sections are mounted in Permount under coverslips and examined (Fig. 3D, C). A section of interest is serially photographed through its depth and then re-embedded for electron microscopy using the acetate sandwich method described in Chaps. 1 and 2.

After mounting the block, the area of interest is photographed again and trimmed. The final trimmed block is again rephotographed so that the location of the appropriate profile can be found at low-power magnification in the ultrathin section. Documentation is crucial if the preparation contains several different Golgi-impregnated cells in the same locality.

During cutting, unstained ultrathin sections are examined for the appearance of profiles. Then subsequent sections are stained with uranyl acetate (Fig. 4A, B). Lead citrate is omitted since fine-grained cobalt-silver deposits would otherwise be lost in the "noise" of lead precipitates.

Golgi-filled profiles appear electron-opaque and obscure cytoplasmic details, except mitochondria. One disadvantage encountered in ultrathin sections is that sometimes the silver chromate precipitate is missing, especially in larger profiles. Substituting potassium or silver dichromate for the water in the cutting trough as suggested by Blackstadt (1970) is supposed to minimize this effect. However, evaporation of water from the chromate solution causes crystals to form on the knife edge, which affect the quality of sections enormously.

A source of misinterpretation is that large cobalt-silver-filled profiles sometimes become preferentially silver-chromate-impregnated. This may occur in patches and results in partial impregnation.

A modification of the method uses the Görcs et al. (1979) silver intensification. We substitute ascorbic acid with 0.2% pyrogallol or 1.7% hydroquinone to give a finer granulation of the precipitate.

Solution A – 3 g sodium acetate
$\qquad$ 1 g AgNO$_3$
$\qquad$ 60 ml conc. acetic acid
$\qquad$ 30 ml 1% (vol/vol) Triton X-100 or
$\qquad$ 0.5 ml DMSO/100 ml developer
$\qquad$ 710 ml distilled water.

---

terisks) gives sufficient contrast for easy identification of the marked profile (see also Figs. 20A, 21 in Chap. 3). The Golgi-impregnated fibre G has lost most of the silver chromate during ultramicrotomy and subsequent staining. **C** High-power electron micrograph of a small Golgi-impregnated profile (double arrowhead). The presynaptic T-shaped ribbon and synaptic vesicles of a contacting unlabelled fibre are clearly resolvable. **D** Electron micrograph showing an HRP-labelled (black asterisk) and a small Golgi-impregnated profile (double arrowhead). The former has sufficient contrast, and the latter has most of the silver chromate preserved. All sections are from C. erythrocephala and are stained with uranyl acetate and lead citrate. Bars **A**, **B** and **D** = 0.5 μm; **C** = 0.1 μm

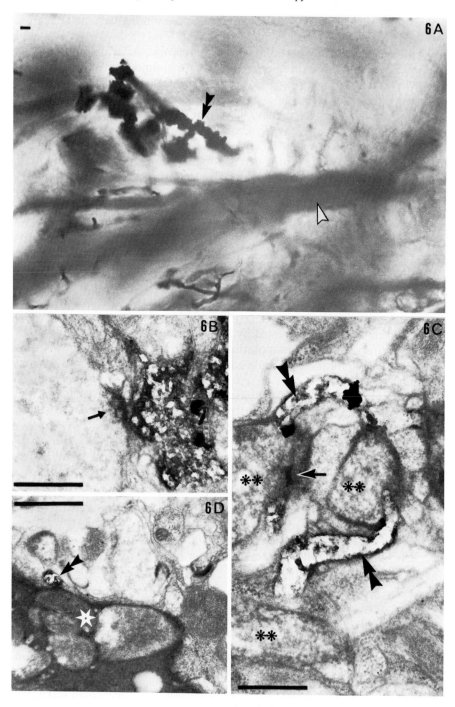

Solution B — 5% sodium tungstate

Solution C — 0.2% pyrogallol or 1.7% hydroquinone

Mix A:B:C: in a ratio of 8:1:1. Incubate at room temperature for 1 h.

Further processing is as for steps 14—22 above (Fig. 3B).

Combined Heme Protein—Golgi Method

This method was introduced by Somogyi et al. (1979). Neurons are back-filled through their cut axons with Boehringer Mannheim Type 1 horserad-ish peroxidase (HRP) which, in contrast to Sigma type VI, achieves less "transneuronal" staining. Lysolecithin is also omitted to eliminate unspecific "transneuronal" staining (see Chap. 3). Complete exclusion of "transneuron-al" uptake is achieved by using cytochrome c (Nässel et al. 1981), but longer uptake periods are needed. Two percent HRP in 0.1 M KCl requires 3—6 h at 4 °C for backfilling 3 mm nerve. About twice as long is needed for cy-tochrome c.

Histochemistry

After filling, the head or body is opened, briefly washed in buffer and then fixed in 2% glutaraldehyde in 0.16 M sodium cacodylate buffer containing 0.1 M sucrose for 4 h at room temperature. Thereafter:

1. Wash tissue in buffer, 3×10 min. Dissect out nervous tissue.
2. Soak in buffer at 4 °C overnight.
3. Wash thoroughly and then soak in 0.04% DAB (3,3'-diaminobenzidine 4-HCl from K & K Laboratories Inc., Plainview, N.Y. or Fluka Chemie, F.R.G.) in cacodylate buffer at between pH 6.5 and 7.9 (CAUTION: BIO-HAZARD). Use DAB in a fume cupboard. Wash all surfaces and instru-

Fig. 6A—C. Cytochrome c and Golgi double marking for electron microscopy (*C. ery-throcephala*). **A** Light micrograph of cytochrome c-filled descending neurons of the visual system (*open arrowhead*) and Golgi-impregnated ocellar fibre terminals (*double arrow-head*). **B** Electron micrograph demonstrating well-preserved Golgi impregnation from the section in **A**. Fine- structural preservation of the surrounding tissue allows identification of presynaptic T-shaped specializations (*arrow*) and synaptic vesicles. **C** Electron micrograph from the preparation shown in **A**. The low concentration of cytochrome c (*black asterisks*) allows the resolution of a presynaptic ribbon pedestal (*arrow*). Two Golgi-impregnated fi-bres (*double arrowheads*) contact two cytochrome c-labelled profiles (*black asterisks*). **D** Electron micrograph showing a Golgi-impregnated fibre (*double arrowhead*) in contact with a heavily cytochrome c-labelled neuron (*white asterisk*). The dark-staining cy-tochrome c—DAB reaction product obscures cytoplasmic details. Sections are stained with uranyl acetate and lead citrate. Bars **A** = 1 μm; **B**—**D** = 1 μm

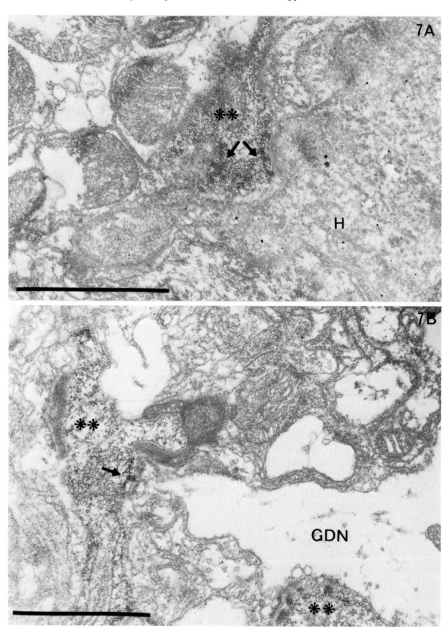

Fig. 7A. Electron micrograph showing double marking with cobalt and HRP (*M. domestica*). Fine granular cobalt-silver precipitate in axon terminals (*asterisks*) of antennal receptors contact the HRP-containing dendritic spine (*H*) of the Giant Descending Neuron. Presynaptic T-shaped ribbons (*arrows*) within the cobalt-labelled terminals are resolvable. The flocculent HRP–DAB reaction product, however, does not show up very well against the surrounding tissue so that its identification is difficult. **B** Control prep-

ments afterwards in sodium hypochlorite. Use protective mask when weighing out the chemical as it is a suspected carcinogen. A lower pH (6.5−6.8) results in heavier staining.

4. Incubate for 1.5 h on a rotator in the dark at room temperature. Prolonged incubation may cause fine-structural damage, however.
5. Add 4 drops 3% H_2O_2 to 10 ml fresh 0.04% DAB in pH 6.5−7.4 cacodylate buffer. Incubate tissue in this for 1.5 h in the dark at room temperature.
6. Wash in pH 7.2 cacodylate buffer 3×10 min, bringing temperature down to 4 °C.
7. Postfix in 1% osmium tetroxide in pH 7.2 cacodylate buffer (without sucrose) at 4 °C.
8. Wash in pH 7.2 cacodylate buffer.
9. Proceed as steps 15−21 to achieve Golgi impregnation.

Soft araldite sections are examined by light microscopy (Figs. 3A, 6A). Selected areas are photographed and the sections re-embedded as described in Chaps. 1, 2. The results are shown in Figs. 5A−D and 6B−D. Other Golgi methods and HRP intensification methods are described in Chap. 3.

Interpretation

A common source of error in the interpretation of backfilled cobalt and heme protein profiles stems from the fact that backfilling results in the uptake of marker by more than one axon and to different extents. Furthermore, neurons containing only trace amounts of marker may be sufficiently contrasted for detection at high resolution but may not be visible in the light microscope (Figs. 3B, 4A, B). For this reason only dark profiles readily seen in the light microscope should be used to substantiate connections between marker-filled and Golgi-impregnated neurons. In all cases uranyl acetate staining of ultrathin sections yields easily identifiable dark and pale profiles.

Double Marking with Cobalt and Heme Protein

In theory, at least, it should be possible to combine the above methods, achieving triple marking. Control experiments, using cytochrome c or HRP backfills and running the DAB−H_2O_2 reaction after sulphiding show that sulphide does not adversely affect the reaction substrate (heme protein).

---

aration, showing cobalt-filled axon terminals (*asterisks*) of antennal fibres contacting a dendritic spine of the unlabelled Giant Descending Neuron (*GDN*). Compare the HRP label and occasional silver grains in the cytoplasm of the filled cell in **A** and the unfilled cytoplasm of GDN in **B**. Sections are uranyl acetate and 1-min lead citrate stained. *Arrow* presynaptic ribbon. Bars **A** and **B** = 1 μm

Subsequent silver intensification gives rise to granular deposits on the re-action product showing it to be a preferential catalyst. On the other hand, backfilling with cobalt, and running the silver intensification after the $DAB-H_2O_2$ reaction results in comparatively poor cobalt intensification. Double-labelled preparations are prepared for electron microscopy by back-filling the Giant Descending Neuron, for example with HRP, while for-wardfilling antennal afferents with $CoCl_2$. After sulphiding and fixing in buffered glutaraldehyde, the tissue is treated with $DAB-H_2O_2$. After the buffer is washed out, the preparation is intensified with glycine or hydro-quinone with silver nitrate in the developer solution (see Chap. 2) osmicat-ed, and embedded in araldite. Neurons can just be discerned in 10 μm sec-tions of the dark brown osmicated brain. Heme protein and cobalt-silver de-posits can be distinguished in the electron microscope. At high resolution it can be seen that $DAB-H_2O_2$ reaction product at HRP sites can act as a cata-lyst for silver reduction. HRP- labelled neurons thus contain sparsely distrib-uted large silver grains in contrast to the fine grained cobalt-silver deposits in cobalt-filled neurons. These compare well with cobalt-filled neurons that have been only silver-intensified. We speculate that cobalt sulphide that is complexed to specific intracellular sites is not affected by H_2O_2 treatment whereas noncomplexed cobalt sulphide may be eluted.

In its present state this combined method can only be used on systems whose connections are already well known, such as antennal afferents sy-napsing onto giant fibres in the fly brain (Fig. 7 A, B). Possibly, lysine- or proline-cobalt complexes (see Chap. 17) will lead to improvements of this method.

Alternative Strategies

Golgi Substitution

Two techniques used in vertebrate neuroanatomy substitute or remove the Golgi precipitate. One substitutes what appears to be a specific reaction product at the inner membrane (see Chap. 1) with lead chromate (Ramon-Moliner and Ferrari 1976) or gold (Fairén et al. 1977). The second method merely dissolves most of the silver chromate with an ammonium solution, leaving an insoluble residue at the inner membrane (Braak and Braak 1982). Interestingly, Ribi's Golgi methods (Chap. 1) and the above substitution methods all indicate a special type of deposit at the cell membrane. Golgi experiments on insects suggest the formation of an insoluble "early" Golgi reaction product as part of the mechanism of impregnation (see Chap. 9, Vol 1). Both of the above methods have been tried by us. The Fairén et al. procedure functions well after Golgi impregnation of "osmium-dichromat-ed" (i.e. Golgi-rapid) tissue and is a useful electron microscopic marker al-lowing visualization of some internal structures. Gold substitution or Ribi's

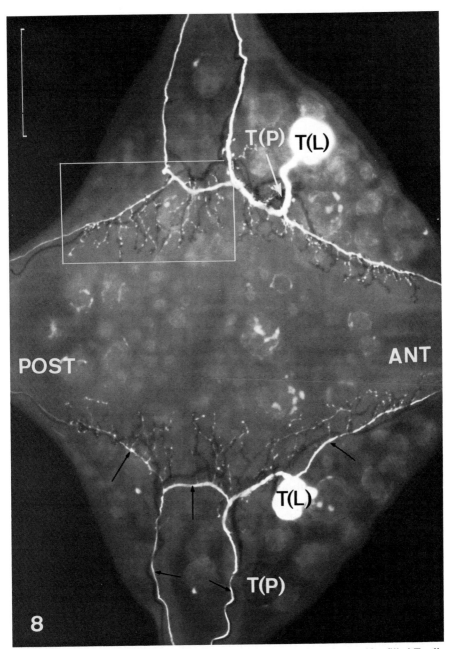

Fig. 8. Dorsal view of ganglion No. 8 of the leech, *Hirudo medicinalis.* Lucifer-filled T cells [*T(L)*] belong to T1 (*top*) and T3 (*bottom*). The darker HRP-filled cells (containing the di-aminobenzidine reaction product) are labelled T(P) (cells T2, *top* and T4, *bottom*). Bar = 100 μm. Compared with many insect neurons the cell bodies shown here are large. *POST* posterior. *ANT* anterior of ganglion. (Photograph by courtesy of Eduardo Macagno)

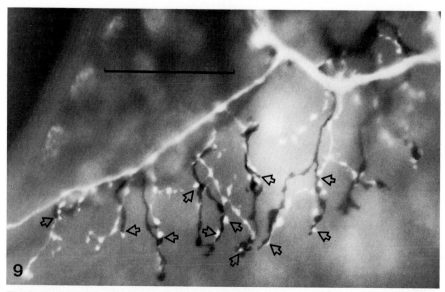

Fig. 9. The inset shown in Fig. 8 is enlarged to illustrate the points of convergence between the T1 (Lucifer-filled) and T2 (HRP-filled) neurons. Each contact is indicated by an *arrow*. Bar = 50 μm. (Photograph by courtesy of Eduardo Macagno)

Golgi modification (Chap. 1) may eventually permit visualization of synaptic sites in cobalt-silver and in Golgi-impregnated neurons. Up to now we are only able to see sites reliably in cobalt- or heme protein-marked cells.

Lucifer Yellow and Heme Protein

Finally, we should mention one double-marking method for light microscopy that, although not yet tried on insects, has been spectacularly successful on the leech central nervous system. This method, invented by De-Riemer and Macagno (1981), combines horseradish peroxidase and Lucifer yellow CH, and is as follows: 20 mg ml⁻¹ HRP with 2 mg ml⁻¹ fast green in 0.2 M KCl is pressure injected from a 30−60 MOhm electrode with pulses of 5−15 psi at the barrel end. HRP is allowed to diffuse through the neuron for 2−3 h after which a second or third cell is filled with 3% Lucifer yellow CH in 0.1 M LiCl using 5−10 nA pulsed negative DC current for 15 min at 1 Hz. Ganglia are placed in Ringer containing 0.5 mg ml⁻¹ diaminobenzidine tetrahydrochloride (CAUTION: powder is reputed to be strongly carcinogenic) until they are infiltrated (5 min) and then a few drops of 3% H₂O₂ are dropped onto the tissue. The reaction is observed under the binocular microscope and is stopped as soon as the background begins to show unspecific reactions (this takes about 2−5 min). Afterwards ganglia are viewed as whole mounts (Fig. 8) and contact points between differently marked neurons can be seen (Fig. 9). This method should be tried on insects that are suitable for whole-mount observation. However, insect ganglia are notoriously impermeable and it may be necessary to treat them with dilute collagenase with a solubilizing agent, such as Triton-X. The double marking would seem ideally suited for combined backfilling.

Acknowledgements. We thank Duncan Byers for much advice concerning the Cajal block silver method and its modifications. We also thank Dick Nässel for advice concerning HRP and cytochrome c filling for double marking with cobalt.

Chapter 7

Lucifer Yellow Histology

Nicholas J. Strausfeld and Harjit Singh Seyan

European Molecular Biology Laboratory
Heidelberg, F. R. G.

David Wohlers and Jonathan P. Bacon

Max-Planck-Institut für Verhaltensphysiologie
Seewiesen, F. R. G.

Introduction

Walter W. Stewart discovered and synthesized the fluorescent dye, Lucifer yellow, based on the commercial dye, brilliant sulphoflavine (Stewart 1978, 1981). Lucifer yellow has since become the most popular and widely used intracellular marker, largely replacing its forerunner, Procion yellow 4RAN, for identifying physiologically recorded neurons.

There are two Lucifer yellow dyes (CH and VS), both 3,6,-disulphonated 4-amino naphthalimides which differ by their substituent on the imide nitrogen (Stewart 1981). Both are listed as commercially available from Sigma Chemicals. Lucifer yellow CH and VS have similar spectral properties, with excitation $\lambda_{max} = 280$ nm and 430 nm, and emission $\lambda_{max} = 540$ nm, conveniently allowing illumination and observation at two widely separate wavelengths. In most respects Stewart has invented the perfect, highly versatile, intracellular marker which, in some circumstances, can also be used as a selective stain of certain neurons (as in the vertebrate retina; Detwiler and Sarthy 1981).

The quantum yield of Lucifer yellow CH is about 0.25 which, in fixed tissue, is about a hundred times better than Procion yellow 4RAN. Lucifer yellow CH fades relatively slowly in blue light, and we have photographed the same neuron for several hours. The CH dye is divalent and negatively charged. Presumably it binds loosely to intracellular structures, not normally crossing the cell membrane, at least within the first 20 min or so after filling. The dye has a low toxicity and moderately affects electrophysiological properties, decreasing spike amplitude. It does not, however, poison neurons, in contrast to Co^{2+}, which accumulates at inner cell membranes and synaptic sites (see Chap. 2). Cobalt presumably poisons cells because it competes with and supposedly blocks calcium receptors (Bassemir and Strausfeld 1983).

The most common application of the dye is to fill cells intracellularly during and after electrophysiological recording. Injection is usually by ion-

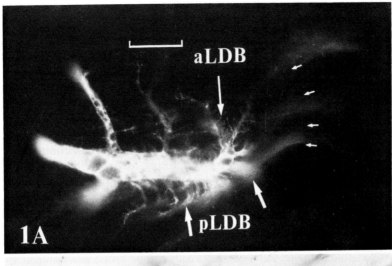

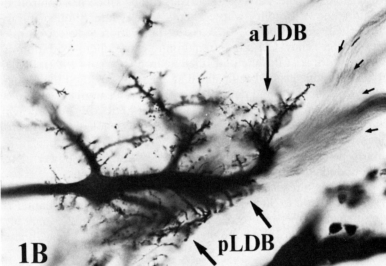

Fig. 1. A 10-μm Spurr's section of a Lucifer-filled Giant Descending Neuron (GDN) in the brain of *Musca domestica*. Anterior dendrites (*a*) from the lateral deutocerebral branch (*LDB*) are not coupled; posterior dendrites (*p*) are dye-coupled to faintly fluorescing profiles collected into three bundles (*small arrows*). This example of weak dye coupling is identical to cobalt coupling between the pLDBs of the GDN and small axons from Col A neurons originating in the lobula (**B**). These share gap junctions with pLDBs (see Chap. 2 and Strausfeld and Bassemir 1983). Bar = 50 μm. **A** taken with Ektachrome 400 copied onto Ilford Pan-F

tophoresis although pressure has been employed on gastropod neurons (Stewart 1978).

Lucifer yellow CH readily moves out of the microelectrode, in contrast to Lucifer yellow VS, which Stewart (1981) reports as sluggish. Lucifer yellow CH is used to study the morphology of adult and developing neurons (Stewart 1978; Glantz and Kirk 1980; Wohlers and Huber 1982; Goodman and Spitzer 1979; Bentley and Keshishian 1982) as well as regenerating cells (Lee 1982) in vertebrates, gastropods, crustaceans and insects. The dye passes between certain neurons, some of which are known to be electrotonically coupled: dye coupling possibly takes place at the gap junctions. Lucifer yellow CH is also used for passive backfilling as described here. In most respects it gives a cleaner fill than cobalt, eliminating background colouration. Its main drawback is that dye coupling from a single neuron into many small cells yields only faint fluorescence which cannot be enhanced, as can cobalt with silver reduction (Fig. 1).

The dye is also used as an ablater. Photosensitization of filled cells causes irreversible damage. Parts of whole cells can be killed in a controlled fashion within a known circuitry whose altered state is later studied intracellularly (Miller and Selverston 1979). A recent innovation for electron microscopy uses Lucifer yellow CH photoexcitation to cause photo-oxidation of 3,3′-diaminobenzidine at the site of a filled neuron (Maranto 1982). Paul Taghert, at Stanford University, has invented an ingenious method that enhances Lucifer yellow deposits and does away with fluorescent microscopy (Taghert et al. 1983, pers. commun.). Peroxidase-tagged antibodies, raised against Lucifer yellow-albumin conjugates, are applied to whole mounts or frozen sections. Later, the peroxidase catalyses oxidation of diaminobenzidine by hydrogen peroxide (see Chap. 3). Doubtlessly, Taghert's powerful technique will be adopted by many researchers.

Much has already been written about the dye and its applications. We have little to add to Stewart's original description and this chapter is merely a resumé of our personal experiences with the dye for functional morphology of insect neurons. We hope that our experience and tips will be useful to others.

Filling from Electrodes

We use a 3% solution of the lithium salt of Lucifer yellow CH (hereafter referred to as LYCH) dissolved in 1% LiCl. The blunt end of the inverted electrode barrel is immersed in the solution and dye is drawn up into the tip by capillary suction in 2−3 min. We use thin-walled borosilicate glass electrodes (Frederick Haer) with an outer diameter of 1.0 mm and an inner diameter of 0.75 mm, containing a fused glass filament. Only the tip is filled. A silver wire from the preamplifier is pushed into the shank to contact the solution. For convenience, the barrel can be backfilled with 3 M LiCl or, for visibility, a coloured electrolyte (e.g. Procion yellow, 2% aqueous solution) using a microsyringe (No. 12, 0.5 mm diameter, ACUFIRM V2A, hypodermic needle).

The shape of the electrode tip is a matter of personal taste. Our LYCH electrodes have a final resistance of 30−80 MOhm. Resting potentials around −40 mV indicated good intracellular penetration with abundant postsynaptic noise at high amplification and spikes between 15 and 80 mV (in "spiking neurons"). After recording, cells are filled by applying hyperpolarizing DC current of 4−8 nA for 5−30 min. Cells were also filled using pulsed negative 1−20 nA DC current (100 ms, pulsed at 5 Hz), but this seems to offer no special advantage.

Passive Back- or Forwardfilling

Backfilling through cut nerves or ganglia employs the same basic method as for cobalt (Chap. 19, Vol. 1). A perfectly tight seal is required between the Lucifer pool and the tissue. The dye travels readily over the tissue surface and has surfactant (detergent) properties. We tamp the vaseline down well and test the pool with distilled water to see if it is tight. Water is removed by a microhypodermic needle and replaced with LYCH. When filling peripheral nerves with LYCH from a blunt electrode (see Chap. 18, Vol. 1), a vaseline seal should be used to prevent spread of dye along the outside of the nerve since it seems to readily cross the blood−brain barrier into neuropil, contributing to background fluorescence and selective uptake by glia cells (see p. 154).

Backfilling is rapid. Dye is distributed through a cell in minutes, even without current (Figs. 2E, F; 9A−C). We backfill from the thorax of *Calliphora* into the brain for 1 h (a distance of 3−4 mm) keeping the animal in a damp chamber at 20 °C. We do not achieve dye coupling in the absence of negative current. Dye coupling occurs between certain neurons using intracellular iontophoresis, negative DC, over several minutes. In Diptera, coupling between neurons includes those that are cobalt-coupled (Fig. 1; Strausfeld et al. 1981). After back- or forwardfilling from electrodes (antennae 30 min, ocelli 40 min) LYCH is withdrawn and the animal is immediately fixed. There is no cobalt-type "diffusion" period (see Fig. 8).

Fixing

Lucifer yellow CH is divalent. It may bind loosely to intracellular structures during filling (Stewart 1978), but in our experience coagulant fixatives that do not employ aldehydes are useless. Most of the dye is lost, as it is in fresh frozen sections (Milde, pers. commun.). Lucifer yellow CH is only properly cross-linked to tissue by aldehydes through its hydrazido group (Stewart

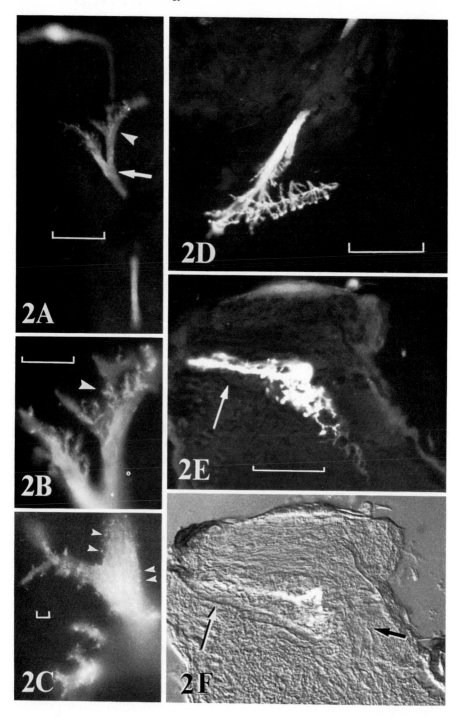

1982), and it is this bond that holds the dye in the cell during dehydration and embedding. Fixatives combining aldehyde and coagulants (e.g. alcohol−acetic acid−formalin; AAF) result in the loss of most of the dye, despite its tolerance to a wide pH range (Stewart 1981), presumably due to inadequate cross-bridging and disruption of membranes and disassembled cytoskeleton.

We fix animals within 1−15 min after iontophoresis or after subsequent recordings. We use pH 6.8−7.2 Millonig (1961) 0.13 M phosphate or 0.1 M TES or 0.16 M cacodylate-buffered 4% commercial formaldehyde (Merck) or 0.16 M cacodylate-buffered 3% paraformaldehyde, sometimes with 1% glutaraldehyde. The latter is acceptable for electron microscopy but gives rise to a high background fluorescence. Light microscopy fixation is in two-stage buffered formaldehyde (1 h at 20 °C). Ganglia are dissected out under buffer and then fixed for a further 6−12 h at 4 °C. We find that a final wash in cacodylate is most efficient for removing excess fixative of any kind. Before dehydration for paraffin embedding, tissue is soaked in 4% formalin in methanol for 2 h (H. S.). One of us (D. W.) prefers to dissect out ganglia and brain from crickets before fixation. The nerve cord is laid out on a Sylgard-covered plastic Petri dish and flooded with an appropriate Ringer (e.g. TES-Ringer). Ringer is then sucked away and the nerve cord, ganglia, brain and peripheral nerves are appropriately arranged, before gently flooding the dish with fixative.

Buffers, Ringers

1. Millonig's Buffer

Solution A: 2.26% $NaH_2PO_2 \cdot H_2O$ in distilled water.
Solution B: 2.52% NaOH in distilled water.

0.13 M Buffer, pH 7.3 is: 41.5 ml A with 8.5 ml B.

Fig. 2. A The most superficial neurons in the highly autofluorescent fly brain are descending neurons postsynaptic to vertical cells. One (type 2 DNVS) is shown in phosphate buffer after 1 h formalin fixation (×16 oil-immersion objective). **B** Detail of the same DNVS in a Spurr's section, ×40 oil-immersion objective from region denoted by arrowhead in **A**, and at ×63 (oil immersion) in **C** from region indicated in **B**. **C** illustrates parallel arrays of minute dendrites covering the dendritic trunk (*small arrowheads*). These structures could not be resolved by cobalt, HRP or Golgi impregnation. Bar **A** = 50 μm; **B** = 10 μm; **C** = 1 μm. **D** Ocellar L-neuron (relay interneuron) ending in the posterior slope of the bee brain. This figure, kindly provided by Dr. Jürgen Milde (EMBL), is a salvaged preparation retrieved by re-embedding an old whole mount in Spurr's sectioned at 15 μm. Bar = 50 μm. **E** Passive forwardfills of severed L-neuron from the ocelli of *Calliphora*. *Arrow* indicates position of vertical cell endings onto which this neuron converges. **F** Combined epi-fluorescent illumination and direct interference contrast illumination to reveal outlines of vertical cell endings (*white arrow*) and axons of DNVS cells (*black arrow* also indicated white at the equivalent position of *Musca* DNVS 2 in **A**). **A**−**C** on Ilford XPI-400, **D**−**F** on Kodak Tri-X (ASA 400). Bar **D, E, F** = 50 μm

2. TES Buffer

Range, pH 6.8−8.2. pK=7.5 at 25 °C (N-Tris[hydromethyl]methyl-2 amino-ethane sulfonic acid) MW 229.2

Adjust pH of 1 M or 0.1 M solution in distilled water by adding conc. NaOH drop by drop which results in a clear solution. When near required pH, dilute the conc. NaOH for final pH adjustment.

3. Cacodylate Buffer

(0.16 M) Sodium dimethylarsonate. Caution: BIOHAZARD

$NaAsO_2(CH_3)_2 \cdot 3 H_2O$ MW 214.0

Dissolve 3.42 g Sodium cacodylate in 100 ml distilled water.
Adjust to pH 7−7.4 with 1 N HCl.

4. TES Ringer (O'Shea and Adams, pers. commun.)

Dissolve in distilled water:

140 mM NaCl	(8.18 g)	
5 mM KCl	(0.37 g)	
7 mM $CaCl_2$	(0.78 g)	
1 mM $MgCl_2$	(0.203 g)	
5 mM TES	(1.146 g)	
4 mM $NaHCO_3$	(0.336 g)	
5 mM Trehalose	(1.71 g)	total mOsm=337

add

100 mM Sucrose   (43.238)

This is an excellent Ringer for a variety of insects.

Whole-Mount Viewing

Lucifer yellow CH in whole unfixed ganglia, viewed with blue light, fades rapidly. Freshly dissected ganglia or tissue that has been fixed a short time is briefly viewed only to verify the presence of a filled neuron. This is possible if the axon projects in a semitransparent connective, or if a cell body is superficially visible. Viewing fixed dipterous tissue is impeded by high background fluorescence, as shown in Fig. 2A−C.

The procedure for whole-mount viewing is as follows: After washing in buffer, fixed tissue is dehydrated and cleared in methyl benzoate, methyl salicylate, terpineol, or in liquid Spurr's low-viscosity embedding medium (Spurr 1969) (Fig. 3A−D). The first three media yield greater transparency

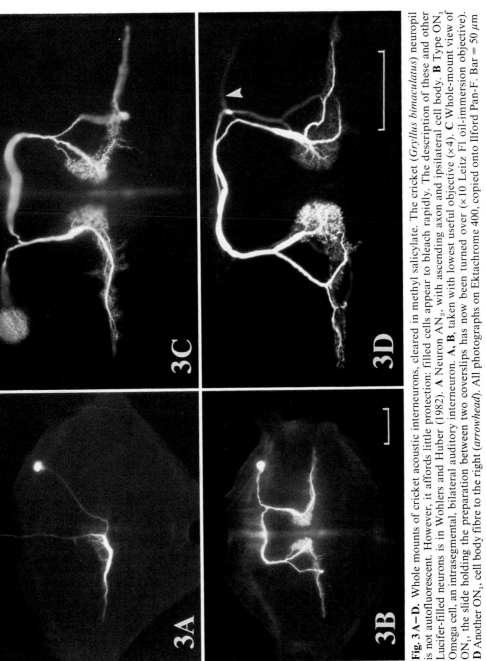

Fig. 3 A–D. Whole mounts of cricket acoustic interneurons, cleared in methyl salicylate. The cricket (*Gryllus bimaculatus*) neuropil is not autofluorescent. However, it affords little protection: filled cells appear to bleach rapidly. The description of these and other Lucifer-filled neurons is in Wohlers and Huber (1982). A Neuron AN₂, with ascending axon and ipsilateral cell body. **B** Type ON₁, Omega cell, an intrasegmental, bilateral auditory interneuron. **A, B**, taken with lowest useful objective (×4). **C** Whole-mount view of ON₁, the slide holding the preparation between two coverslips has now been turned over (×10 Leitz Fl oil-immersion objective). **D** Another ON₁, cell body fibre to the right (*arrowhead*). All photographs on Ektachrome 400, copied onto Ilford Pan-F. Bar = 50 μm

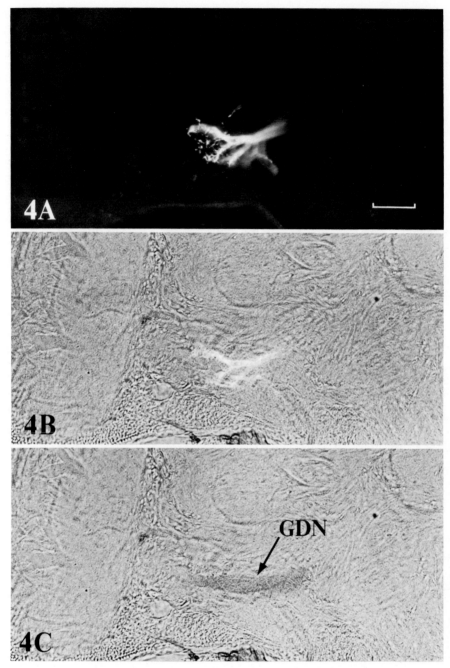

Fig. 4. A Combined illumination showing a unimodal descending neuron (of *Musca*) in the lateral deutocerebrum (10-μm Spurr's section: background fluorescence reduced by cacodylate wash). The cell shares part of a dendritic domain with the GDN and belongs to a

but must be washed out of tissue before embedding. It is most convenient to view Spurr's-infiltrated ganglia during the embedding procedure. Dehydration is through propanol and then into xylol before going into salicylate or benzoate (xylol is omitted when using terpineol) or into acetone before Spurr's. For viewing we use a special slide designed by Dr. Les Williams consisting of a steel plate (normal slide size; $75 \times 26 \times 2$ mm) with a centre hole 10 mm in diameter, closed on one side by a glass coverslip stuck to the metal with Merck Entellan. Tissue is supported on the glass, flooded with clearing medium, and a coverslip holding a drop of medium is dropped over the hole. The preparation can be viewed from both sides with a $\times 25$ oil-immersion objective (Fig. 3 B, C). A convenient long-working-distance oil-immersion lens is the Leitz Ks $\times 22$ (numerical aperture 0.65; working distance $0-2.3$ mm). Steel slides of various thickness cope with various sizes of ganglia.

Whole-mount viewing can resolve the entire neuron and if the tissue has low autofluorescence, as is the case with some Orthoptera (Fig. 3), the results can be stunning. Once polymerized, however, a Spurr's whole mount is completely opaque and alarmingly fluorescent. We know some unfortunates who have thrown away their hardened preparations. Don't. Sectioned, the block will reveal a beautifully glowing neuron (Fig. 5 C, D).

Embedding and Sectioning

Spurr's Embedding Routine (see Figs. 1 A; 2 A−C; 4 A−C; 5 A−F; 6; 9 A, B)

1. Fix tissue in buffered formalin for 1 h.
2. Wash in buffer, dissect out ganglia,
   or dissect out unfixed ganglia onto Sylgard, under Ringer. Drain Ringer and flood with fixative.
3. Fix for a further 6−12 h at 4 °C.
4. Wash for 1 h with several changes of cacodylate buffer.
5. Bring through graded alcohols (propanol is cheap and efficient for light microscopy) until 98% for 10 min each.
6. Wash 2×10 min in dried absolute alcohol (propanol).
7. Immerse 2×10 min in dried acetone.
8. Immerse in 1:1, 1:2 and 1:4 acetone: Spurr's for 10 min each.
9. Immerse in Spurr's 1 h at room temperature. If necessary, view as a whole mount.

---

uniquely identifiable cluster of descending neurons (Strausfeld and Bacon 1983). **B** Epifluorescent illumination with direct illumination through stepped-down condensor. **C** The same, directly illuminated without epi-fluorescence. The GDN was presumably damaged during penetration, hence its photon opacity due to early retrograde degeneration. All photos taken with $\times 16$ oil-immersion objective onto Ilford XPI-400 at 800 ASA. Bar = 50 μm

10. Put drops of Spurr's on a Sylgard-covered glass Petri dish. Arrange ganglia under the binocular microscope.
11. Cut the caps off BEEM capsules. Stick a row of capsules into a strip of plasticine. Fill each capsule to the brim with Spurr's. Invert each capsule over each drop of Spurr's containing ganglia (the liquid will not leak out). Also make two or three "dummy" blocks to test hardness later before cutting. If the blocks are too soft, they can be polymerized for a few more hours. Wear rubber gloves when handling Spurr's.
12. Polymerize blocks at 70 °C for 12–16 h.
13. Cool slowly. This is a critical step. Plastic that has cooled rapidly develops enormous internal stress forces. Sections can crack, break apart and cause other kinds of stress forces in the experimenter when the prize cell fragments to smithereens.
14. Cut 10–12 μm sections on a sliding microtome. The smallest and most sensitive is made by Jung (F. R. G.) and can be operated by pushing the sledge between finger and thumb. If the block is too hard, section under an infrared lamp (Chap. 1). However, the best sections come from soft blocks cut in a cool room (≤ 15 °C).
15. Arrange sections in rows on tiny drops of water, resting the slide on a cool steel plate (4 °C) to prevent evaporation. When all the sections are cut and oriented, put the slide onto a hot-plate at 60 °C. After the water has completely evaporated, draw a line of Gurr's Fluoromount on a coverslip and place it over the sections while the slide is still on the hot-plate. The mounting medium penetrates the Spurr's and air bubbles are driven to the edge of the coverslip. Weigh down the coverslip for 2 h at room temperature to flatten sections. Clean off excess Fluoromount.

Spurr's has a pot life of 3–4 days at 4 °C. We find that Spurr's at least 6-h-old seems to give rise to less autofluorescence for whole-mount viewing than a freshly made batch. Background fluorescence of sections is unacceptable if the block is polymerized too long or at too high a temperature.

Spurr's for Lucifer yellow embedding is made up as follows:

Vinylcyclohexandioxide (ERL 4206)	10.0 g
Diepoxide (DER 736)	6.8 g
Nonenyl succinic anhydride (NSA)	26.0 g
2,4,5 Tris(dimethylaminomethyl)phenol (DMAE or DMP)	0.4 g

DMP is toxic and smells revoltingly of decaying fish. Mix components under the fume cupboard and wear rubber gloves.

Paraffin Embedding Routine (see Figs. 2 E, F; 9 D)

1. Fix tissue in buffered formalin (p. 137) 1–12 h.
2. Postfix in 4% formalin in absolute methanol (up to 1 h).
3. Dissect out brains and ganglia under 70% alcohol.

4. Dehydrate through graded alcohols.
5. Wash 2×20 min in dried absolute ethanol.
6. Rinse in benzene (BIOHAZARD: Use only under fume hood and with appropriate waste disposal containers)
   or
   rinse in xylene.
7. Infiltrate with paraffin wax (or substitute, e.g. Tissue Prep; Fisher Scientific) for 1 h at 60 °C.
8. Infiltrate again in fresh paraffin, or substitute, for 1 h at 60 °C.
9. Block out in aluminum foil cake-tins on a hot-plate (cheap and convenient). Let cool, trim.
10. Cut 10-μm serial ribbons.
11. Place ribbons on a strip of distilled water drawn on a slide from a microsyringe. Do not use adhesives as these contribute to background fluorescence.
12. Warm slide a few degrees under melting point of wax to stretch sections. Drain away water with the cut edge of a filter paper. Blot sections with fluff-free lens tissue.
13. Let dry 2 h at just below melting point of wax.
14. Place slide, section upwards, over a waste trough. Gently flood with xylene to dissolve away wax, rocking the slide from side to side. Repeat several times until wax is washed out.
15. Drain off xylene. Put on a coverslip holding a strip of Gurr's Fluoromount.
16. Let set for 2 h before observing.

Microscopy

The basic filter set-up used for viewing with a Zeiss microscope is described by Stewart (1978). We use a Leitz Ortholux photomicroscope equipped with an automatic camera that integrates from the whole field or from a central spot. The microscope has Leitz interference phase contrast optics (NPL Fluotar ×25/0.55 ICT and NPL Fluotar ×40/0.70 ICT, both dry lenses) and three additional Zeiss objectives, all oil-immersion: Plan Neofluar ×16/0.8; Plan Neofluar ×25/0.8 and a Planapochromat ×40/1.0. The ×25 can be also used for water and glycerol immersion. Epifluorescent illumination is provided by a Leitz Ploemopak, housing an Osram 100 W mercury lamp (HBO 200), which has an emission peak at 435 nm suitable for Lucifer yellow excitation. Infrared heat absorption and a low-pass red filter interrupt the beam which is directed through one of four alternative filter blocks arranged as a carousel above the objective revolver. We employ the following filter blocks: the Leitz L2 (high-pass excitation filter 450–490 nm, barrier filter with cut-off at 520–525 nm), the E3 (436 nm and 490 nm, respectively), and the G2 (350–460 nm and 515 nm, respectively). With L2, the background is dark

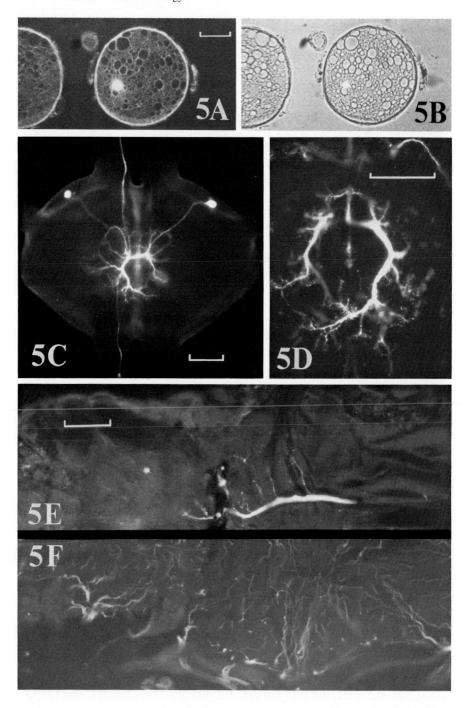

green; the filled cell, brilliant blue-green. With E3 the background is green; the filled cell, brilliant yellow-green and the tracheae, yellow-green. With G2 background is yellow-green; the tracheae, yellowish; and the filled cell, brilliant yellow. An appropriate block combination is used to resolve only tracheae when a cell is suspected of being weakly dye-coupled. Figure 5E, F compares tracheal autofluorescence with LYCH fluorescence (A2 block alternating with the L2 block: see p. 154.).

Trivial though it seems, transparent sections have to be found on the stage. Visibility is provided by low-power interference phase contrast. We home in on the area suspected of containing part of the injected cell and then quickly go through the whole series using low-power epifluorescence to check the cell's extent before taking serial photographs at ×25. High-power enlargements are left to the last since intensely focussed blue light rapidly bleaches the dye.

The Leitz Ortholux is equipped with intermediate magnifications of ×2, ×2.5 and ×3.2. The Ploemopak block further magnifies by ×1.25. Using the ×25 Zeiss Plan Neofluar, final magnification on film is ×78 (with ×2.5 intermediate magnification) and for the Zeiss ×40 Planapochromatic it is ×125. The theoretical limit of resolution is $0.4\,\mu m$ and $0.2\,\mu m$ in each case, but smaller diameter fibres show up crisply because they emit, rather than absorb light. The two higher magnifications embrace an area on the slide of $300 \times 450\,\mu m$ and $190 \times 280\,\mu m$, respectively.

Photography

Chapter 15, Volume 1 described the ideal film for fluorescent microscopy as one whose resolution (lines per millimeter or Lpm) recorded the maximum number of data points (1.3×10^7 and 2×10^7 at the two magnifications above). The resolution of most films is less than the desired Lpm, but since theoretical values assume perfect optics and zero flare, an optimal film would be wasted. Excellent prints can be made using quite fast black and white film such as Kodak Tri-X (ASA 400), developed soft (Fig. 2D). Kodak high-speed Ektachrome 400 for colour slides can be used for serial reconstructions

---

Fig. 5. A 40-μm Spurr's transverse section through neck connective of *Gryllus bimaculatus* showing cross section of the ascending neuron AN$_2$. Combined epi-fluorescent axon illumination and direct illumination with closed-down condenser iris brings out contrast of other axon profiles (**B**). Bar = 50 μm. **C** Low power whole mount of the acoustic relay neuron TN. **D** After apparent bleaching TN was embedded and sectioned in Spurr's. The cell is rotated through 180 °C. Section thickness is 30 μm. Note the low background autofluorescence achieved in cricket neuropil. **A–D** on Ektachrome 400. **E** and **F** Multimodal descending neuron terminal in the fused mesothoracic ganglion of *Musca* (10-μm Spurr's: Ks ×22 oil-immersion objective; filter block G2). **F** shows the other side of the ganglion (A2 filter block excitation λ 270–300 nm, barrier filter cut-off at 410–580 nm) to show tracheal autofluorescence. Bar = 50 μm

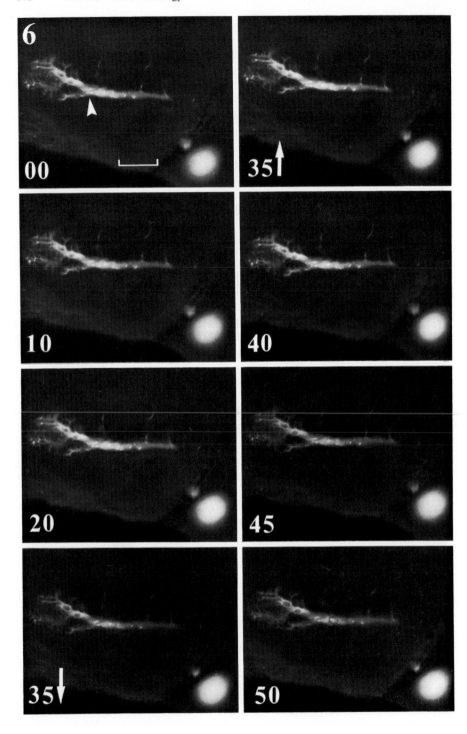

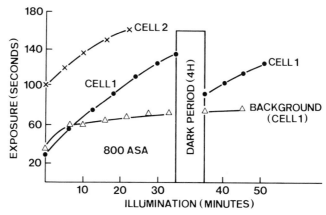

Fig. 7. Exposure (s) plotted against continuous illumination (excitation λ 450–490 nm). Two profiles were photographed in adjacent sections, one at ASA 800 (cell in Fig. 6; *cell 1*) the other at ASA 400 (*cell 2*) using spot measurements as indicated in Fig. 6. Background emission was measured from the integrated whole field on a third section. Background bleaching achieves a plateau after 7 min. There is no recovery of background.

and as master prints for fine-grain black and white copies: We use Ilford Pan F (ASA 50) developed soft in Tetanol Emofin two-stage developer (Figs 1 A; 3A–D; 5A–D). Two newer films are Ilford XPI-400 (Figs. 2 B, C; 6) and Agfa Vario-XL (Figs. 2 E, F; 4; 5 E, F). We use the latter at a range of between ASA 125 and ASA 800, but it can be pushed to ASA 1600. Its black and white negatives are developed by AP70/C 41 colour processing. If this is done in the laboratory, each film should be shot at a uniform ASA rating (speed) because processing times are adjusted according to speed. If the film is sent away for Vericolour processing, the commercial laboratory should be informed at what speed the film was shot. Most automatic cameras can be set for integrated light measurement over whole field or for a spot measurement. We use emitted light from a strongly fluorescing part of the neuron to determine exposure time and then photograph off-automatic. Flare can be reduced at the expense of background detail by shooting ASA 400 film at ASA 800.

Fig. 6. A series of photographs of a dendritic tree in the fly deutocerebrum. The neuron was first exposed for 2 h at excitation λ 450–440 nm and then stored in the dark for 1 year (20–28 °C). It was then photographed using spot measurement of emitted light (*arrowhead*), interrupted by a 4-h dark period after 35 min (*arrow down*). The cell appears to fade within the first 35 min, but after the dark period (*arrow up*) it recovered some fluorescence. It was almost invisible to the human eye after 45 min. Ilford XPI-400. Bar = 50 μm

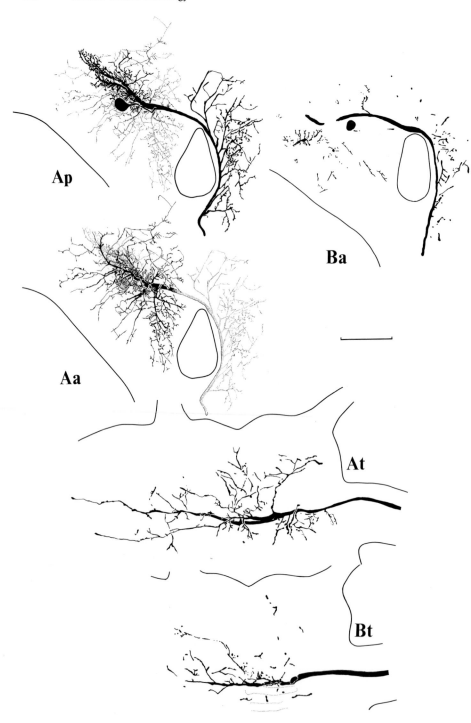

Fading

1. In Tissue

If tissue is not fixed immediately after the experiment, dye is lost from the cell. Possibly this is a special property of insect nerve cells. Stewart (1978) reported retention in gastropod neurons. For controls we filled cells and left the animal in a damp chamber for 10, 20, 40, 60 or 120 min before fixation. The last component of the neuron to show retention is the cell body. Figure 8 compares a complex multimodal neuron in the deutocerebrum of *Musca* fixed immediately after filling (A) and another example of the same cell type fixed 40 min after filling (B). Incidentally, Procion yellow requires a long diffusion time (12 h at 4 °C before fixation) before it is distributed at high enough concentrations to be visible throughout the cell.

    Fading occurs in whole mounts (e.g. in methyl salicylate) and during the washing-out procedure before embedding. Viewing in liquid Spurr's and then polymerizing and section cutting is the safest routine if cells are to be properly studied.

2. In the Section

Prolonged illumination at excitation λ 435 nm bleaches out the dye. However, this is not as rapid as commonly supposed. The cell in Fig. 6 was exposed for a total of 3 h. At the time of writing (2 years after Fig. 6 was made), it still fluoresced. Background, however, bleaches rapidly, particularly in wax sections. We tested fading by setting the automatic camera on point measurement of an axon, and exposed repeatedly at ASA 400 or 800, measuring the exposure times. These are plotted against the total illumination period (Fig. 7). Contrast between cell and background reaches a maximum after a few minutes. Background fading then rapidly reaches a plateau without any later recovery of autofluorescence. However, if the preparation is put in the dark for some hours, the first exposure of the neuron taken after the dark period is of shorter duration than the last exposure before the dark period (Figs. 6, 7). This partial recovery of LYCH fluorescence is extremely useful for obtaining as much information as possible out of a preparation. Although a cell seems subjectively bleached, it is still worth taking optical-section photographs. Long exposure times amplify the residual emission as a stored image on the negative. We have salvaged several weak fills that, to the adapted human eye, seemed hopelessly lost.

Fig. 8. Reconstructions of two (presumably) identical neurons (*Musca*) fixed at different times after iontophoresis of LYCH. Cell *A* [seen from posterior (*Ap*) and anterior (*Aa*) in the brain], and terminal in the thorax (*At*) is a complexly branched descending neuron whose full extent is seen in tissue fixed 2–4 min after iontophoresis. Cell *B* was in tissue fixed after a "diffusion" period of 40 min following iontophoresis. Bar = 100 μm

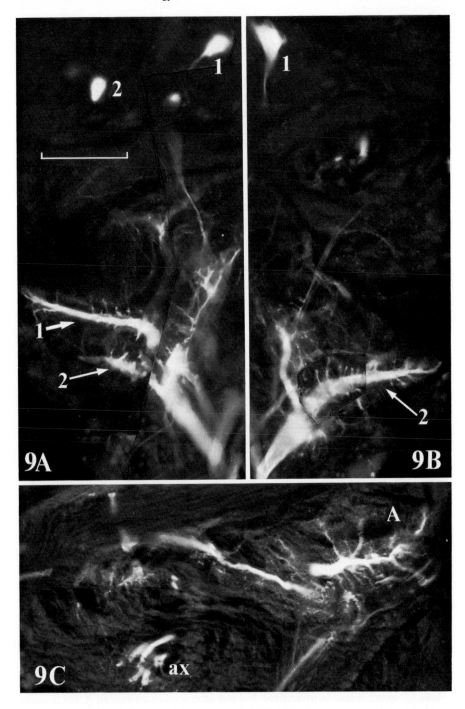

Reconstructions

Drawings cannot be made with a camera lucida because of bleaching. We photograph the cell every $3-4\,\mu$m (optical sections) onto colour slides and project these via a surface-glazed mirror onto paper or a graphics tablet linked to a computer (see Chap. 8). It seems obvious, but surprisingly few people realize the value of coloured pencils which "remember" what was in front and what was behind. Use green for outlines and a different colour for each set of optical sections representing a certain number of 10-μm sections. The cell in Fig. 8 needed 78 slides and one colour would have led to confusion. Drawings are blacked-in in one direction, breaking lines that underlie others to give an impression of depth. Computer reconstructions should be presented as stereopairs (Chaps. 8, 9, this vol), isometric views (Chap. 8, this vol.) or as anaglyphs (Fig. 10).

Geography

Filled profiles should be related to brain regions, tracts and, when possible, to other identifiable cell profiles (Fig. 9). Although Leitz Fluotar objectives are not ideal for epi-illumination, they can combine interference phase contrast with epifluorescent illumination as shown in Figs. 2 E, F; 5 A, B. We usually take these photographs before any others while the cell is at its brightest. If interference phase contrast is not available, an easy trick is to rake down the condenser slightly and close down the condenser iris. This is said to be (and is) a cardinal sin and every student is warned against it for obvious reasons. But if the result can be compared with a reduced silver preparation, the margins of neuropils, tracts and even certain large neurons can be related to the fluorescing neuron (Fig. 4A−C).

Storage

No one likes to have to remain in the laboratory after a long day's recording, waiting until the small hours for brains to dehydrate or infiltrate. Table 1 summarizes stages where the preparation can (or cannot) be left overnight or longer.

◄───

Fig. 9. A, B Backfills of two DNVSs (types *1* and *2*); *Calliphora* (cf. type 2 DNVS in *Musca*; Fig. 2A). Lucifer efficiently shows these neurons whereas silver intensification usually obscures cobalt-filled DNVSs because surface silver deposits lie almost in the same plane as the lateral dendrites. 10-μm Spurr's sections photographed on Agfa Vario-XL at ASA 800. Bar = 50 μm. **C** Backfilled descending neurons; *Calliphora. A* is possibly a homologue of cell A (*Musca*) in Fig. 8. Axon cross sections (*ax*) illustrate good preservation of spatial relationships in this paraffin-embedded preparation. Note background detail. Photographed on Agfa Vario-XL at ASA 400. Bar = 50 μm

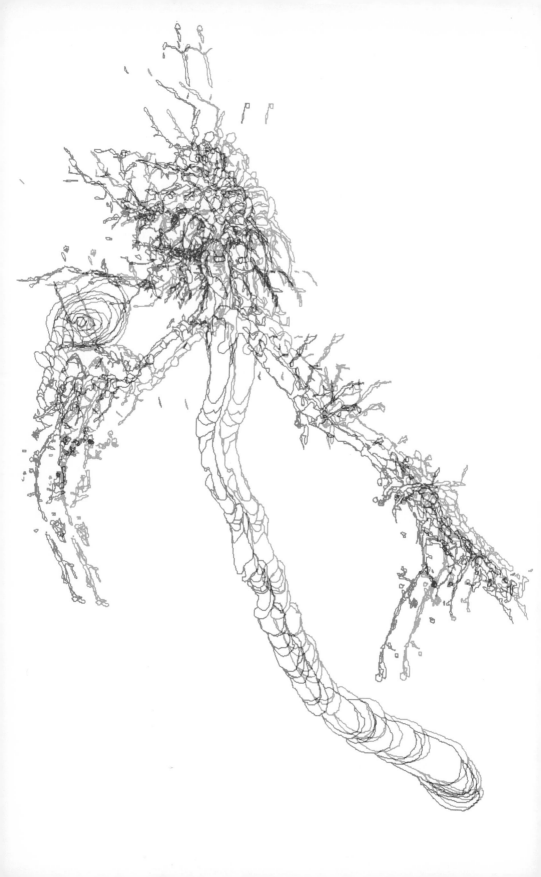

Table 1. Storage of Lucifer yellow (LYCH) filled neurons:

Preceding Step	Storage medium	Temperature	Time
Buffered formaldehyde (1 h primary fixation)	Phosphate or cacodylate buffer	4 °C (dark)	Maximum 4 h
Buffered formaldehyde (1 h primary fixation)	Buffered formaldehyde (secondary fixation)	4 °C (dark)	Up to 12 h
Buffered formaldehyde (6 – 12 h secondary fixation)	4% Formaldehyde in absolute methanol	Room temperature (dark)	1 – 2 h
Buffered formaldehyde (secondary fixation)	Phosphate or cacodylate buffer	4 °C (dark)	Up to 48 h
70% alcohol	DON'T	–	–
Dehydrated into liquid Spurr's	Liquid Spurr's	4 °C (dark)	Up to 4 days (useful for transport)
Polymerized Spurr's	Uncut block	Up to 28 °C	At least 3 years
Paraffin wax	Uncut block	Up to 20 °C	Maximum 3 days
Polymerized Spurr's Paraffin blocks	Sections under Fluoromount	20 °C (dark)	At least 3 years

Fig. 10. Anaglyph of Lucifer-filled and computer-reconstructed Giant Descending Neuron in the brain of *Musca domestica* (Bacon and Strausfeld, in prep.), reconstructed from 74 optical sections, photographed at every 4 μm onto Ektachrome 400 at ASA 800, using ×25 oil-immersion objective. Sections were oriented and digitized as explained in Chap. 8, Fig. 4. The size of the axon at its widest is 20 μm. View through Kodak (Rochester, NY 14650, U.S.A.) wratten filters; No. 25 (red, left eye) and No. 61 (green, right eye)

Artefacts

1. Long filling times seem to cause local damage to the cell at the point of penetration, although the cell may appear normal elsewhere. The membrane appears torn and a halo of extracellular LYCH surrounds the axon. Lucifer yellow will also be incorporated into sheathing glia cells.

2. Extracellular leakage can occur when filling nerve bundles. Glia nuclei readily accumulate LYCH as does the perineural sheath. Extracellular iontophoresis from electrodes into cricket ganglia shows local uptake by glia cells but not by neurons.

3. Glancing penetrations. The advancing electrode continually leaks Lucifer yellow. Several cells may be penetrated for periods ranging from a few seconds to some minutes before a stable recording is achieved. Presumably all penetrated cells receive some LYCH which dissipates by the time a stable cell is recorded and filled. The cell that has been filled for the longest time can usually be distinguished from glancing penetrations by its brightness alone. However, pale fills from glancing penetrations are critical when interpreting dye coupling. When two cells are apparently contiguous but their axons share the same height in a connective, they cannot be assumed to be dye-coupled. Only when two cells converge far from the recording site (and ideally arise from contralateral neuropils), can dye coupling be claimed with assurance.

4. Mechanical tension. Cells cannot always be filled over long distances. Although we do not know why, imperfect fills, distal or proximal to the electrode, can occur when the connective is stretched over the indifferent electrode (spoon) or some other kind of mechanical support. The stretched axon is presumably compressed, locally blocking dye diffusion. The best place to obtain stable recordings from long-axoned neurons is at their point of entry into a ganglion.

5. Extraneous fluorescence. Native pigments can ruin an otherwise good preparation. The retina should, if possible, be detached from the brain after a 1-h fixation if tissue is to be stored in fixative or buffer for longer than 6 h.

6. Autofluorescence. Unfilled neurons and glia yield a green background fluorescence (auto-, or tissue fluorescence). Large axons appear dark, however. Trachea usually fluoresces at excitation $\lambda\ 340-430$ nm with an emission at $\lambda\ 520$ nm. If the LYCH concentration in a cell is low, we use a G2 filter block (high-pass excitation at $350-460$ nm, low-pass barrier at 515 nm) since this shortens exposure times. However, when G2 is used, tracheae can be mistaken for weakly dye-coupled processes and it is advisable to check if they are genuine Lucifer-filled elements by switching over to A2 which lets through only tracheal autofluorescence (Fig. 5 E, F).

Conclusions

Stewart's Lucifer yellow has already persuaded even the most reluctant electrophysiologist to go intracellular. The dye is a boon, but at the same time it can be dangerously seductive because it so conveniently provides instant feedback about the shape of a recorded neuron. However, a drawing of a filled neuron displayed alongside electrophysiological data does not comprise much more than the message "I was there". On its own, any single neuron tells us rather little about the functional organization of the central nervous system. However, we hope that this chapter will encourage the physiologist to look more closely at the organization of the neuropil in which the cells project and make their connections.

Acknowledgements. We thank Walter Stewart for the gifts of Lucifer yellow. We thank Franz Huber for hosting us (J. P. B., D. W., N. J. S.) at various times in his institute. We affectionately remember the aid of Andechser Dunkles, and the "Rondelfliegen."

Chapter 8

Portraying the Third Dimension in Neuroanatomy

Peter T. Speck and Nicholas J. Strausfeld

European Molecular Biology Laboratory
Heidelberg, F. R. G.

Introduction

Reconstruction of biological objects should, ideally, demonstrate the significance of structures in terms of their dynamic roles in a functional system, such as a cell membrane or a brain. This is the main concern of neuroanatomical illustration, and it is equivalent to graphical modelling in molecular biology. Indeed, three-dimensional manipulation by a computer need not distinguish a molecule from a neuron.

This chapter describes the design and some of the achievements of a computerized neurographics system which was initiated in 1975 with the encouragement and advice of Drs. John White and Sydney Brenner at the MRC Laboratory in Cambridge where a system had already been developed for reconstructing the central nervous system of the nematode *Coenorhabditis elegans.*

Visualization of surface textures, the recognition of frontness and viewing in depth are encountered and solved by our visual system in everyday life. However, it is surprising how difficult it is to recognize the relevant data structure and to formulate an algorithm that will enable a machine to perform a fraction of the same functions. From this point of view, designing computer graphics for neuroanatomy not only involves methods for storage and retrieval but also poses questions about basic concepts in experiencing the viewing process. "What do I see?" is rephrased to "How can I make a simple statement describing what I see?" Recognizing the necessary and sufficient elements by which to describe a structure fully is the essence of computer graphics. The neuroanatomist is obliged to operate within precisely defined constraints.

Why Computer Graphics in Neuroanatomy?

Computer graphics serves two purposes. One is to overcome the dimensional constraints of the page. The other is to investigate strategies for quantifying complex branching structures. Since the first has to be implemented before the second can be achieved, most of this chapter is devoted to outlining the way we re-assemble and display structures from serial sections. The results are illustrated by examples of normal and experimentally altered neuropils, cobalt-coupled systems, electron micrographs of synaptic systems and dye-filled neurons. Other techniques, such as grey-tone enhancement, predictions of two-dimensional patterns and skeletonization are dealt with briefly at the end of this chapter.

For simple images with planar surfaces, reconstructions and rotations can be done by hand on graph paper in much the same way as that employed by an architect drawing an elevation. However, curved surfaces and branches defeat even the best draughtsman and in order to display them from viewpoints other than that occupied by the microscopist, quite complex computations have to be used. For the topologist, neurons are simply bags like any other cell. But although the neuron starts its life as a simple bag, during development it undergoes many topographical distortions. Its final three-dimensional branching pattern contributes to local electrical events and, in part, determines its integrative properties.

The study of form and function is indivisible and the most appropriate way to describe form is to illustrate it. This is the power of computer graphics: to convey a complex message in the most direct way possible (Levinthal and Ware 1972).

Designing the System

In their unholy alliance, the computer scientist and the neuroanatomist share one basic idea: that a complex three-dimensional object can be described as a set of two-dimensional structures (Fig. 1). The anatomist has to create sec-

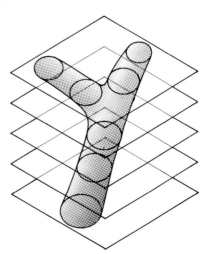

Fig. 1. A branching object described as a sequence of two-dimensional sections

tions to resolve a nerve cell. The computer scientist has to operate on three-dimensional structures in terms of two-dimensional graphs since most physical display devices operate in two-dimensional space. This is the starting point for designing a simple interactive system that couples the observer to the machine. All other requirements are subsequently specified by both members of the alliance, arriving at a consensus that is optimal for the naive user and that allows further additions and refinements by the designer.

For the user, it is important that the program is lucid and that the strategies used are basically simple. Feedback between the user and the machine is achieved by implementing a simple set of instructions (the menu) which are sufficient for all operations during graphical input. This first stage involves a dialogue between the user and the machine and is called the "interactive" input.

Computations that manipulate the stored image (the noninteractive stage) employ a separate set of instructions initiated after the user answers an operational questionnaire presented on the screen.

Audiovisual signals are used during the interactive phase for error elimination, correction and continuity. The preliminary graphical information relayed back from the computer to the screen reflects the structural representation of the information held in core memory.

The Equipment

Operational simplicity must be achieved if a graphics system is to be run on a modest general purpose computer (e.g. storage requirements not exceeding

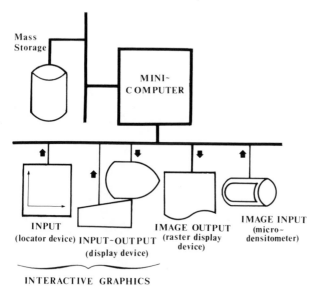

Mass Storage

MINI-COMPUTER

INPUT
(locator device) INPUT-OUTPUT
(display device)

IMAGE OUTPUT
(raster display device)

IMAGE INPUT
(micro-densitometer)

INTERACTIVE GRAPHICS

Fig. 2. The NEU hardware components. The micro-densitometer is an extra preprocessing device not linked to the interactive computer graphics

64 Kbytes in main memory). The external memory, which leads to a page-wise organization of the secondary memory structure, requires a minimum of 256 Kbytes. The basic hardware is a Kontron MOP 1 magnetostrictive $35\times35\,cm^2$ tablet equipped with a signal stylus interfaced to a 192 KW NORD-10S, with two CDC 66 Mbyte disc drives. Interactive visualization is with a Tektronix 4014 storage display oscilloscope and paper prints are made with a Versatec 1200 A high-resolution electrostatic printer-plotter. Data traffic between the terminal and computer per session is about 1 Mbyte with a permanent disc file space requirement not exceeding 10 Mbytes. Completed data is archived on tape.

The NEU System

The Programs

The neurographics (NEU) system consists of a set of programs to collect, model and display spatial information. Most of the programs are written in PASCAL (Jensen and Wirth 1978; Wirth 1976). This is a highly flexible language enabling underlying ideas to be explicit in individual algorithms. These are declared within the program itself (see p. 165).

In this chapter we describe colloquially the rationale underlying the design of programs without actually going into their technical details. The main program used for data acquisition (CLOVER) employs a hierarchical structure whose components are subroutines with high-level specificity. For potential designers of a computer graphics system the program reference manual and user's manual are available from the European Molecular Biology Laboratory (Speck 1981 a, b).

Modest systems require splitting the representational task into an interactive and noninteractive part. The latter's purpose is to provide a high-quality graphical representation (hard copy print-out) of the acquired structure from an arbitary viewpoint. In designing the basic framework, vector representation was employed for interactive operations (drawing in profiles with feedback of the entered image). Matrix representation was employed during the noninteractive rendering.

Interactive Operations

These are no more complex than pointing a stylus at specific areas of the tablet (the menu or instruction area) and then using the stylus to draw outlines of sections onto paper attached to the tablet. The stylus houses a receiver coil for detecting electromagnetic waves transmitted through wires beneath the tablet's surface. Typing instructions are kept to a minimum, and ideally these are only necessary for logging in and out of the computer.

Using the Kontron system, flexibility is increased by two possible modes. One is used to create the profile data set and send coordinate instructions representing polygons of the projected image to core memory. In this mode when the drawing pen of the digitizer is set on the tablet's surface and moved around, a continuous stream of coordinates (presently up to 50 s^{-1}) is sent to the host computer. The other mode is used for operating in the menu. The simple act of touching a certain zone of the tablet, representing elements of the "menu", will perform updates, deletions, jumps in the sequence of frames and other necessary operations. The basic set-up, interfaced to a Nord computer, is shown in Fig. 2. Additional peripherals, such as an electrostatic printer-plotter and an optronics microdensitometer require separate programs, but their operation is basically simple.

Before the first profile is entered, the operator defines the frame perimeter, or view box, which is the operational area on the tablet. This contains the projected image. The view box will fill the screen of the control display even if it occupies only a fraction of the tablet. The final representation is thus enlarged or reduced depending on the size of the original projection within the defined working area of the tablet.

The second operation that precedes any drawing is setting the proper distance between successive frames. Assuming that the intersection distance is known, then the scale in the plane of section is employed to draw a line representing the distance between sections. This sets equidistant spacings of successive stored-frame images for the final rendition. An error at this stage

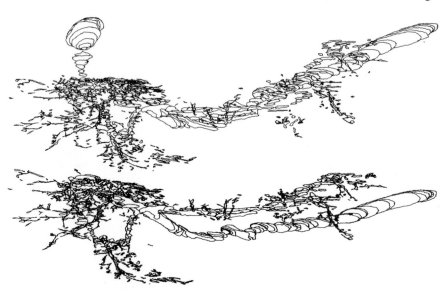

Fig. 3. Incorrect frame spacing can give rise to what is, subjectively, a plausible but incorrect view (*upper rendition*). *Lower rendition* shows a view of the Giant Descending Neuron of *Musca domestica* (see also Fig. 9) where sections are separated by 4 μm, not 8 μm as above

carries distortion in the Z-axis. Sections will be too close or too far apart with obvious consequences in hidden-line removals when the image is rotated (Fig. 3).

Alignment of Sections

Some reconstruction systems employ sophisticated methods for pre-aligning photographs before they are traced into the computer (Stevens et al. 1980). This is usually done by matching successive film frames by optical align-

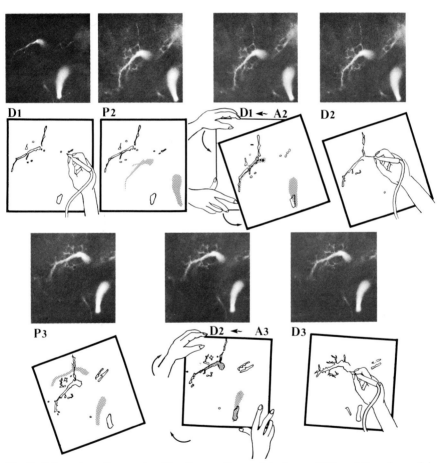

Fig. 4. Orientation of successive film frames. Frame 1 is drawn (*D1*) and then frame 2 is projected onto the tablet (*P2: stippled*). Frame 2 is next aligned by hand over the stylus trace of *D1* (*D1 ← A2*). The ballpoint stylus trace of D1 is removed, and a new sheet of paper receives the stylus trace of the drawing of frame 2 (*D2*). Then the third frame is projected (*P3: stippled on tablet*) onto the tracing of the second. The third projection is aligned by moving the tablet (*D2 ← A3*) and so on. The final result from this projection series is shown as an anaglyph in Fig. 10, Chap. 7

A

B

C

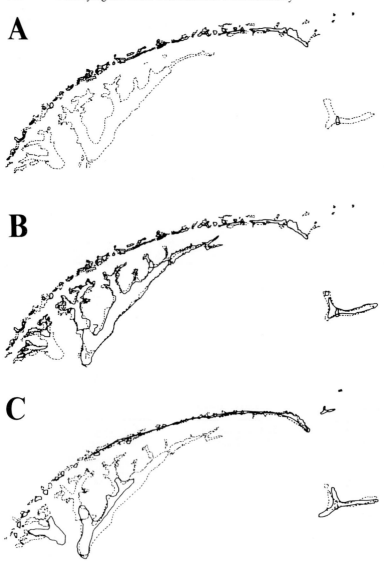

Fig. 5. Each tracing is relayed as a set of coordinate values which are displayed "on-line" on the storage oscilloscope. The previous frame is shown as a dotted outline. In **A** film frame 1 (*dotted*) of a sex-specific neuron in the male fly is superimposed by outlines of the next section (film frame 2). The completed set of outlines (**B**) is shown in the next frame (**C**) as a set of dots which now underlies the traced outlines of frame 3

ment (stroboscopically or by pseudostereopairs) and then rephotographing the aligned frames onto a continuous strip of 35-mm positive film.

We have dispensed with pre-alignment and do the matching directly, during the interactive phase, while tracing into the computer. Each film frame is mounted in a slide holder and projected onto the tablet. In the case of high-contrast colour positives or black/white negatives (Lucifer yellow fills or cobalt fills), it is sufficient to draw the profile on paper stuck to the tablet. This is a temporary record of the profile outlines of the initial film frame (No. 0) and is simultaneously entered into the computer. The next film frame down (−1) or up (+1) is brought into register with the drawn outline simply by moving the magnetostrictive tablet holding the attached drawing until this coincides with the projected image (Fig. 4). Alignment is by triangulation and can be done for ten or so frames before changing the paper. The close correspondence of the reconstructions to neurons seen in whole mounts demonstrates the accuracy and reliability of this method (see also Chap. 9).

Electron micrographs consist of many similar profiles without distinctive landmarks unless they include cobalt or Golgi or heme protein reaction products (see Chaps. 1−3). Alignment of unmarked sections requires that each electron microscope negative is matched to its own and to the foregoing positive in the series. The positive of the first negative is attached to the tablet, and tablet-with-positive is aligned with the first negative. After the profiles are entered, the second negative is aligned with the first positive by moving the tablet this way and that. The first positive is removed, and the second positive is aligned with the projected image of the second negative and then stuck to the tablet. This procedure is repeated through the series. Each new film frame that is entered appears on a graphical display as a solid line superimposed on a dotted line of the previous frame (Fig. 5). Alignment can be immediately checked and, if necessary, the misaligned frame can be deleted from the file page to avoid parallax errors.

Interactive Profile Acquisition

A Tektronix 4014 storage tube display is used to monitor graphical input and to provide audiovisual feedback for interactive data acquisition. The device physically retains the graphical information sent from the central minicomputer and can "write-through" the stored drawing, thus allowing the dynamic display of a limited number of vectors. This feature is used to indicate the current pen position of the digitizing tablet either by a "crosshair" or a "rubber-band" indicator on the screen.

The acquisition of profiles relies upon a program called CLOVER, divided into the following command modules which deal with depth control. These cause the current content that is displayed on the screen after a frame has been drawn to be appended to the data base, and the plane of the next frame to be moved up or down to receive the next input.

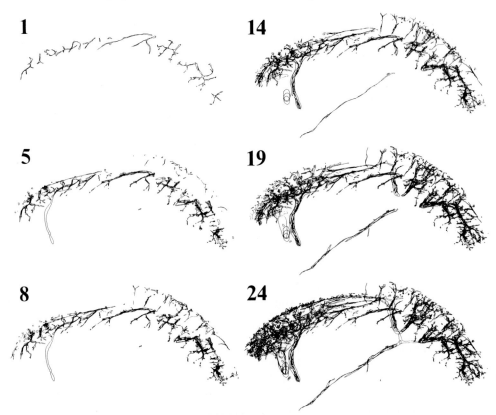

Fig. 6. An image is built up frame by frame. This shows superimposed profiles drawn from frames *1, 5, 19,* leading to the final rendition of three male-specific cells of the fly *Calliphora.* Here the cells are shown as "open" profiles and have not been treated as hidden lines. Some rotations are shown in Fig. 11

The command UP moves the drawing plane up to the next frame which is numbered in sequence. The command DOWN does the reverse. The command SELECT moves the drawing plane from the present frame to one that is defined by the operator on the keyboard (one of the few keyed-in commands). The command TOP moves the drawing frame to a new frame that surmounts all others previously employed. The command BOTTOM does the reverse.

The previous frame, which is the frame below when moving up and the frame above when moving down, is displayed in dotted lines so that alignment with the new frame can be checked (Fig. 5).

CLOVER allows any frame to be revisited and new profiles to be added to it while alignment can be checked either against the next lower or upper frame. All profiles drawn after the last command are kept in temporary store. Any command other than ERASE transfers the profiles entered after the previous command from temporary store to the graphical data base. This data transfer can be initiated explicitly with a SAVE command, freeing temporary store occupied by numerous or elongated profiles for new profiles on the same frame.

The ERASE command wipes out all new profiles entered since the last command and leaves old profiles untouched. If there are only old profiles in the current frame, CLOVER asks how many profiles the user wants removed and wipes out the given number of profiles in the reverse order in which they were entered. If both old and new profiles have to be removed, one ERASE command will delete all new frames and a second ERASE command will remove as many frames as the user desires.

A RETURN command returns control to the operating system and terminates the CLOVER program. Restarting CLOVER sets it up in drawing mode on a new frame on top of the frames already entered. Highly complex structures can thus be composed (Fig. 6).

Graphical Data Structures

Data structures for building large graphical data bases in the NEU system use a relational model treating line drawings as graphs.

Basic constituents of this model are *points* with their spatial location attached and *lines* forming a binary relation within the points. In addition, points can carry integer *marks*. Thus the model forms a *labelled digraph* (Harary 1969), the points of which are labelled with their coordinates in three-dimensional space and occasionally with an integer. A point can be connected to an arbitrary number of neighbouring points. This is realized by maintaining a dynamic list of lines for each point. The basic data types of the model expressed in PASCAL notation are:

```
point
= record
    head of line list: line;
    x, y, z: integer
    end;
line
= record
    neighbour: point;
    next: line
    end;
mark
= record
    badge: integer;
    end;
```

MARKS   **POINTS**   **LINES**   **Fig. 7.** Elements of the NEU
                                                          graphical data base (see text)

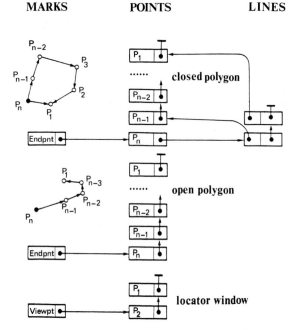

These building blocks are designed to construct descriptions of complex graphical objects. However, the representation is entirely modular, express-ing topological simplicity. This is shown in Fig. 7 which depicts the elements of a *graphical data base* used in the NEU system. Description of *open* and *closed polygons* and a *locator window* (on the tablet) are represented in basic modules of the relational model. Note that the notion of a polygon is in-troduced with a unique mark (endpt) anchoring the endpoint of a polygonal chain. Open and closed polygons are distinguished by the linear and circular arrangement of their line pointers. For points only connected to their im-mediate predecessor (in store) no explicit line element is created, instead the head of line list entry of such a point holds a reference to a unique line element (see p. 10 in Speck 1981 c).

Collections of these polygons listed in a common file compose a spatial object as a collection of outlines on consecutive cross sections. Such a data collection is called an *object graph*.

Drawing a profile results in its simultaneous representation in a core memory "page". When this is saturated, it is transferred to a disc and a new page is made available. The overall memory structure is analogous to a re-cord in a log book. Core memory holds only the information of the currently "open" page while disc storage holds all the remaining pages, including a copy of the current page before its transfer into core memory.

New graphical information is always appended at the end of the graphi-cal data base regardless of its positional content. This means, for example, that deletion of profiles does not result in removing information pertaining to them and then reshuffling all consecutive data. Rather, deletion tags the relevant records as being "invisible."

In this conservative fashion the data base is kept relatively static while maintaining flexibility for the user program by letting it view the entire structure through a "window" page in core. This is filled and replaced by a set of memory management programs written in a systems-programming language similar to "C" (Kernighan and Ritchie 1978).

Noninteractive Operations

Selecting the View

The subsequent operations performed on the stored image run automatically after an appropriate interactive questionnaire has been called and answered by the user. Noninteractive operations are entirely concerned with surmounting the constraints imposed by two-dimensionality. Their final output is onto the printer-plotter.

Dimensional constraints require that a three-dimensional image is obtained and mapped into two-dimensional space. This must be performed so that it still conveys information about the third dimension by means of "projection mapping." This can be achieved in a number of ways, but the most reasonable method is to apply an ergonomic approach whereby the user has an intuitive understanding of how the view is obtained. For this we have used the concept of a "synthetic camera" which is also used in modern graphics standard (see Graphics Standard Planning Committee 1979).

The Synthetic Camera

This is an imaginary entity that, by means of its orientations in object space, creates a parallel projection of the reconstructed image. It establishes correspondence between vectors in picture space and vectors in object space. The viewing direction, which is in object space, is determined as normal to the viewing plane that holds the projected view. A "view right" is an additional device, also defined in object space, whose projection onto the viewing plane defines the direction of the left margin of the print out.

The vector/matrix operations are based on the notion of homogeneous coordinates (Rogers 1976). A kernel procedure composes a transformation matrix to map the original object coordinates under given viewing parameters into *normalized device coordinates* (for glossary of terms see Yen 1981). These coordinates (u, v, n-system) represent horizontal $u = (1, 0, 0)^T$, vertical $v = (0, 1, 0)^T$, and depth expansion $n = (0, 0, 1)^T$ of an imaginary three-dimensional viewing display.

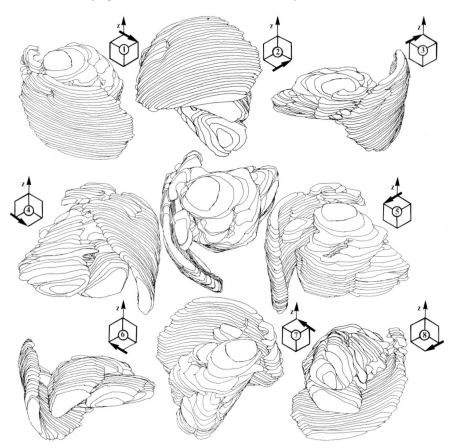

Fig. 8. Eight isometric views of a developmentally abnormal medulla, lobula and lobula plate of the fly optic lobe. The orientation of the original input is at centre and is composed of 30 sections. The viewing angles are projections from the centre to the vertices of an imaginary cube enclosing the original input. There are eight "corner" views (input reconstructions by courtesy of D. R. Nässel)

Fig. 9. Eight views of a reconstruction of the Giant Descending Neuron of the fly (dendritic tree only). The original input orientation of 70 sections is shown centre. The eight views encompass a good all-round view of a cell filled with Lucifer yellow (Courtesy of J. P. Bacon) and any view can be observed as a stereopair

Viewing parameters are:

 centre of projection
 view right direction
 viewing plane normal defined in object space

and,

 depth of the viewing volume
 centre of the viewing volume
 height and width of the viewing volume

referring to the u, v, n-system.

The viewing transform is set up as a coordinate transformation from object space into the u, v, n-system.

The *centre of projection*, or eyepoint, marks the origin of the u, v, n-system in object space. For orthographic projections the *projection direction* coincides with the *viewing plane normal*. Only the latter needs to be specified. The *view right vector* projects onto the u-axis.

The operator is asked to specify the desired viewpoint rather than actual matrix terms (which are indeed used internally during the numerical evaluation of a mapping). A map of the reconstituted structure is available from any viewing direction. However, for simplicity the user is restricted to only ten possible aspects, any of which can be given as a single picture or as a stereopair. The operator must imagine that the reconstructed object lies in a cube with the last entered profile lying uppermost and parallel to the plane of the cube's upper surface. Lines drawn from the centre of the cube out through its eight vertices are the viewing directions. The ninth view is the stack of profiles, oriented in the same way as they are in the original frames, as seen from above; and the tenth, as seen from below. The eight isometric views, and the top and bottom views, are displayed in parallel projection. Enlargements are shown as percentages of the original top view. All views are therefore subject to real measurements.

The ten views are sufficient to give an "all-round" impression of a relatively simple or a highly complex object (Figs. 8, 9), Depth perception is achieved if the operator calls for a stereopair of any of the ten views (see Chap. 9), tilting the image to give a 7° retinal disparity. Ten stereopairs of ten views is a poor man's "real-time" rotation. Some examples of stereopairs are shown in Figs. 15–17.

Rendering the Reconstruction

The main reason for switching from vector to raster data representation was the availability of a brute force algorithm for hidden-line removal, which was one of our essential requirements. Wire figures, whose profiles are transparent, cannot simply provide information about near surfaces even though,

Fig. 10. Comparison of non-hidden lines (wire-frame) and hidden-line renditions. The first gives rise to a "Neckar-cube"-like optical illusion. The surface features of the second are unambiguous. (Optic lobe neuropils of the fly). (Original reconstruction courtesy of D. R. Nässel)

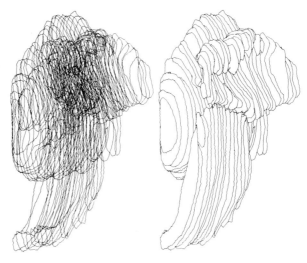

Fig. 11. Wire-frame figures function well for branched structures of limited depth and complexity (see Chap. 9) particularly when viewed as stereopairs. However, complex neurons occupying great depths give rise to confusing images when rotated through more than 5−10 °. In the *second* picture the original rendition (*left*) is X-axis rotated towards the viewer by 10 ° and around its Z-axis also by 10 ° (moving the lower edge nearer and the upper edge farther away). The *right-hand view* shows rotation around the Z-axis 10 ° in the opposite direction and, in addition, a Y-axis rotation of 10 °

viewed as stereopairs, they have distinct advantages for viewing through several cells when these are overlying each other (see Chap. 9). However, it is often impossible to make head or tail of the final rendition, particularly in the case of neuropil volumes or groups of neurons comprising 50 or more frames. Figures 10 and 11 illustrate this rather nicely.

Hidden-line removal algorithms operate either in object space or in image space. In the first, every pair of subdivisions (patches) of the surface of every profile in every frame is compared. Patches that coincide are removed from the appropriate frame representing behindness. To function accurately without eroding (aliasing) the structure, the subdivisions of the surfaces have to be extremely small. The operation consumes large amounts of computer time. A less cumbersome method is the "brute force" operation in image space (Sutherland et al. 1974). In this the final picture points are compared with each other through the reconstituted image. One may imagine that the wire-frame, three-dimensional image is intersected by many closely packed girders forming a tight two-dimensional matrix holding Z-values corresponding to the X- and Y-values of their indices. Only points that occupy one matrix element in depth are compared, and the point nearest the imaginary "camera" is retained in the final print-out (Fig. 12). The resolution of the final rendering can be chosen according to the desired image quality (Fig. 13) and the amount of computer time available. The time taken for such hidden-line operations is linearly proportional to the complexity of the structure.

Smoothing the Image

Various methods exist for smoothing jitteriness and minor errors of tracing during the interactive data acquisition. However, these are costly in time and can be neglected in our application. During transformation of the vectorial representation to quantified raster representation, jitter is eliminated as is scatter due to foreshortening of lines after projection.

The subdivisions of points in vectorial space are so much finer than subdivisions in screen or raster space that the conversion from one to the other is a smoothing process in its own right.

Types of Rendition, Speed and Other Attributes

Two types of rendering are defined during the interactive acquisition phase. Closed polygons trigger a different operation than open polygons since the former are filled with surface elements. Thus, closed polygons are subject to the hidden-line operation during the noninteractive rendering phase. Open profiles cannot trigger hidden-line operations and are rendered as transparent polygons or "wire-frame" figures. The experienced user can draw a

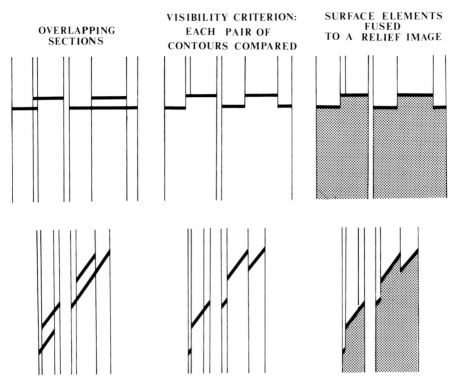

Fig. 12. The principle of the hidden-line algorithm. In the *top row* the hidden-line operation is performed on the original top-view stacking. In the *lower row* the same operation is performed but after rotation. Further explanation in the text

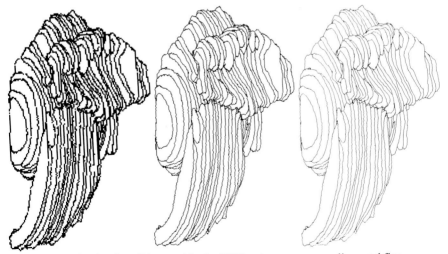

Fig. 13. Three levels of rendition used in the NEU system: coarse, medium and fine

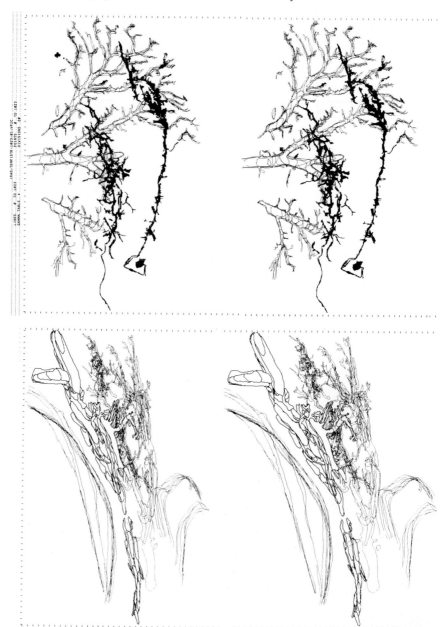

Fig. 14. Closed and open profiles used in the NEU system. The *top rendition* is a stereopair of two "vertical type 2" amacrine cells (*black closed profiles*) whose processes extend from the caudal layers of the lobula plate to the frontal layers containing the horizontal cells. Here parts of the north horizontal cell (see Chap. 9) are shown as open profiles (*dotted lines*). The *lower stereopair* shows terminal branches of the north and equatorial horizontal cells of the lobula plate of *Calliphora*, impregnated by the Golgi method, terminating onto

Fig. 15. Stereoview of "top view" of blacked-in Giant Descending Neuron dendrites (Lucifer yellow filled by J. P. Bacon. For comparison see the anaglyph in Fig. 10, Chap. 7)

wire-frame "closed" profile by describing first one of its edges and then the other, avoiding closure of a continuously drawn perimeter. Open profiles are printed as a dotted line derived from one of three possible grey levels as described below.

After acquisition, the computer instructs the user to answer a list of questions that pertain to the final rendition. One of these is the selection of picture (raster) points and grey levels. In the present system there are only two levels of representation in the final output: black points or nothing (1 or 0). The problem encountered in drawing thin lines or grey shading is the same as that in newsprint. Continuous shades have to be represented by dot patterns which are entities smaller than the picture points themselves. In the present system two sizes of picture points are used (picture elements or pixels) consisting of 4×4 or 2×2 subdivisions (nibs) giving medium and fine lines, respectively. Faint lines used for wire-frame profiles employ large pixels with two nibs each. Surface profiles of hidden-line (solid) figures employ either 4×4 or 2×2 nibs of the medium and fine pixels. Blacked-in pro-

certain dendrites of descending neurons (*open profiles*) that were backfilled with cobalt chloride and intensified (see Chap. 6). Closed profiles do not obscure open profiles. In this rendition open lines also depict the lateral edge of the posterior brain and the oesophageal foramen (*lower right*)

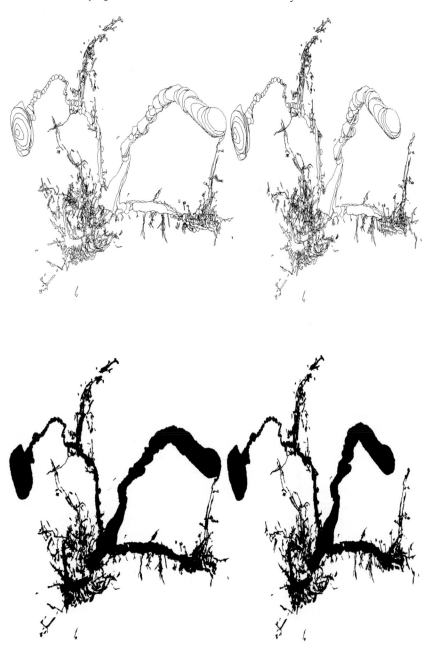

Fig. 16. Two pairs of stereoviews of the same Giant Descending Neuron as in Fig. 15 shown as "corner" views (cf. Fig. 9). The top stereopair is unfilled, with hidden lines, whereas the other is shown black (filled in). There is no loss of depth perception. The ocular disparity in Fig. 15 is about $\pm 3.5°$. Here it is $\pm 5°$. Even this hyperstereoscopy allows good cortical fusion of the two images

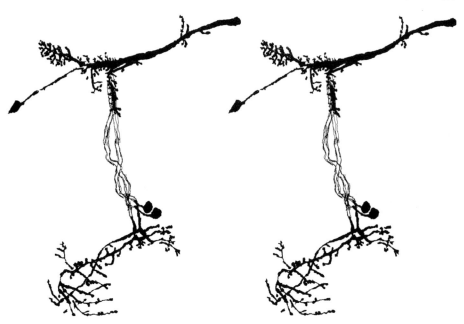

Fig. 17. Filled-in closed profiles and transparent (wire-frame) open profiles allow optimal resolution of contacts between cells. In this figure two vertical cells in the lobula plate of *Drosophila* (VS cells No. 7 and 8) are shown filled transsynaptically from both sides of a dendritic branch of a vertical descending neuron (*top*) (Preparation by courtesy of H. S. Seyan)

files can be achieved by printing surface elements at the highest grey level. Usually, however, solid renditions have surface elements at the lowest grey level (0×0 nibs in a 2×2 pixel). In other words, they are white. The various options available are shown in Figs.14–18.

Final Remarks

The present system is a useful aid for visualizing neuroanatomical structures including groups of neurons and for interpreting electron microscopical serial sections (Fig. 18). Probably its only limitation is storage space and the ability of the user to sort out tangles of cross sections belonging to many different neurons within complex (unstratified) neuropil.

It would be desirable to use the system for prediction of cell arrangements (Fig. 19) and quantitative analysis. One example is plotting-in synaptic distributions and keeping a permanent record of the measurable aspects of cell morphology so that they can be compared and statistically analysed. One method is to make simple stick diagrams of nerve cells, recording synaptic densities, widths and lengths of processes and so forth. Programs have

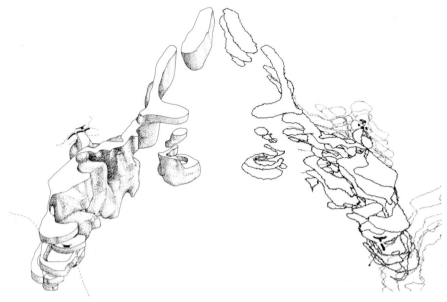

Fig. 18. Hidden line and open profiles can be combined usefully in electron microscopy reconstructions. *Right* rotated view of a small-field terminal synapsing onto a vertical cell (*dotted profile lower right*). The T cell (closed profiles with hidden lines) is also post-synaptic with at least two other profiles to a single element (*middle right*). An artist's rendition of the synaptic connections is shown to the left. (Courtesy of U.K. Bassemir)

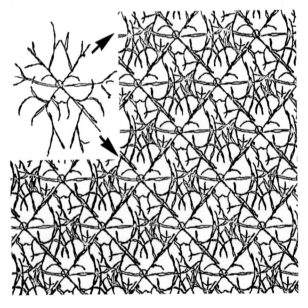

Fig. 19. A "spread-out" display of a reconstructed planar dendritic tree (*upper left*) showing its "ideal" dendritic networks within a planar neuropil. Neurons are spaced out according to the retinotopic mosaic, whose X- and Y-axes are *arrowed*

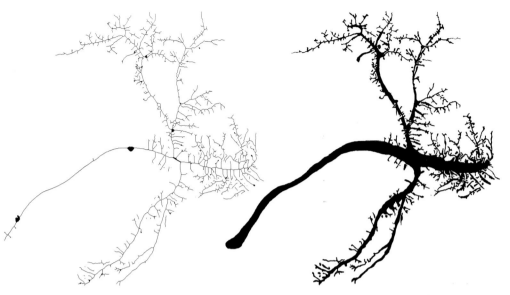

Fig. 20. Line diagrams (*left*) can be made from a two-dimensional "collapsed" image (*right*) and can be used for plotting out additional data such as spine density or connections

been devised (Speck 1981 c) that can render a branched structure as a stick structure (Fig. 20). Scanning densitometers, which first operate on the exclusion of defined grey levels, can exclude vague outlines of neurons, extracting only those profiles exactly in focus. This creates a highly contrasted image for "stick" figures (Fig. 21). This type of pattern enhancement is a useful adjunct to preprocessing of information before its entry into the computer, particularly in histological preparations where there is a large amount of ambiguous background structure and subjective bias. Serial sections of mass-impregnated cells, where there is a high noise level, are particularly amenable to this approach as shown in Figs. 21 and 22.

Finally, a look into the future. Since 1975 there has been an industrial revolution in circuit design, the refrain "Chips with everything" referring to quite another form of cookery. We may soon expect to see "off-the-shelf" interactive graphics systems that are embodied in personal desk-top computers performing the kinds of operations summarized here. When this happens, the neuroanatomist is recommended to abandon prejudice and use

Fig. 21. In-focus and out-of-focus images (**1**) are subjected to optronics scanning techniques which eliminate all but specified grey levels to enhance contrast (**2–3**). A further program to eliminate all but the continuous lines eventually eliminates all profiles except those in the plane of focus (**4–6**). (Mouse visual cortex: Cobalt chloride diffusion into pyramidal cells; preparations in Fig. 21 and 22 by courtesy of M. Obermayer)

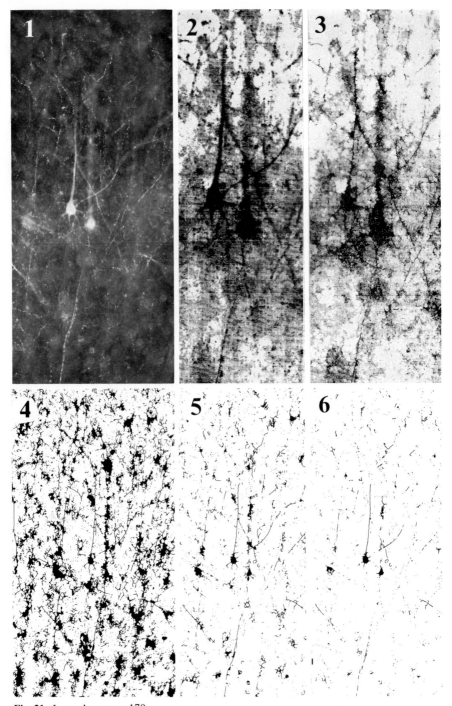

Fig. 21. Legend on page 179

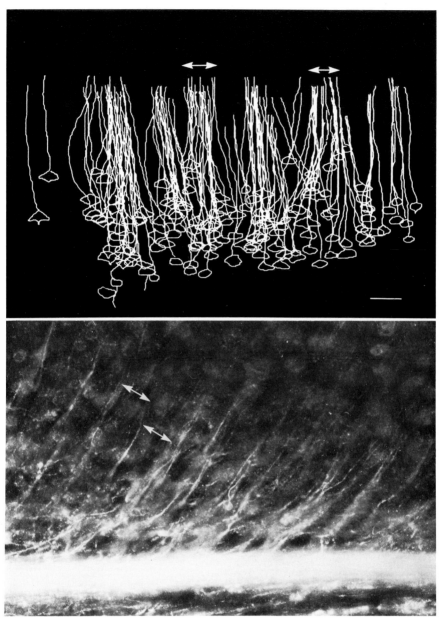

Fig. 22. Serial reconstructions and rotation (*upper picture*) of mass cobalt-filled mouse pyramidal cell and apical dendrites showing the existence of small columnar-like islets (width 50−60 μm) whose lateral dimensions correspond to the distances separating afferent fibres filled by cobalt injection into the geniculate (*lower micrograph*)

these aids if the funds are available. If nothing more, computer graphics provides a surprising insight into ways of seeing.

Addendum

Since this book went to press "NEUREC", a BASIC program for 3D-serial reconstruction (employing stereopairs), has been implemented on a micro-computer. The total cost of about $ 5000 includes the desk-top computer, disc-drives, digital plotter tablet (linear resolution of 0.05 mm) and a printer plotter. The program is available from the cited authors, presently at the Department of Zoology, University of Göttingen, F.R.G. (see: Gras H and Killman F (1984) NEUREC − a program package for 3D-reconstruction from serial sections using a microcomputer. Computer Programs in Bio-medicine (Elsevier): In press).

Acknowledgements. We thank Dr. Dennis Iverson for his help, Dr. Peter Mobbs for advice and John Stanger for excellent photographic assistence.

Three-Dimensional Reconstruction and Stereoscopic Display of Neurons in the Fly Visual System

Roland Hengstenberg, Heinrich Bülthoff
and Bärbel Hengstenberg

Max-Planck-Institut für biologische Kybernetik
Tübingen, F.R.G.

Introduction

Nerve cells are complicated anisomorphic bodies, intertwined amongst thousands of others. Their structural complexity is apparently crucial for their function and, therefore, the function of the nervous system as a whole. However, it is notoriously difficult to visualize the three-dimensional structure of nerve cells and their spatial relationships within different brain regions (His 1887).

We present a straightforward and simple procedure for three-dimensional display of structures from serial sections. We use the human brain for the most difficult part of the work; recognizing contours, aligning sections and perceiving depth by stereopsis. The boring and laborious part of the work (labelling, noting, handling and storage of large amounts of data) is left to a minicomputer with standard peripheral equipment. In its present form our procedure does not require sophisticated subprograms. These can, however, be implemented if required. General aspects of computer-aided neural reconstruction and solutions of special problems are described by Gaunt and Gaunt (1978), Lindsay (1977), Macagno et al. (1979), Sobel et al. (1980) and Ware and LoPresti (1975); see also Chap. 8, this Volume.

Procedure

Our procedure, initiated in 1975, is based upon manual reconstruction of stained neurons, consisting of the following steps: (1) intracellular penetration and recording; (2) fluorescent dye injection (Procion yellow M4-RAN; Stretton and Kravitz 1968, or Lucifer yellow CH; Stewart 1978); (3) histological processing (fixation, dehydration, infiltration with paraffin and serial sectioning at $7-10 \mu m$); (4) fluorescence photography on Kodak EHB 135 at a magnification of $40\times-60\times$ with high aperture fluorescent microscope lenses (Zeiss, Universal microscope, incident light illuminator III RS,

Neofluar 16/0.5 W-Oil, 25/0.8 W-Oil; (5) sequential projection and manual alignment of slides on a digitizer tablet (Fig. 1); (6) drawing and simultaneous digitizing of contours with a data acquisition program (HISDIG); and (7) transformation and reproduction of digitized structures with a transformation program (HISTRA). Steps 1–4 have previously been published in detail (Hengstenberg and Hengstenberg 1980).

Our procedure should be usable without special knowledge of computers; it exploits the operator's image processing capabilities for contour extraction, section alignment and depth perception by stereopsis. Data acquisition is ergonomic in that structures can be digitized as they would be drawn, including their fine details. During drawing, the operator is not loaded with additional computer chores. Resolution is preserved and alignment between sections utilizes the present state of the reconstruction rather than only the preceding section. Data and program structures allow later implementation of other more sophisticated evaluation routines. We do not use expensive hardware such as image processors for real-time image transformation and display, nor do we use highly sophisticated software for reconnecting cut fibres, hidden-line removal, shading or plane-surfaced polyhedra (see Chap. 8 this Vol.). Instead we explore how much of an "ideal" three-dimensional reconstruction of neurons can be achieved by high resolution stereograms of the raw data, and image processing by the human brain in stereopsis.

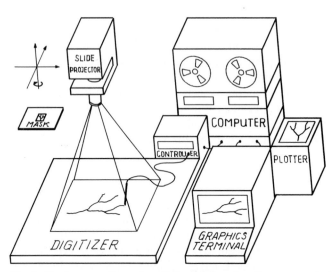

Fig. 1. Hardware configuration. Colour slides of sections are projected at about 600× total magnification by an adjustable slide projector onto a large digitizer tablet, covered with drawing paper. Contours are drawn, and simultaneously digitized, fed to a minicomputer and displayed on a graphics terminal which also serves to control acquisition and reproduction programs. Final graphs are produced by an electrostatic printer/plotter

Hardware Configuration

Hardware (Fig. 1) consists of a slide projector (24×26 mm), mounted on an adjustable gimbal. Slides are aligned by an adjustable mask and projected onto a large digitizer tablet (Summagraphics ID-RS232; 70×120 cm) whose surface is covered with drawing paper. Contours are drawn with a colour pen while being digitized (30 x,y-values per s). Contour data are fed (via a microprocessor-controlled buffer memory, 32 Kbyte) to a general purpose minicomputer (Digital Equipment, PDP 11/34), and are preprocessed and stored on magnetic disc (CDC 9762; 80 Mbyte) or tape (Kennedy 9000; 9 tracks, 800 bpi). Digitized contours are plotted on-line on a storage tube graphic display (Tektronix 4006) next to the digitizer, so that the operator can see and perform necessary corrections in real time. The terminal also serves to control the data acquisition program HISDIG. Stored data are processed by an evaluation program HISTRA and plotted on an electrostatic printer/plotter (Gould 5200; 20×30 cm; 80 dots cm^{-1}).

The Data Acquisition Program HISDIG

Data acquisition is in principle similar to manual drawing. Its major features are outlined below.

Data Specification

The first part of HISDIG specifies the data file (filename, storage location, field size, thickness of sections etc.). It further accepts a list of 20 identification codes (ID) for structures to be handled separately (e.g. neuropil surface, fibre tracts, stained and unstained neurons etc.). A resolution increment (RI), specified for each identification code, determines fine (stained cells) or coarse (brain outline) digitization. This has the advantage of saving memory space and reconstruction time for contours with low information content. HISDIG is now ready to accept data and asks for the first section and ID-code.

Digitization of Contours

While the projected image is being traced, the digitizer continuously samples the pen location and yields data points (x, y) with a resolution of 0.1 mm at a rate of 30 points s^{-1}. Together, with the section number and thickness, these coordinates define a point in space P (x, y, z). Subsequent data points are stored by the computer only if they are more than one resolution-increment (RI) away. By appropriate choice of RI for each of the structures to be digitized, an adequate resolution can be maintained for coarse and fine structures without accumulation of redundant data.

Data points separated by more than two RIs are interpreted as the end of one and the beginning of a new contour, i.e. a new profile of the same structure. To reconstruct contours, subsequent data points are joined by a line, as long as their distance D < 2RI. When D > 2RI, no connecting line is drawn, and profiles remain separate. The advantage of this is that many complicated profiles may be digitized without having to converse with the computer. This means that outlines can be drawn in any sequence, even arbitrary segments of one profile. This is practical since few people can draw lines accurately in every direction. Unlike automatic contour tracing systems (see Sobel et al. 1980) our semi-automatic system needs neither expensive hardware nor sophisticated software.

Alignment of Sequential Sections

The most crucial step in serial section reconstruction is alignment of sequential sections. For this we use the most fancy, fast and foolproof image processor (FFFIP) which is presently available: the operator's brain (see also Chap. 8, this Vol.).

We do not usually start reconstruction with the first or last section of the series, but begin with a middle section containing many stained profiles and having the largest brain area. This achieves optimal placement of the reconstruction, with respect to field boundaries. The wealth of detail in this initial section facilitates alignment of subsequent sections, especially if the stained profiles (or other prominent landmarks) are scattered. It is then very easy for an experienced operator to align the first few sections perfectly, even though the reconstruction is still quite fragmentary. At this stage, it is also useful to draw auxiliary structures that are not to be digitized to aid alignment. Later, the alignment of a section is quite easy because it matches not only the adjacent section, but all previously drawn structures. This minimizes the risk of cumulative errors such as linear shift or "helical distortion". As luck has it, sections of important preparations tend to be technically imperfect, i.e. they may contain folds, scratches, tears or bits of dirt. At worst, a section may even be missing. Even using the whole reconstruction as the reference frame for alignment one cannot eliminate such flaws, but it obviously helps the operator to make appropriate corrections.

The upper- and lowermost sections of a series usually contain only fragments of a stained cell and often very little of its environment. With an almost completely drawn reconstruction, it is usually no problem to fit in small fragments.

Correction and Storage of Data

Data are automatically saved on magnetic disc, and data acquisition can be interrupted whenever all contours of one ID-code are digitized. Should the computer break down, all data, except for those of the ID-code presently being digitized, are saved. Such an emergency, or if the computer is needed for other purposes, does not require that the work is repeated.

If during a reconstruction, certain data are incorrect (e.g. inappropriate matching of profiles), that section and ID-code can be called and digitized again. The incorrect data are automatically erased.

Data generated by HISDIG are stored on magnetic disc for fast access until the desired reconstructions and graphs are completed. Then they are compressed and stored on magnetic tape. They can be updated and compared with other preparations. On average a preparation requires < 1000 blocks of 256 words. Our largest data file is < 3000 blocks.

Field Size and Resolution

The present set-up was developed for our particular research on the fly's brain (Hengstenberg et al. 1982). It is, however, generally applicable. Other magnifications of microscope or projection yield various trade-offs between field size and resolution. The figures in Table 1 emphasize that the resolution of the fluorescence microscope is preserved throughout the reconstruction procedure and is still present in the stored data (Table 1, rows a−d). This is done by using a large digitizing tablet, which, at an overall magnification of about $600 \times$, allows one to draw and digitize profiles of < 1 μm diameter. When reconstructions are plotted or printed, considerable loss of resolution occurs due to the limited resolution of these processes (Table 1, row e). This is pronounced when the whole field of 1200×1800 μm

Table 1. Field size and resolution

		Width	Length	Resolution	Frames per section	Data per section
a	Largest section size	1200 μm	1800 μm	–	1	–
b	Field (Neofluar X 16/0.5)	600 μm	900 μm	0.55 μm	4	7.1×10^6
	Colour film	24 mm	36 mm	15 μm	4	15.4×10^6
c	Field (Neofluar 25/0.8)	400 μm	600 μm	0.35 μm	9	17.6×10^6
	Colour film	24 mm	36 mm	15 μm	9	34.7×10^6
d	Digitizer tablet	700 mm	1050 mm	0.10 mm	1	73.5×10^6
e	Gould plotter	190 mm	280 mm	0.15 mm	1	2.4×10^6
	Printed stereograms	50 mm	75 mm	0.10 mm	1	0.4×10^6

is displayed. However, smaller fields of $500 \times 700\ \mu$m can be plotted or fields of $200 \times 300\ \mu$m can be printed without loss of resolution. It is therefore necessary either to compromise between field size and resolution (Figs. 3, and 5–11) or to divide plots into survey and detail pictures (Fig. 4).

The Reproduction Program HISTRA

Digitized data are reproduced by a second program specifying: (a) digitized structures to be reproduced or omitted, as selected by their ID-code; (b) part of the preparation (length, width, depth) to be evaluated (three- dimensional window); (c) shift along x-, y-, z-axes; (d) rotation in space about x-, y-, z-axes; (e) linear scaling along individual axes; (f) perspective, i.e. choice of viewing distance; (g) choice of two-dimensional reproduction (shadowcast from a specified viewpoint) or three-dimensional reproduction by computation of stereo pairs; (h) type of contour line for different structures (e.g. light/heavy or dotted, broken, solid or colour), and the plotting device (graphic terminal or plotter).

HISTRA further accommodates a variety of subprograms for quantitative evaluations (contour length, profile area, cell volume etc.). A particularly important aspect of data transformation is computation of stereo pairs for viewing digitized structures in three dimensions.

Stereoscopic Vision

Euclid (280 BC) stated that "stereopsis" (Greek: solid sight) is caused by the simultaneous impression of two dissimilar images of the same object in the two eyes (Okoshi 1976). This dissimilarity is illustrated in Fig. 2a for three points of an object. When the eyes are fixed on the cross, it will be focussed by the lenses onto corresponding locations in the foveae of the two eyes. Other points in the fixation plane (horopter) will be focussed onto corresponding but laterally displaced locations. However, points behind or in front of the fixation plane are projected onto noncorresponding ("disparate") locations (Fig. 2a). This disparity (amongst several other cues) is used by the stereoprocessor in our brain to compute the depth of an object relative to the plane of fixation, where the disparity is zero. This trigonometric calculation is simple if there are only a few objects in the surround, but it becomes most difficult for many objects as in a natural scene, or in some of the stereograms presented in this article. In those cases, the brain must first solve the "correspondence problem," i.e. it has to work out which image points in the two eyes belong to which object point. This problem is clearly illustrated by Frisby (1979); Marr and Poggio (1976, 1979) suggest how the brain may solve it.

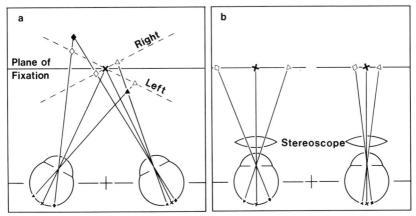

Fig. 2a, b. Stereoscopic vision. **a** Projection of points in space (♦, ×, ●) onto the two retinae, while fixating an object. Notice that the fixated point (×) is focussed onto corresponding locations of the two retinae, whereas points in front or behind the plane of fixation are projected onto noncorresponding ("disparate") locations. The disparity between left and right retinal image is used by the brain to perceive depth. The oblique lines (*Left, Right*) denote those planes into which the object points are projected to generate the two half-images of a stereogram (◇, ×, △). **b** When a stereogram is viewed with a stereoscope, the projected points in the half-image are focussed onto the same (disparate) locations of the two retinae as when looking at object points in different depths. Consequently such image points are perceived to have the same depth as the real object points

The construction of stereograms mimics the optical transformation of points in space, as they would be seen by two laterally separated eyes. This may be achieved by tilting the object or the assembly of data points P (x, y, z) by + 2.5°, and projecting all points into a plane normal to the line of sight. This yields the "half-image" for one eye and repeating the procedure with − 2.5°, the half-image for the other eye. A disparity angle of ± 2.5° is equivalent to a fixation distance of 75 cm. A nonlinear disparity correction (Glenn and Burke 1981) is in our opinion not necessary.

The two half-images have to be presented independently to the two eyes by a stereoscope. The brain will then solve the correspondence problem and thereafter compute the appropriate depth for all pairs of image points (Fig. 2b). When the sign of the disparity angle is reversed either by computation or by exchanging the two half-images, the depth sensation reverses also, and objects may be viewed "from the other side" (Figs. 6, 7).

Stereoscopes

Stereoscopic vision requires that the two half-images are independently presented to the two eyes. This can be achieved in different ways:

a) *Anaglyphs:* red and green half-images are printed in register and viewed through red/green spectacles so that each eye sees only one of the stereo pair. Anaglyphs can be viewed directly and printed if colour facilities are available (see Fig. 10. Chap. 7. this Vol.).

b) *Mirror stereoscopes* of 220 mm base are best suited for direct inspection and stereometry of large-format stereograms (20 × 30 cm each half-image) such as are obtained directly from the plotter. Mirror stereoscopes are available from C. Zeiss, Oberkochen, Germany, mod N2; Balzers AG, Lichtenstein, mod SB180.

c) *Lens stereoscopes* are useful for small stereograms of 65 mm basis as in Figs. 4—12. They are available from C. Zeiss, mod TS4, order no. 516404; or from Edmund Scientific Corp. Barrington, N.J., USA, order no. 42118. Note: The addition of convex lenses considerably improves the depth perception because the virtual images of the stereo pair are now located several meters ahead and therefore accommodation of the eyes becomes less effective as a cue for depth perception.

d) *Improvisation of a lens stereoscope:* mount two lenses of 90—150 mm focal length (ca. 30 mm diameter) 55—65 mm apart, so that their optical axes are parallel. Mount them about 1.1 focal length above the stereogram, so that both planes and the baselines are parallel. Stereopsis may also be achieved without a stereoscope by uncoupling of convergence and accommodation of the eyes.

e) *Slide projection* of anaglyphs is possible by two aligned projectors and red/green filters; colour matching may, however, be critical. Stereograms may be similarly projected through crossed polarizers onto a metallic screen and viewed through polarizer spectacles.

Stereopsis for Beginners

The stereoscope must be parallel with respect to the baseline of the stereogram. It may be shifted up/down or left/right to scrutinize details. Tilting, however, creates vertical disparities which are "unfamiliar to our brain" and which spoil stereopsis straightaway. Stereopsis is usually not achieved immediately (Julesz 1971) and may take up to 10 min to achieve. The intensity of depth perception increases gradually during this time, and many details become visible only towards the end of this period. This slow and gradual increase of depth perception is particularly conspicuous when looking at the stereogram for the first time. Fixate patiently a central area and view relaxedly. Observers with glasses should wear them.

Examples of Displays and Stereopairs

The directional preference of certain wide-field movement-sensitive interneurons in the fly visual system is correlated with their location within the neuropil: horizontal cells (HS) are in the anterior surface layer, and vertical cells (VS) are in the posterior surface layer of the lobula plate (Hausen 1981). It was, however, not clear whether or not this correlation holds also for other types of cells. Furthermore it was not clear how neurons with apparently "oblique" preferred directions (Hausen 1976a, b; Eckert and Bishop 1978; Hengstenberg 1981) are arranged in the neuropil. We therefore combined intracellular studies on the directional specificity of neurons with three-dimensional reconstruction to clarify the spatial organization of the lobula plate. For this it was not only necessary to reconstruct the stained cells "floating in vacuo," but also to show their spatial relationship to adjacent identifiable neurons and to the neuropil boundaries. At the same time, already available knowledge about many of these neurons could be utilized to develop and evaluate this procedure.

The examples of stereopairs (Figs. 4−12) ought to be viewed with a lens stereoscope. For readers with an impairment of stereopsis the text describes what can be seen. It should be kept in mind that bookprinting can seriously reduce resolution (cf. Table 1).

Selective Display

Selective staining procedures often show a preference for particular neurons or cell classes. Neurons which are refractory to such procedures will therefore remain concealed. Reconstructions from unspecifically stained semithin serial sections may be used to reveal the presence, shape and location of such cells. Figure 3 shows two plots, where 27 comparatively large tangential neurons in the lobula plate of the blowfly were reconstructed from 85 sections of 3 μm thickness. Contrary to all other reconstructions shown later, the cells were not reconstructed on a single sheet of drawing paper but on a stack of prealigned foils. Plotting all neurons simultaneously (Fig. 3a) yields a chaos which cannot be disentangled. By plotting only a few neurons, (as selected by their identification codes) at a time, and by employing various combinations, it is possible to reveal the structure of single neurons, their location in the neuropil, and their relative position with respect to other identifiable cells (Fig. 3b). By this procedure, it was also possible to confirm the existence of a particular neuron (VS1) in the blowfly (Hengstenberg et al. 1982), which had previously been described in the housefly (Pierantoni 1976), but had never showed up in cobalt mass-impregnations of VS-cells (Hausen et al. 1980).

Survey and Detail

Nerve fibres are usually long and thin and may have fine arborizations in different, widely separated neuropils. It is impossible to display such neurons in toto and with sufficient detail. Our scheme of digitizing with high

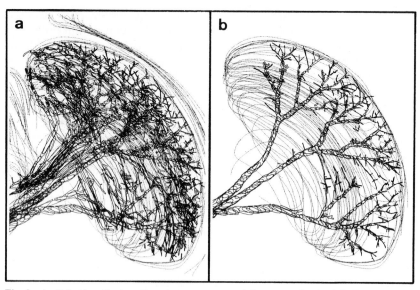

Fig. 3a, b. Selective display. Twenty-seven tangential neurons in the lobula plate were reconstructed from 85 unspecifically stained semithin sections in order to map the number, position and structure of neurons accessible for microelectrode penetration. **a** Displaying all reconstructed cells is a useless chaos. Displaying only a few at a time, as selected by their name (Identification Code), and in various combinations, allows recognition of identifiable neurons, clarification of their relative positions, and statements about the existence, shape and position of neurons which have so far not been impaled. **b** shows as an example the three horizontal cells HSN, HSE and HSS as selectively displayed from the data file of **a**. Field width 480 μm. Neurons in Figs. 3–11 are from *Calliphora erythrocephala*

resolution on a large tablet allows low resolution survey plots with data compression and recovery, when necessary, of high resolution by magnification of particular regions. This is not to be confused with photographic enlargement, where just the scale is blown up. Figure 4 shows a heterolateral projection neuron (H4) of the fly visual system with three distinct arborizations: one in the lobula plate, another in the ipsilateral protocerebrum, and a third in the contralateral posterior slope. Figure 4a shows the whole neuron in a generalized neuropil outline, which was digitized along the largest circumference of different neuropils (cf. Fig. 3). Figure 4b is a stereogram of the contralateral axonal arborization, magnified $\times 2.2$. The axon is about 5 μm in diameter where it enters the plot, and the small collaterals are about 1 μm or less in diameter. The contour jitter of about 0.2 μm demonstrates the resolution limit of digitization, in part due to the drawer's tremor, and in part due to the resolution limit of the digitizer tablet (0.2 μm). The stereogram reveals that the telodendritic arborization extends through a volume of about $200 \times 220 \times 150$ μm and where each single collateral is located in this volume. If another neuron passing through this region had been stained, a

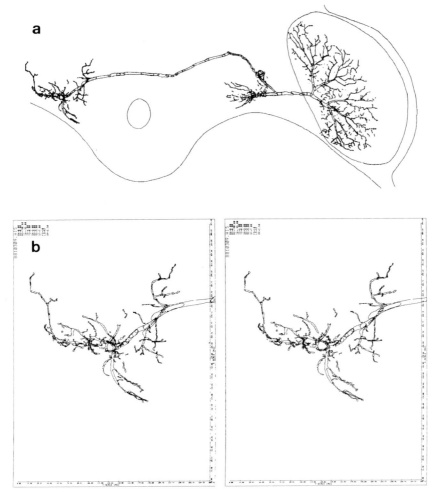

Fig. 4a, b. Survey and detail. **a** Survey picture of the heterolateral projection neuron H4, as seen from behind. The cell has three discrete arborizations, one in the right lobula plate, another in the ipsilateral protocerebrum, and a third in the contralateral posterior slope. **b** Stereogram of the contralateral, presumably telodendritic arborization of H4, seen from behind. Field width: **a** 1250 μm; **b** 285 μm. This high-resolution stereogram was generated by the window function of the reproduction program HISTRA, retrieving the full information about a small area from the data file of the whole cell. The axon is about 5 μm, most collaterals are about 1 μm in diameter

fair estimate could have been made about the proximity of the two neurons, and if collaterals were possibly involved in contacts. It should be kept in mind that selection of the area to be plotted is a function of the reproduction program HISTRA, operating upon stored data. There is no need to do the tedious reconstruction again.

Depth Interpolation by Stereopsis

We compute and display stereograms from incoherent contour data. This means that contours in one section are compressed into a plane and subsequent sections are represented by a series of such "contour-planes," each separated by an empty space of section thickness. One would therefore expect to see in the stereograms piles of irregular squiggles, separated from one another. This is in fact true for coarse structures like neuropil outlines. Reconstructed neurons, however, are remarkably coherent and smooth. Apparently our stereoprocessor interpolates between different contour planes if there is no doubt about the local correspondence. This fortunate property allows us to do without any advanced subroutines to reconnect cut profiles or to suppress "hidden contours". Figure 5 shows the dendritic arborization of the tangential neuron H2 and the outlines of the right lobula plate, as seen from the front. H2 responds selectively to horizontal back-to-front movement in the ipsilateral visual field. The arborization is essentially flat and closely apposed to the anterior surface of the neuropil. This is best seen in the upper part of Figure 5, where sufficient contour lines of the lobula plate are present.

Functional Stratification

Previous studies on the fly visual nervous system (Eckert 1979; Hausen 1976a, b, 1981; Hengstenberg et al. 1982; Pierantoni 1976; Strausfeld 1976) have shown that the location of tangential cells in the lobula plate is correlated with their characteristic response properties. Cells which respond best to horizontal movements in the visual field are located in the anterior surface layer of the lobula plate, and cells which respond best to vertical movements are located in the posterior surface layer. Figure 6 illustrates this correlation for two types of large movement-sensitive tangential cells. The upper part of Fig. 6 shows a stained horizontal cell (HSN) digitized together with the unstained profiles (fluorescence shadows) of horizontal (HS) and vertical cells (VS). It shows further the outlines of the right lobula plate neuropil, as seen from behind. The stained HSN and unstained HS profiles appear to lie deeper, i.e. they are closer to the anterior surface. The unstained VS-profiles "float" on top, close to the posterior surface of the lobula plate. The lower part of Fig. 6 illustrates the reverse situation: a giant vertical cell VS4, in the left lobula plate is viewed anteriorly. Like the VS-shadows in Fig. 6a, its main branches are closely apposed to the posterior surface of the lobula plate, and the small dendritic branches extend mostly towards the observer, i.e. into the neuropil. Such reconstructions of single stained neurons together with profiles of unstained but identifiable cells and the neuropil boundaries allow us to generalize the correlation between location in the neuropil and functional specificity. The horizontal movement-sensitive neuron H2 (Fig. 5) and several other tangential cells (not shown) obey this rule.

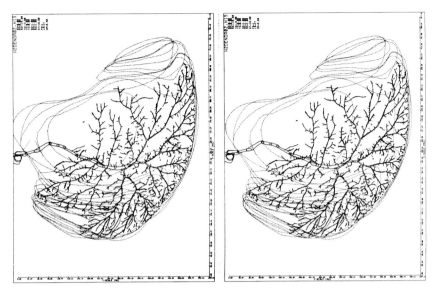

Fig. 5. Depth interpolation by stereopsis. Dendritic arborization of interneuron H2 in the right lobula plate, as seen from anterior. The bulge of the lobula plate towards the right indicates the large fibre tracts leaving the neuropil. Field width: 620 μm. Notice the coherent and smooth appearance of the arborization, even though no subprograms for reconnection of cut profiles or hidden line removal have been applied

Synaptic Contact Layers

The correspondence between the physiological specificity of neurons and their location within the neuropil, as stated so far, may be misleading if only the location of the main branches is considered. This is due to the fact that the vast majority of synaptic contacts is found on the outermost dendritic branchlets (Pierantoni 1976; Hausen et al. 1980; Bishop and Bishop 1981). In HS- and VS-cells these branchlets extend more or less distinctly towards one another (Fig. 6) and may in principle form either distinct "horizontal" and "vertical" synaptic contact layers, or they may approach and penetrate one another to form a mixed layer. By sequential penetration, dye injection and three-dimensional reconstruction, this question can be unambiguously answered. Figure 7 shows two stereograms of the horizontal cell HSE (Figs. 3, 6) and of the vertical cell VS2; in the upper one, as seen from posterior, and in the lower, as seen from anterior. As in previous figures, the main branches are seen to lie in the anterior and posterior surface layers of the lobula plate, and the small dendrites extend towards the interior of the neuropil. In the region of overlap the small dendrites of both cells form discrete layers separated by an empty space. Not a single dendritic branchlet of either neuron invades "foreign territory." This finding is consistent with results

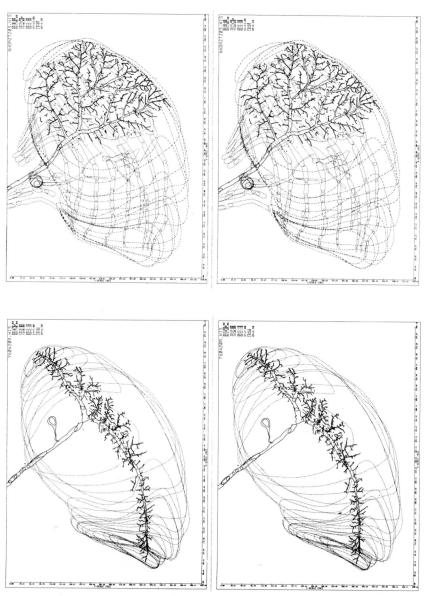

Fig. 6. Functional stratification. *Top* Dendritic arborization of neuron HSN, together with unstained profiles of HS-, and VS-cells, and the outlines of the right lobula plate; seen from posterior. *Bottom* Dendritic arborization of neuron VS4 in the left lobula plate, seen from anterior. Field width: *top* 450 μm; *bottom* 550 μm. The lobula plate appears like a thick-walled transparent bowl; stained neurons float inside the "wall;" vertical cells are apposed to the posterior (outer) surface of the neuropil and horizontal cells are apposed to the anterior (inner) surface. Compare Fig. 5: H2, which also responds specifically to horizontal movements

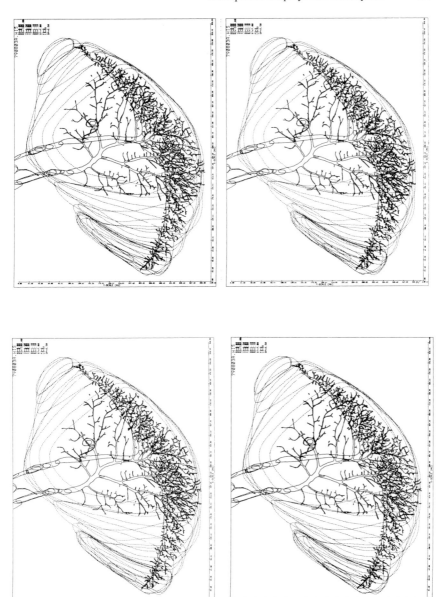

Fig. 7. Synaptic contact layers. A horizontal cell HSE and a vertical cell VS2 stained and reconstructed from the same fly. *Upper stereogram,* is seen from posterior; *lower* is seen from anterior. Field width 420 μm. Notice the location of main branches, and the perfect separation of small dendritic branchlets of the two neurons. This corresponds to directional specificities to wide-field movements in the visual field. Compare also the different attractiveness of structural features in the two stereograms

from cobalt mass-impregnations (Strausfeld and Obermayer 1976; Hausen et al. 1980). The conclusion that there are discrete synaptic contact layers, specified by their directional specificity was recently corroborated by ^{3}H-2-deoxyglucose labelling under specific stimulation and subsequent high-resolution autoradiography (see Chap. 11). Figure 7 also illustrates how much more information can be retrieved from a stereogram than from the same graph viewed two-dimensionally. It shows further that even in this case of dramatic overlap, hidden-line removal is not necessary to perceive the structural relationships between the two neurons.

The two stereograms of Fig. 7 are identical except for the sign of disparity. Therefore they are excessively redundant and one would not expect them to provide different insights. It is, however, interesting to compare the two stereograms and to observe how attention is attracted to different structural features when looking from anterior or posterior into the same structures. Apparently, attention is mainly guided to objects in the foreground.

Bistratified Vertical Cells

Some of the tangential neurons in the lobula plate do not exhibit a clear-cut preference for either horizontal or vertical movements in the visual field

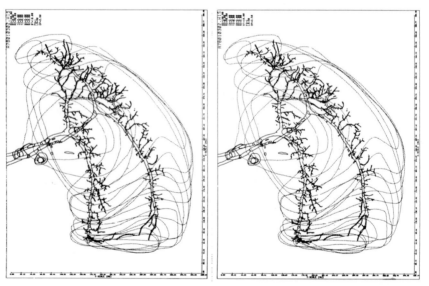

Fig. 8. Bistratified vertical cells. Two vertical cells, VS4 and VS8 from the same fly, as seen from anterior. Stereo effect overemphasized by computation with 1.5 × real section thickness. Field width: 420 μm. VS4 (*right*) is entirely confined to the posterior layer of the lobula plate, and responds exclusively to vertical movement. VS8 (*left*) is bistratified: its ventral dendrites are in the same layer as VS4, but the dorsal dendrites invade the anterior layer of horizontal cells (cf. Figs. 5–7). Local movement stimuli in respective parts of the visual field reveal that local directional specificity correlates with dendritic stratification

(Eckert 1979; Hausen 1976 b, 1981; Hengstenberg 1981). The vertical cells respond in different ways to horizontal and vertical movement. Three-dimensional reconstruction of such neurons shows that the dendritic arborization of VS1 and VS6—VS10 is bistratified. Figure 8 shows an example of this arrangement. As in Fig. 7, two neurons, VS4 and VS8, were successively impaled in the same animal, studied physiologically and injected with Procion yellow. VS4 is exclusively excited by downward movement, whereas VS8 responds to downward movement and to front-to-back movements. The former response dominates in the ventral half of the visual field and the latter in the dorsal half. The stereogram (Fig. 8) shows clearly that VS4 is monostratified and resides entirely in the posterior surface layer of the lobula plate. VS8, however, is not confined to one neuropil layer: in the ventral part of the lobula plate it resides in the posterior layer but in the dorsal half it occupies the frontal surface layer, as if it were a horizontal cell. Here again the small dendritic branches apparently invade discrete layers of the neuropil presumably because appropriate input information is available only there.

From the retinotopic mapping of the visual field into the lobula plate, combined with the local directional preference of a neuron, flight manoeuvers of the fly can be inferred that would maximally excite such neurons. VS8 is expected to respond best during pitch movements about the transverse axis of the fly.

Intermediately Located Neurons

The examples of neurons presented so far (Figs. 5—8) seem to suggest that entire tangential cells or parts of them are located either in the anterior or in the posterior layer of the lobula plate. There are, however, other neurons which occupy intermediate positions (Hausen 1981). Figure 9 shows as an example the arborization of H4 in the lobula plate (Fig. 4). It is displayed together with the fluorescence shadows of HS-, and VS-cells. With some scrutiny, its intermediate position between the two cell classes can be recognized. Interestingly, its small branchlets extend both ways, anteriorly as well as posteriorly, and the cell responds well to oblique movements from postero-superior to antero-inferior in the ipsilateral visual field. Even though the spatial arrangement of H4 differs in principle from the previous examples, the correlation between its structure and function suggests that directionally specified input elements are organized in discrete layers.

Sequence of Neurons

A central problem of neuroanatomy is that of synaptic connectivity. Which neurons converge upon which subsequent ones? Figure 10 shows again two

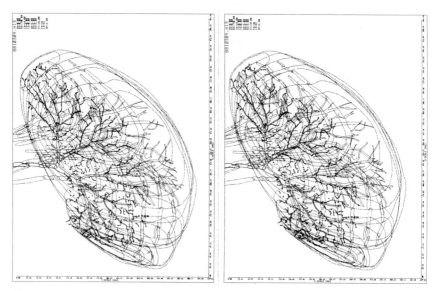

Fig. 9. Intermediately located neuron. Arborization of H4 (cf. Fig. 4) in the right lobula plate seen from posterior. Field width 430 μm. The axon and major branches lie between the fluorescence shadows of HS- and VS-cells. Small branches extend both ways: anteriorly into the layer of horizontal cells, and posteriorly into the layer of vertical cells. It responds best to oblique pattern movement

sequentially stained neurons of the visual system of the fly: a vertical cell VS9 (top), and a descending neuron YDN (bottom). Already the two-dimensional representation and even more so the respective stereogram (Fig. 10) show that the telodendritic arborization of VS9 and one of the two major dendritic branches of YDN are intimately intertwined. The second dendritic branch of YDN approaches large fibres of the ocellar system. Thus stereograms suggest sites of synaptic contact between branches of different neurons. They also draw attention to branches that seem to end blindly indicating additional inputs or outputs of neurons.

Neurons of Irregular Shape

The previous examples (Figs. 3—9) were all tangential neurons in the lobula plate of the blowfly. This neuropil has a highly ordered arrangement of retinotopic input elements (Strausfeld 1976). It is furthermore fairly shallow and essentially flat. It is therefore possible to select a plane of sectioning (frontal), where shadowcast reconstructions of the stained neurons represent reasonably well the structure of the neurons. The same would be true for Purkinje cells in the vertebrate cerebellum. In such cases the effort and cost of three-dimensional reconstruction may favour manual shadowcast reconstructions.

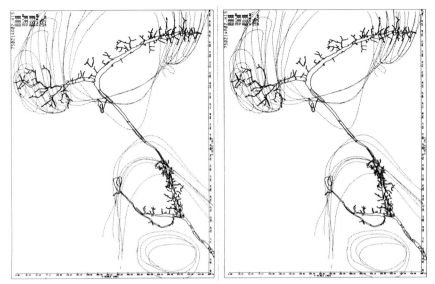

Fig. 10. Sequence of neurons. A vertical cell (VS9), and a descending neuron (YDN) were sequentially impaled and stained. The stereogram is rotated by 90 degrees (dorsal to the left) to utilize the print format optimally, and is seen from posterior. Field width: 500 μm. The terminal arborization of VS9 is in close contact with one dendritic branch of YDN, whereas the other approaches large fibres of the ocellar system (not shown). The axon of YDN, projecting into the mesothoracic neuropil is omitted. The close apposition of the two neurons suggests synaptic convergence but is of course not sufficient to prove contiguity. Stereograms of this kind show however those axon collaterals and dendritic branches which do not contact one another, thereby indicating additional inputs and outputs

The majority of neurons are however irregular in shape (e.g. stellate, granule and pyramidal cells of the vertebrate cerebral cortex, or interneurons and motor neurons of invertebrate ganglia). For these neurons it is impossible to define a unique plane of sectioning (e.g. a unique direction of view) representing the neuron in full detail. Single two-dimensional reconstructions of such neurons are necessarily ambiguous with respect to the orientation of their arborizations, and their three-dimensional structure can only be partly deduced by tedious orthogonal projections (Murphey 1973).

A large proportion of central interneurons connect particular neuropil areas, leaving others out. Single branches of such neurons may pass through well-known fibre tracts or less obvious routes. It is quite obvious that three-dimensional reconstruction and stereoscopic inspection of such neurons are well-suited for studying such complex arrangements. Figure 11 gives an example of such a neuron with irregular shape, extending into several different neuropil areas of the fly's protocerebrum. It is a descending neuron (DN3) whose axon leaves the brain about normal to the plane of view, close to the oesophageal canal, and then projects through the cervical connective to end bilaterally in the pro- and mesothoracic neuropils (Hengstenberg and

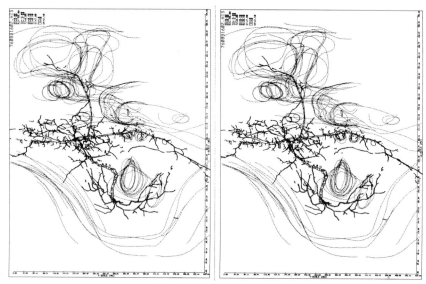

Fig. 11. Neurons of irregular shape. Dendritic arborization of a descending neuron DN3, seen from posterior. Field width 560 μm. Six major dendritic branches and the axon originate from a common branch point in the left lateral protocerebrum. Two project to the left and right optic foci through the posterior optic tract, one ascends towards the optic tubercle, bypassing the central complex and corpus pedunculatum, three smaller ones remain close to their origin in the lateral protocerebrum. In the uppermost section, the axon gives off a few collaterals which project bilaterally into the subesophageal ganglion. The axon projects then through the cervical connective bilaterally into the pro- and mesothoracic neuromeres (omitted). Obviously such complicated arborizations are best represented by stereograms

Hengstenberg 1980; Strausfeld 1976 plate 7.26 B). Figure 11 shows the dendritic arborization of DN3, and some of the protocerebral neuropil areas, as seen posteriorly. Near the cut end of the axon two collaterals arise and spread ipsi- and contralaterally into the optic foci of the posterior slope. Further up, the axon narrows down and divides into at least six major dendritic branches. Two prominent ones project either way through the posterior optic tract to lateral (presumably visual) parts of the protocerebrum. A third branch ascends ipsilaterally and projects to the optic tubercle, bypassing the central complex and the corpus pedunculatum, except for a small branch at the surface of the calyx. The remaining three main branches are confined to the ipsilateral protocerebrum; their precise destination is unknown.

　　Possibly, the complex structure of DN3 may correlate with quite complex (largely unknown) response properties: the cell is silent in the resting state and fires 1−2 spikes with fixed delay when surprised by bright flashes, but this response habituates very rapidly. No response was obtained with pattern movement of any kind. The dendritic structure of this neuron, especially its dendritic connections with several different central neuropils, suggests that sensory modalities other than visual may contribute to the characteristic response of DN3, and that the apparently low responsiveness with purely visual stimulation may be due to the lack of adequate stimulation.

Comparison of the stereogram with the previous two-dimensional reconstruction of the same neuron (Hengstenberg and Hengstenberg 1980) shows how much more can be learned about the structure of neurons by use of three-dimensional reconstruction and stereoscopy.

Further Applications

The examples demonstrated in this article (Figs. 3–11) have been taken from current research on visual interneurons in the fly's brain. Obviously the very same procedure can be used for other problems, where three-dimensional data of arbitrary structures is available in the form of serial sections or serial frames.

a) Electron Microscopy. The same procedure should be applicable for EM serial sections, because the ratio between section thickness and resolution (500 Å/25 Å = 20) is about the same as that of the paraffin sections used for this article (10 μm/0.5 μm = 20). We expect in particular that stereograms with transparent surfaces will be useful to display synaptic complexes etc.

b) Brain Architecture. A further application will be to reconstruct from unspecifically stained serial sections of the fly's brain the outlines of major neuropil areas and fibre tracts. By plotting such outlines faintly, the three-dimensional reconstruction can be kept quite "transparent" and even multiply enclosed and irregularly shaped neuropils can be depicted in their natural spatial relationships. We expect stereograms to help conceptualization of brain architecture and introducing newcomers to this field.

c) Skeleton and Musculature. It is notoriously difficult to visualize how insects move their head, wings, legs, etc. because skeleton and joints are complicated three-dimensional structures, and the muscles usually extend in different directions. Here too, "transparent" stereograms could help to visualize what movement ensues from the contraction of one, or a group of muscles. We plan to reconstruct the neck musculature of the blowfly in order to find out which of the many small muscles in this region (Sandemann and Markl 1980) generate which of the different head movements (Hengstenberg and Sandemann 1982).

d) Free Flight Trajectories. Using slightly modified data acquisition and display programs, the trajectories of free flying insects can be reconstructed in three dimensions from high-speed films (Bülthoff et al. 1980). Stereograms allow the visualization of the flight path, the change of flight velocity and of body posture for different flight manoeuvers (Bülthoff et al. 1980), for landing on a target (Poggio and Reichardt 1981; Wagner 1981), or for a male chasing a female fly (Wehrhahn et al. 1982). It is particularly interesting to compare in three dimensions the flight path of a real fly with the trajectory predicted by a mathematical model (Fig. 12). This way the accuracy of the model's theoretical framework is easily revealed.

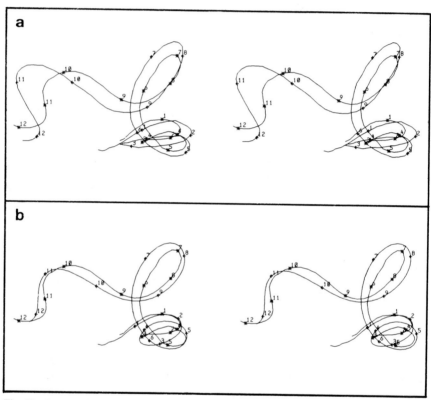

Fig. 12a, b. Free flight trajectories. Stereograms of real and simulated flight paths of two male houseflies, one chasing the other. **a** Flight paths of real flies as reconstructed from a high speed film (80 frames s⁻¹), according to Bülthoff et al. (1980). Corresponding points of the two trajectories are numbered at 100 ms intervals. **b** Flight path of the real, leading fly, and the simulated chasing fly according to an algorithm described by Poggio and Reichardt (1981). The model computes from the 3D-coordinates of the leading fly the future torque, lift and thrust of the chasing male, translates them into angular-, vertical-, and forward velocities, and integrates these into a three-dimensional trajectory. Notice that the "artificial fly", as well as the real one, can accurately follow sharp turns and loopings of the leading fly

Concluding Remarks

Our intention was to develop a simple, computer-assisted method for re-constructing and displaying neural structures in three dimensions using standard equipment. We employed simple procedures transparent enough for people with little experience with computers to use them straight away. We also wanted to find out how much of an "ideal" three-dimensional reconstruction could be achieved by computation of stereograms using only contour data but no sophisticated data processing. Our procedure was developed for analyzing dye-injected visual interneurons of the fly. Its handiness and performance were tested against the usual manual reconstruction procedures, and against results achieved by cobalt mass-fillings of neurons. We have now entirely abandoned manual reconstruction in favour of the computer-assisted procedure.

Control of the acquisition program HISDIG is self-explanatory and fool-proof. The reproduction program HISTRA can be handled equally simply. It allows various useful transformations, such as computation of stereograms. These operations are controlled by a plain language dialogue.

With thick sections our use of contour data prohibits image rotation by 90 degrees because the low resolution in the z-direction (given by the section thickness) yields only poor reconstructions. With thinner sections ($1-3\ \mu$m) this drawback becomes less prominent, and rotation up to 60 degrees may be possible (E. Buchner, unpubl.). However, in many cases stereograms eliminate the need for orthogonal projections. At high resolution, they show clearly and in fine detail the three-dimensional structure of neurons and their spatial relationships with other structures. Somewhat to our surprise there seems to be no need to reconnect cut fibres explicitly or to remove hidden contours by special subprograms. Apparently the superb image processor in the observer's brain performs these operations rapidly and reliably.

We now consider our procedure sufficiently developed to be distributed. Operation can be learned within one day. The same system can be realized de novo at an estimated cost of less than US $ 15,000. If a minicomputer or even a microprocessor with suitable peripherals is available, only the digitizer tablet needs to be added. Readers wanting to set up a 3D-reconstruction system may contact us for advice and more detailed information.

Finally, we believe that the heuristic value of stereoscopy for neuroanatomical studies has been deplorably underestimated. The possibility of displaying, together, many complex shapes, combined with our own outstanding abilities of form and depth perception, enormously help the analysis of neural structures.

Acknowledgements. We wish to thank Prof. K. G. Götz, Drs. R. Cook, E. Buchner, K. Hausen, Mr. J. Pagel and U. Wandel for critical advice and help during this work. May 1982

Chapter 10
Laser Microsurgery for the Study of Behaviour and Neural Development of Flies

GAD GEIGER, DICK R. NÄSSEL
and HARJIT SINGH SEYAN
European Molecular Biology Laboratory
Heidelberg, F.R.G.

Introduction

Since laser microbeam irradiation was first used in experimental cell biology over 20 years ago (Bessis et al. 1962; Townes 1962), it has been extensively employed in numerous modes in many cellular systems (reviewed by Berns et al. 1981). Many applications of laser micro-irradiation are suitable for studies of subcellular physiology, microdissection and surgery. The design of the laser microbeam unit is principally similar for all these purposes although some parameters can vary. Lasers include: ruby lasers (Bessis et al. 1962), N_2-lasers (Isenberg et al. 1976), argon ion lasers (Berns 1971; Cremer et al. 1974), neodynium YAG lasers (Berns et al. 1981), and flashtube pumped dye lasers (Sulston and White 1980; Geiger and Nässel 1981, 1982; Berns et al. 1981). The wavelength, pulse duration, power and spot size can vary, depending on the desired application of the instrument. Since our purpose was to make ablations deep in the tissue without causing damage to other cells in the ray path, we chose a pulse dye laser where some of the parameters could be adjusted.

This chapter discusses irradiation and ablation of cell groups for the study of development and behaviour of insects. Our aim has been to correlate structure and function by selectively ablating and eliminating identified nerve cells in higher-order visual neuropil of the fly (*Musca domestica*) and then to study visual orientation behaviour. The laser ablations were performed on larval animals to eliminate precursors of certain behaviourally relevant nerve cells. The procedure of localizing these precursors in post-embryonic developing brains is fairly complex and is described in one section of this chapter, summarizing studies published elsewhere (Geiger and Nässel 1981). Eliminating selected nerve cells also results in valuable data that can be used to describe developmental mechanisms (Nässel and Geiger 1983; Nässel et al. 1983). These data and results from experiments in other invertebrates are discussed.

The Laser Microbeam Unit

The laser surgery unit described here is similar in principle to other systems (Berns 1972; Sulston and White 1980), based on one designed by John White (Sulston and White 1980; White, pers. commun.), but is operated in a mode that enables us to make ablations deep in the tissue. To achieve localized ablations in depth the following conditions were met:

a) Sufficient power of the laser to destroy the target cells.
b) Avoidance of tissue damage above or in the neighbourhood of target cells.
c) Minimal damage to enzymes due to interactions with stray light.
d) An adjustable and versatile unit applicable to different tissues, sizes and depths of lesions to determine appropriate lesion parameters for different purposes.

The Dye Laser: Energy and Power Considerations

We used an Electro Photonics Model 43 flashtube-pumped pulse dye laser which has a pulse duration of 1.3 μs and can be used with a wide range of dyes. We chose the organic dye Coumarin 2 which operates at wavelengths of 450–470 nm, depending on concentration. At these wavelengths a compromise is achieved between transparency of the tissue and minimal interference with enzymatic reactions by laser photons. Tissue is more transparent at shorter wavelengths whereas at longer wavelengths (within 400–800 nm) interactions with enzymes are less likely to occur.

The second parameter considered was the power of the laser beam. Since we intended to make lesions at about 0.2–0.7 mm from the tissue surface, it was clear that high power was needed to destroy the desired cells. To destroy single cells in one pulse the destruction sphere (spot) could not be too small and the power of the laser beam at the focal plane had to exceed 10^{10}–10^{12} W/cm². At 10^{10}–10^{12} W/cm² (450–470 nm) a normally transparent substance becomes opaque due to the nonlinear properties of the dielectric coefficient of the substance at high power (Ready 1978).

When tuning wavelength and power, it is important to consider that for any given power the longer the wavelength, the bigger the lesion. Destruction is, however, less controlled in the longer wavelengths (cells tend to "explode"). Therefore, short wavelengths and high power were preferred within the limits of the compromise between tissue transparency and enzyme destruction.

Optics and Alignment of the Laser

The laser is incorporated into the ray path of a Zeiss Standard 18 microscope with Nomarsky optics (Figs. 1, 2). The laser beam enters the micro-

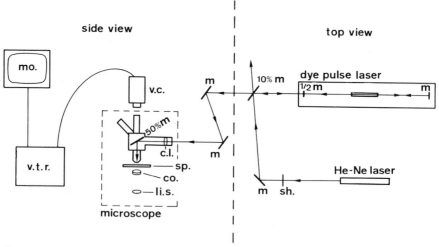

Fig. 1. Schematic drawing of the laser microbeam surgery unit. The dye pulse laser beam is aligned with the He−Ne laser. The two are incorporated in the microscope ray path with the aid of mirrors (*m*) and a collimating lens (*c.l.*). The specimen is put on the microscope stage (*sp.*). Ablation control and aiming was made with the aid of a video system (*v.c.* video camera; *v.t.r.* time-lapse video tape recorder; *mo.* monitor). *li.s.* light source; *co.* condenser; *sh*, shutter. Note that the *right* is a top view and the *left* a side view

scope from the side. The microscope was modified by incorporating a pellicle (partially reflecting thin mirror) above the objective lens to bring the laser beam into its path. Before entering the microscope the laser beam is collimated by a lens (50 or 100 mm focal length depending on the lesion size desired) and passes through an iris aperture (J. White, pers. commun.). The collimating lens brings the laser beam into focus at the image plane of the objective lens and allows us to adjust the lesion site.

Highest possible numerical aperture (N.A.) lenses were used to achieve ablations restricted in depth without damaging cells above the target cells. Changeable lenses were used (Zeiss Planachromat; $\times 16$, N.A. 0.35; $\times 40$, N.A. 0.65; $\times 100$ oil immersion, N.A. 1.25) to achieve different sizes of lesions. We most commonly used a $\times 40$ lens, however, which at a depth of $100-300 \, \mu m$ in the tissue resulted in single lesions (spot size) $2-3 \, \mu m$ in diameter. This magnification allowed optimal N.A. and optical resolution for our experiments. Using the $\times 100$ lens obtains a much smaller spot size (less than $0.5 \, \mu m$).

An axially reflected He−Ne laser beam emerging from the back mirror of the dye laser was used to align the dye laser with the optics of the microscope. The beam of the He−Ne laser was brought to coincide with the axial pathway of the dye laser with the aid of a partially reflecting thin mirror.

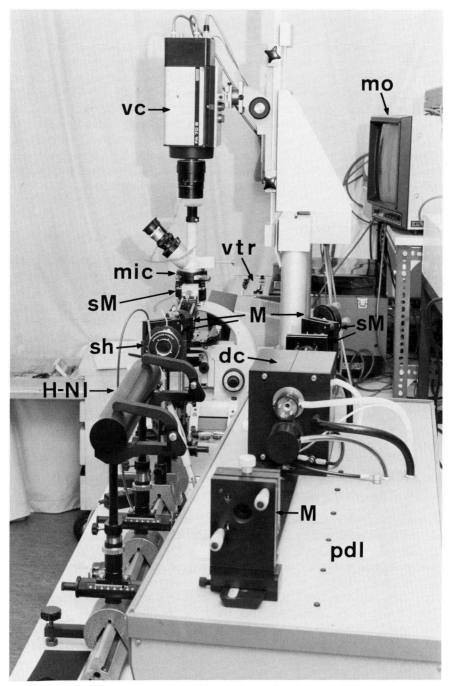

Fig. 2. Laser microbeam surgery unit. Abbreviations as in Fig. 1. In addition: *pdl* pulse dye laser; *dc* dye cell, *H-Nl* He−Ne laser; *sM* semi mirror; *mic* microscope)

Videocontrol and Recording of Ablations

The laser beam ablations were indirectly observed through a closed video system. This was more safe and practical than viewing directly through the microscope. A videocamera (Grundig FA70B) connected to a time-lapse video recorder (National 8030) and to a monitor (National WV-5400) was used for observation and to control focussing the depth of the lesion site. Documentation of each ablation was with a time-lapse video recorder. A second video camera and mixer was used for adding auxiliary information to the video recordings. For instance, a number could be assigned to each experiment recorded. With the time-lapse facility lesions could afterwards be analysed frame by frame for measurements and photography.

Procedure of Laser Surgery

In simple cases a specific embryonic cell in tissue (or peripheral cell in an adult) can be irradiated and destroyed. Ablations of identified precursors or peripheral sensory cells require little effort in determining the extent of the lesion because these structures can be observed in vivo through the micro-scope (see e.g. Sulston and White 1980; Edwards et al. 1981). Although our system is also designed to ablate superficial structures and axonal tracts, we emphasize here its special use for ablating groups of developing precursor cells within the postembryonic brain of the fly. It can, of course, be used on embryonic tissue as well.

The brain of the freshly hatched fly larva consists of a large number of undifferentiated cells of roughly the same size and shape. Therefore, the choice of target cells is necessarily arbitrary at first, but is progressively tuned by trial and error to achieve the desired effect. We first describe making lesions and go on to describe how we achieve reproducible structural changes by correlating the alterations of adult brain morphology with carefully recorded ablation parameters.

Synchronized Larval Development. Larvae of *Musca domestica* were synchronized by collecting eggs from single females. These were kept under observation and larvae were collected 2 h after the first hatching. The larvae were reared on a mixture of bran, quark cheese and Nipagin, at 29 °C and 55% humidity. Larvae of desired age (see later) were put on microscope slides and anaesthetized with ether for 10−15 s. (Control experiments showed that ether anaesthesia did not affect brain postembryonic development.) A trough of vaseline was made around the larva, filled with fly saline (Case 1957) and covered with a thin (0.17 mm) coverslip. This preparation was then brought into the laser ray path.

Aiming at Target Cells. Using the microscope stage to orient the larva, the target area was brought into the center of a crosshair, marking the axis of the

Fig. 3. Videograph of the 28-h-old larval central nervous system. A laser lesion is visible at the cross-hair (*arrow*)

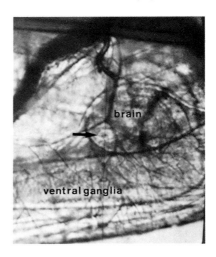

Fig. 4. Schematic drawings of the anterior side and top view of a 28-h-old larva. The cephalopharyngeal skeleton (*c.s.*) can be seen anterior (*a*) to the central nervous system (*c.n.s.*) The salivary glands (*s.g.*) and muscle layers (*m.l.*) are shaded. The drawings at the *bottom* depict the two views of the central nervous system (enlarged). Lesions causing elimination of HS and VS cells on one side of the adult are marked as are coordinates assigned to larval brain hemispheres. Initial lesions are made by aiming the laser beam from the side (illustration on the *right*). The coordinate location of this lesion was determined from the fate-correlation described in the text. Next, the coordinate location after a 90° rotation of the central nervous system (and the larva) is calculated, resulting in the same frequency of lack of the HS and VS cells (*left*)

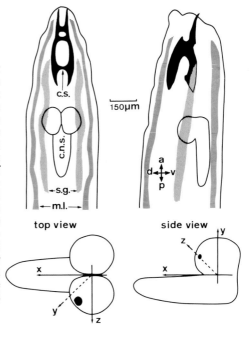

laser beam (Fig. 3). At this point the focal plane and the centre of the beam's destruction sphere are in the same plane.

Brain hemispheres of the first through early third instar larvae are roughly spherical and are symmetrical enough to assign aiming coordinates (Fig. 4). Lesion depth was controlled by reading the μm scale on the microscope's fine focussing knob. The aiming coordinates enabled reproducible lesions resulting in similar alterations of adult morphology. The ablation (caused by damage and evaporation of the cells by local heating) was followed on the monitor (Fig. 3). A lesion caused by several laser pulses is shown in a brain hemisphere of a second instar fly larva (Fig. 5). As can be seen in Fig. 6, the target cells lie deep inside the larva, beneath cuticle, muscles and other tissues through which the beam passes without causing damage.

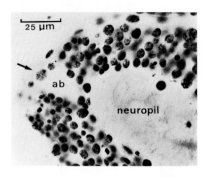

Fig. 5. Silver-stained frontal section of a larval brain hemisphere, fixed 1 h after ablation. At this stages (28 h development) the larval brain consists of a cortex of undifferentiated imaginal cells and a neuropil core. The lesion (*ab*) (diameter approx. 20 μm corresponding to 2–3 cell diameters) is inside tissue and a few intact cells can be seen on the surface of the brain. The laser beam penetrated the tissue approximately in the direction of the *arrow*

Rearing Experimental Flies. After laser surgery, larvae were reared individually in plastic bottles with food mixture at 29 °C and 55% humidity. This relatively high temperature was most appropriate for survival and fast development (West 1951). Rearing individual larvae resulted in the same rate of survival as rearing larvae in bulk. Mortality due to the actual laser surgery is probably negligible since lesions were fairly small and the extent of the lesion well-localized. Furthermore, deficiencies in the optic lobes (our only target) do not appear to be lethal (cf. Power 1943; Greenspan et al. 1980).

Behavioural Tests

Since one aim was to eliminate identified nerve cells to study their role in visual orientation behaviour, treated flies were appropriately tested a few days after eclosion. Tests included visual fixation of a single stripe, torque and thrust responses to a single moving stripe and to moving gratings. These are described in detail later.

Histological Procedures

After behavioural tests, fly heads were cut off into fixative with a razor blade, exposing the posterior brain surface, and were processed for either (a) reduced silver staining, (b) semithin sectioning of glutaraldehyde-fixed tissue postfixed in 2% osmium, or (c) cobalt filling of neuron assemblies.

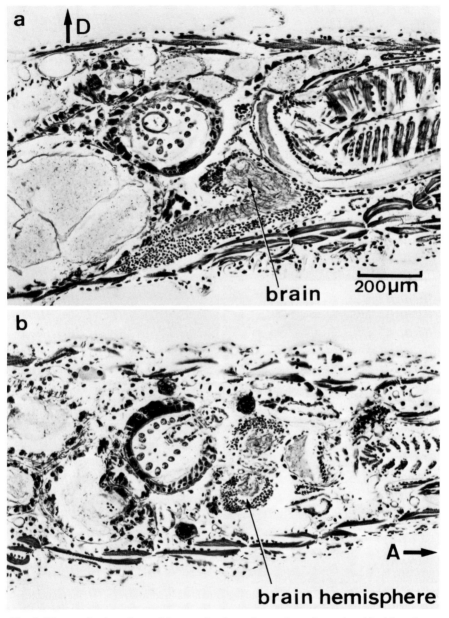

Fig. 6. Silver-stained sections of front ends of two larvae [seen from the side (**a**) and top, (**b**)] to demonstrate cuticle thickness and muscle layers which the laser beam must penetrate to reach brain targets. D = dorsal, A = anterior

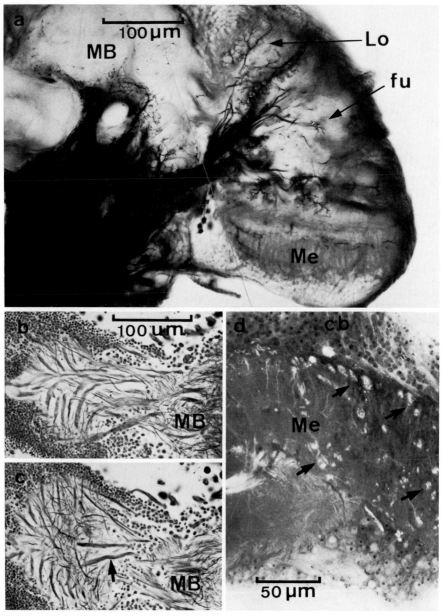

Fig. 7a–d. Examples of the three histological methods used to screen for anatomical defects in adult brains. *MB* midbrain; *Lo* lobula; *Me* medulla. **a** Cobalt backfilling from a severed thoracic ganglion, showing transneuronally stained optic lobe neurons that are disarranged due to a larval ablation. Note the fusion (*fu*) of the lobula plate and part of the lobula. **b** and **c** Examples of Bodian silver-stained sections of the intact lobula plate. Note the large axons of HS (*arrow*) and VS cells. **d** Semithin section (3 μm) stained with toluidine blue. The medulla is abnormal, penetrated by numerous axon bundles (at *arrows*) whose irregularities are due to laser ablation in the larva. *cb* medulla cell bodies

Reduced silver staining was found to be the most informative, giving a detailed overall picture of neuropil structure, cell body locations and fibre projections (Fig. 7). Some individual cells can even be identified from their shape and location. We mainly use Clark's (1973) modified Bodian protargol method (Bodian 1937). Rowell's reduced silver method (1963) was also tried but gave less satisfactory staining of identifiable cells. AAF (alcohol-acetic acid-formaldehyde) was used as a fixative for both methods.

Semithin araldite sections (2 μm) give good resolution of some structures but are less useful for viewing overall neuropil and small fibres (Fig. 7). This method is not used routinely but has potential for studies of early development.

Cobalt filling (Figs. 7, 9) is useful for studying details of single cells or whole sets of certain neurons (Pitman et al. 1972; Strausfeld and Obermayer 1976; Strausfeld and Hausen 1977). Filling assemblies of optic lobe neurons is done by immersing the cut thoracic ganglion in a pool of 5% $CoCl_2$. Cobalt ions pass transneuronally from descending neurons, linking the brain to the thoracic ganglia, into certain optic lobe interneurons (Strausfeld and Obermayer 1976; see also Vol. 1). This method could be especially useful for screening alterations or elimination of certain nerve cells [given that the cell type is one that normally fills transneuronally (Hausen and Strausfeld 1980; Strausfeld and Bassemir, 1983)]. This method does not, however, fill the majority of cell types.

Histological Analysis

The adult visual system of the fly is characterized by its orderly arrangements of receptors and nerve cells (Fig. 8A). The optic lobes, under the compound retina, consist of successive retinotopic neuropils: the lamina, the medulla and the divided lobula complex (lobula and lobula plate). Their neurons are arranged in columns and their stratified neuropils contain many larger nerve cells, the processes of which cover the whole or specific parts of the retinotopic mosaic (Strausfeld 1976; Strausfeld and Hausen 1977; Strausfeld and Nässel 1980). These features make the visual system of the fly especially suitable for studying perturbed development since even small structural alterations can be detected.

Alterations of fibre patterns, deletions of certain cells or gross rearrangements or deficiencies are recorded conventionally using micrographs and tracings that are compared with normal anatomy (Figs. 8, 9). One aid for display and analysis of gross changes in shape, volume or orientation is computer graphics, compiling data from serial sections (see Chaps. 8 and 9). A few computer reconstructions of disarranged optic lobes are shown in Fig. 14.

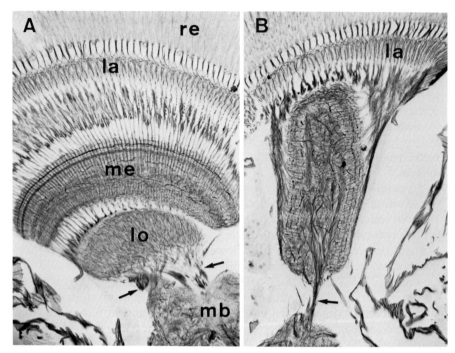

Fig. 8A, B. Silver-stained (Rowell's method) frontal sections of normal (**A**) and ablated (**B**) optic lobes of the same adult fly depicting retina (*re*), lamina (*la*), medulla (*me*), lobula (*lo*) and midbrain (*mb*). The ablation caused elimination of the lobula complex and a folded medulla which received inputs from the lamina on both sides. Note the difference in connections (*arrows*) to the midbrain and the lack of a separately structured lobula complex in (**B**)

Finding the Appropriate Target for Ablations: Fate-Correlation

In embryonic tissues of, e.g. nematodes, leeches and locusts, the number of primordial cells is small and their locations and identities (cell lines) known. The tissue itself is "less three-dimensional" and cell elimination is relatively straightforward. The fate of the ablated cell can be directly observed and recorded histologically.

In later (postembryonic) developmental stages and in more complex organisms the number of precursor cells is large; their identities, obscure; and their location, often deep in the tissue. (The reasons for making ablations in first and second instar larvae rather than embryos will be described in the next section.) In flies, larval brain hemispheres (containing optic lobe Anlagen) contain a fairly homogenous mass of undifferentiated cells that can be seen by Nomarsky optics. Choosing initial lesion sites is thus somewhat arbitrary, the only rough guidelines being earlier descriptions of the locations of optic lobe neuron precursors (Meinerzhagen 1973).

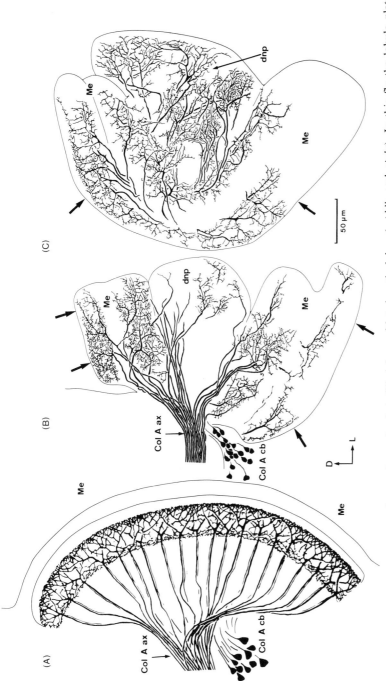

Fig. 9. Tracings from two consecutive sections of cobalt-filled laser-ablated optic lobes (*middle* and *right*). In this fly the lobula plate was eliminated and many lobula cells formed a disorganized neuropil (*dnp*), whereas others invaded medulla neuropil (*Me*), which is split into several neuropil regions. The medulla's surface faces the lateral protocerebrum rather than the inner surface of the lamina, as in normal flies. Normal Col A cells (*left*) in unablated flies are arranged as orderly pallisades in the lobula. In this example of an ablated fly the dendritic morphology of Col A cells is altered, whereas their cell bodies (*Col A cb*) seem to be situated normally and their axons (*Col A ax*) terminate in the appropriate region of the lateral midbrain (see Strausfeld and Nässel 1980; Strausfeld 1983). In the ablated fly, axons from the lamina enter the medulla at the sides marked with *arrows*. In normal flies, axons enter the distal surface of the medulla only

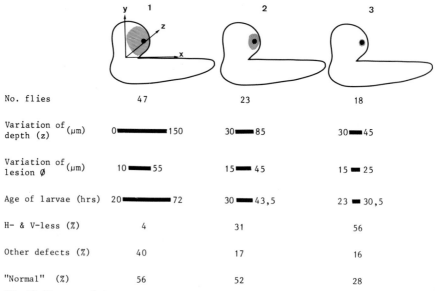

No. flies	47	23	18
Variation of depth (z) (µm)	0 ■■■■ 150	30 ■■ 85	30 ■ 45
Variation of lesion Ø (µm)	10 ■■ 55	15 ■ 45	15 ■ 25
Age of larvae (hrs)	20 ■■■■ 72	30 ■■ 43,5	23 ■ 30,5
H- & V-less (%)	4	31	56
Other defects (%)	40	17	16
"Normal" (%)	56	52	28

Fig. 10. Fate-correlation. Three steps towards a specific alteration of HS and VS cells in the brain of the adult fly. Laser ablations were made in the larval brain as shown schematically above. *Black area* shows the location (and its standard deviation) at which elimination of HS and VS cells in the adult was achieved. *Shading* indicates the area in which ablations were made in each step. Correspondingly, other parameters are shown and correlated with the rate of success in eliminating the HS and VS cells (i.e. HS- and VS-less). Note that in the first step each ablation parameter was used randomly within the constraints described in the text

In the first experiments, lesions varied greatly in location, depth and size, and we used different ages of larval development (Fig. 10). Lesions were specifically made to eliminate several larval brain cells by repeatedly pulsing in adjacent sites. We took care not to damage tracheae, eye-antennal imaginal discs or optic stalks. All variations of parameters used in making these lesions were recorded (see p. 208). After behavioural tests and histological treatment, these parameters were correlated with anatomical alterations of the adult brain. The resulting multiparameter map (fate-correlation) was used to refine lesion parameters to achieve a desired anatomical abnormality in the adult. By statistics, we excluded uninteresting alterations and aimed at the interesting ones (Fig. 10). Incidental anatomical and behavioural abnormalities that occurred during this "refinement" process shed some light on developmental mechanisms (see p. 220).

Anatomical-Behavioural Correlations of Laser-Eliminated Lobula Plate Neurons

The fate-correlation map enabled us to selectively and reproducibly eliminate two sets of large identified neurons in the third visual neuropil (lobula
plate; Fig. 11) on one side of the brain and to study their role in visual orientation behaviour. These neurons (HS and VS cells) have horizontally and
vertically arranged dendrites, respectively. They are directionally selective
motion-sensitive neurons, believed to play a role in the control of optomotor
turning and thrust responses (Pierantoni 1974; Dvorak et al. 1975; Hausen
1976a, b, 1981; Eckert and Bishop 1978; Hengstenberg 1977, 1981).

Ablating cells in adult flies proved technically difficult. We therefore
ablated HS and VS neuron precursors in pre-third instar larvae before onset
of lobula plate differentiation and after early first instar because larvae are
optically denser in the first few hours after hatching. Quantitative fate-correlation which determined parameters for elimination of HS and VS cells is
described in Fig. 10. Three tests recorded visual orientation behaviour
(Geiger and Nässel 1981):

1. The fly's torque (tendency to turn) and thrust (tendency to change flight
   speed) in response to a single stripe moving transiently clockwise and (after an interval) counterclockwise around the fly.
2. The fly's ability to visually fixate a single black stripe.
3. The fly's torque and thrust in response to monocular stimulation with a
   grating moving progressively (from front to back) and regressively (from
   back to front) around the fly.

The apparatus used for these experiments is described in Fig. 12. The
symmetry of the stimulus conditions and unilateral elimination of HS and
VS cells give us the advantage of having an experimental and a control side
in the same animal.

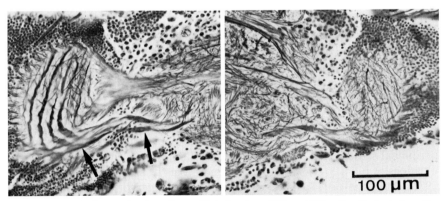

Fig. 11. Frontal silver-stained sections of lobula plates HS and VS cells eliminated on the
right side. VS cells are seen on the normal side (*left, arrows*)

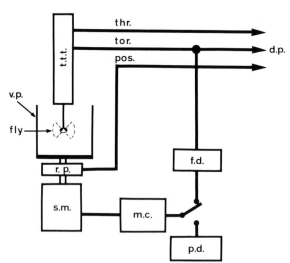

Fig. 12. Apparatus for testing visual orientation behaviour in flight. *t.t.t.* torque and thrust transducer; *v.p.* visual panorama; *r.p.* ring potentiometer; *pos.* information of stimulus position; *s.m.* servomotor; *m.c.* motor control; *p.d.* pattern drive; *f.d.* electronic flight dynamic simulator; *d.p.* data processing of torque (*tor.*), thrust (*thr.*) and position (*pos*) (Geiger et al. 1981)

The results of these tests show that the response to a single stripe was similar on both sides of the fly. The responses to moving gratings were, however, reduced and had different temporal properties on the side lacking the HS and VS cells. From this we concluded that single-object motion detection is processed mainly by cells other than HS and VS cells, and that HS and VS cells serve as visual stabilizers and wide-field background motion detectors (for more details see Geiger and Nässel 1981, 1982). Using the same approach, other cell types can be eliminated and further behavioural tests made for a more detailed understanding of visual information processing.

Aspects of Neuronal Development

Laser microsurgery is also an appropriate instrument for studying neuronal development, being employed to study cell lineages, the role of pioneer fibres in the formation of neural connections, and interactions between developing neurons. The latter was done using UV-irradiation but can be performed with a pulse dye laser unit as well.

Sulston and White (1980) used selective laser ablations of identified cells in early larvae of the nematode *Caenorhabditis elegans* to study regulation and cell autonomy during postembryonic development. The organization of peripheral sensory nerves from the cerci in locusts was observed after laser ablations of pioneer cells in the cercal rudiments of embryos (Edwards et al. 1981). Macagno (1977, 1978) eliminated photoreceptor cells of the small crustacean *Daphnia* to study the role of sensory innervation in the organization and differentiation of central neurons.

Several other aspects of neural development can be studied using laser ablations of single neuron precursors or groups of these, for example: (a) the role of the CNS in the control (interaction) of differentiation of peripheral sensory cells, (b) the control by the sensory cells of cell proliferation and cell death in the CNS, (c) the role that afferent and efferent interactions play in the differentiation of interneurons and (d) regulation of cell number in CNS after ablations [for recent reviews on invertebrate neural development, see Meinertzhagen (1973); Bate (1978); Anderson et al. (1980); Stent (1981); Goodman (1982)].

In our own experiments with *Musca* we made ablations of neuronal precursors in the larvae only. Our interest was in the visual interneurons deep in the optic lobes, and we did not ablate sensory cells of the retina. Although our aim was to eliminate specific sets of nerve cells, we obtained incidental results that, with further studies, may possibly shed some light on neuronal development (see Nässel and Geiger 1983; Nässel et al. 1983). Our preliminary findings are summarized as follows. Laser ablations resulted in five main types of alterations of the optic lobes (see Figs. 7, 8, 9, 13, 14): (1) Alteration of the gross orientation of neuropils, (2) alterations of fibre patterns in and between neuropils, (3) fragmentation of neuropils (which have normal organization in other respects), (4) elimination of parts or whole neuropils and (5) elimination of identified cells. These alterations are either due to the direct elimination of nerve cell precursors or to indirect elimination or disruption of other cells or tissues.

We shall briefly present three types of alterations achieved after laser ablation of CNS precursors since these shed some light on the control of differentiation of visual interneurons (Nässel et al. 1983). These experiments employ removal of the precursors of central target neurons to study differentiation of their more peripheral input neurons. In one set of experiments the entire optic lobes were eliminated except for parts of the lamina. The retina in these animals was normal in appearance. Fragments of the lamina seemed structurally normal in that they received retinotopic inputs from the lamina. Neither the lamina nor the retina had any fibre connections to more central neuropils (Fig. 13D). It therefore seems that the differentiation and maintenance of the lamina neurons is controlled by the retinal input and that efferent input is not crucial. In another experiment, precursors of the entire lobula complex were ablated. The lamina differentiated normally on the ablated side. The medulla appeared to have differentiated normally with a columnar and layered organization, but it was folded along its central edge (Fig. 8). The only connections between the medulla and the midbrain were those formed by the axons of the serpentine layer as in normal flies (Strausfeld 1976). Thus, many of the medulla neurons that would normally have projected to the lobula complex in this animal differentiated without central targets and their axons did not leave the medulla neuropil. Another ablation, which resulted in the absence of the medulla and lobula complex in adults, gave rise to fibres connecting the lamina with even more central neuropil

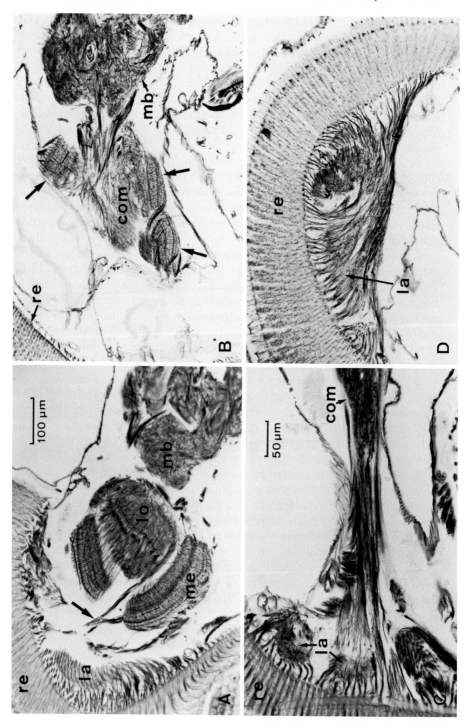

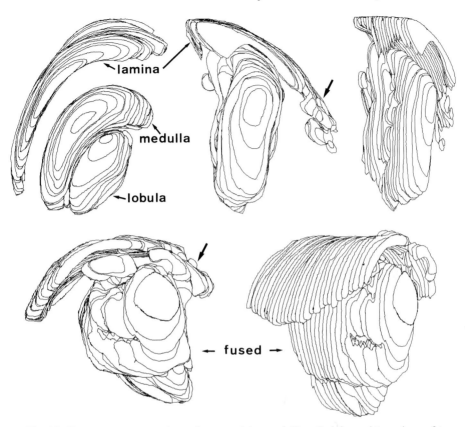

Fig. 14. Computer reconstructions of a normal (*upper left*) optic lobe and two views of two altered optic lobes. *Arrow* points to fragmented lamina. The alteration in the *upper right* is similar to the one shown in Fig. 8 B

Fig. 13 A–D. Reduced silver- stained sections of variously altered adult brains. **A** Medulla divided into two parts with its connections to the lamina disarranged (*arrow*). **B** Disarranged and fragmented optic lobes. Note several medulla-like neuropil regions (at *arrows*) and a fused neuropil mass (*com*) in the centre. The lamina, which is normal, cannot be seen in this section. Bar as in (**A**) **C** Axons from the lamina appear to connect directly to a small neuropil region (*com*) that possibly represents a reduced and fused lobula complex. **D** In this fly no optic lobes were seen on the ablated side except for parts of the lamina. For further abbreviations see Fig. 8. Bar as in (**C**)

(Fig. 13C). It therefore seems that the lamina and many of the medulla neurons can differentiate without forming their appropriate central connections or they can form connections to central neurons that are not their normal targets (Nässel et al. 1983).

Studies of mutants of *Drosophila* (Power 1943) and *Musca* (Nässel and Geiger, 1983) with reduced retinae show a corresponding reduction of the lamina, a smaller reduction of the medulla and an almost normal lobula complex. Surgical removal of the retinal primordia yield similar results (See Meinerzhagen 1973; Nässel and Sivasubramanian 1983).

Discussion

One advantage of the laser microsurgery unit described here is its ability to make precise and reproducible ablations of selected cells deep in the animal without causing other damage. This extends the use of laser ablations to larger tissues since it is possible to adapt the wavelength and power of the laser and the optics of the unit to meet specific requirements and conditions. We operate the laser at λ 450−470 nm rather than at UV to restrict cell damage (which can be monitored by video immediately after the ablation) and to minimize the effect of stray light (such as its interactions with enzymes and other proteins). Using the present set-up we performed control experiments (at the above wavelength) on *Drosophila* embryos to study the effect of laser irradiation. We found that to achieve an alteration in the larval integument pattern an immediately visible alteration had to be produced in the embryo precursor cells (leading to cell death), whereas only bleaching cells had no visible effect on development. Using UV-irradiation, however, a graded effect on development was observed where constant spot size yielded defects of variable sizes (Lohs−Schardin et al. 1979). In this context another control experiment should be mentioned: *Musca* larvae were taken through ether anaesthesia and all other steps except laser irradiation. No visible alterations of the adult nervous system were observed.

In summary, we have demonstrated that controlled laser ablations deep in developing tissue, with subsequent fate-correlation, can achieve reproducible elimination of identified nerve cells, or alteration or elimination of whole neuropils. This approach seems promising for further studies of the functional roles of identified nerve cells and the analysis of postembryonic development.

Chapter 11

Anatomical Localization of Functional Activity in Flies Using ^{3}H-2-Deoxy-D-Glucose

Erich Buchner and Sigrid Buchner

Max-Planck-Institut für biologische Kybernetik
Tübingen, F.R.G.

Introduction

Functional analysis of complex insect neuropil by electrophysiological techniques suffers from specific limitations. Because many cell somata are not invaded by electrical signals (Hengstenberg and Hengstenberg 1980), axons or dendrites have to be impaled. The majority of these are, however, too small to be routinely accessible, and, in general, more than two or three cells can rarely be recorded at the same time. Neuroanatomical techniques, on the other hand, resolve even the finest processes of a number of cells. Thus, a technique that would "fixate" the state of physiological activity in nervous tissue and visualize it along with its structure would greatly aid the investigation of local neuronal circuits even though it would be ill-suited for investigating temporal characteristics of nervous activity.

The first functional mapping technique widely applied in vertebrates, at a regional rather than cellular level, was developed by L. Sokoloff and his colleagues (Sokoloff et al. 1977; Sokoloff 1978). This "radioactive deoxyglucose technique" has also been applied at the cellular level to vertebrate retina and to molluscan ganglia, employing high-resolution light microscope autoradiography (Basinger et al. 1979; Sejnowski et al. 1980). This chapter describes the state of the deoxyglucose technique as applied to insect brain at a regional level (Buchner et al. 1979; Buchner and Buchner 1980, 1983). Recent developments to extend the technique to the cellular level in insect nervous system by electron microscope autoradiography are briefly described (Buchner and Buchner 1981, 1982).

Electrical activity in neurons increases the cell's demand for high-energy phosphates, which in vertebrate brain appear to be provided mainly by the metabolic breakdown of glucose (reviewed by Crone 1978). In insects, however, brain metabolism appears to be less uniform. Some species (e.g. *Calliphora erythrocephala*, *Apis mellifica*) exclusively or predominantly utilize carbohydrates. Others (e.g. *Locusta migratoria*, *Manduca sexta*, *Bombyx mori*) show a comparatively high capacity for metabolic breakdown of fatty acids, and apparently make use of it, as was shown by measurements of respiratory quotients on isolated cerebral ganglia (Kern et al. 1980). The fact that glucose levels in nervous tissue are almost as high as in the hemolymph (Treherne 1974) suggests the existence of a facilitated transport mechanism across the blood−brain barrier and the neuronal cell membrane. In vertebrates such transport systems are known to carry D-glucose and, when artificially supplied, carry its analogues, 2-deoxy-D-glucose and 3-O-methyl-D-glucose, at comparable rates. Intracellular glucose as well as deoxyglucose are phosphorylated by hexokinase. This enzyme does not, however, accept 3-O-methylglucose as a substrate. Glucose-6-phosphate is further metabolized by the glycolytic and tricarboxylic pathway to yield carbon dioxide and water, and thus is rapidly lost from the tissue. Deoxyglucose-6-phosphate (DG-6-P), however, cannot be converted to fructose-6-phosphate nor can it enter the pentose phosphate pathway. Thus DG-6-P is essentially trapped intracellularly and accumulates at a rate indicative of the rate of glucose uptake. Glucose uptake in turn correlates with physiological activity (Mata et al. 1980). The common transport of the three compounds, phosphorylation of deoxyglucose and glucose, and the breakdown of glucose-6-phosphate are schematically depicted in Fig. 1. When labelled deoxyglucose is employed, the accumulation of DG-6-P can be localized in sectioned tissue by autoradiography, provided that postmortem diffusion of the (water-soluble) sugar derivatives is prevented. With incubation periods of 30 min or more, neither 3-O-methylglucose nor glucose or any of its derivatives accumulates in a similar fashion (Sokoloff et al. 1977; Hawkins et al. 1979). If, in an animal of unknown brain metabolism, preferential accumulation of label in active nervous tissue is obtained by the radioactive deoxyglucose

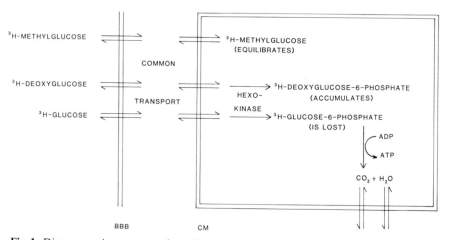

Fig. 1. Diagrammatic representation of transport and metabolism of glucose, deoxyglucose and 3-O-methylglucose. *BBB* blood−brain barrier; *CM* cellular membrane

technique, but is not found when labelled 3-O-methylglucose or glucose are used, then transport mechanisms and glycolytic enzymes in this species are likely to be similar to vertebrates.

Essentials of the Technique

Rather than repeating the detailed steps of the deoxyglucose technique, as presently applied to the fruit fly *Drosophila melanogaster* (Buchner and Buchner 1980), we shall discuss a number of points that may be useful when the method is extended to other insect species.

Application of Radioactive Deoxyglucose (DG)

Tritiated deoxyglucose has been used exclusively by the authors because spatial resolution of ^{3}H autoradiography is superior to ^{14}C by a factor of 5 or more, depending on section thickness (Rogers 1973). Although feeding the compound to very small insects may be technically simpler, injection of it into the hemocoel may shorten incubation time and yields more reproducible results. As rather high doses are needed for high-resolution autoradiography (see below), commercially available ^{3}H-deoxyglucose solutions (Amersham TRK.383) should be condensed by freeze-drying and instant reconstitution with the desired amount of water. (Freeze-drying of 0.1−0.2 ml of aqueous solutions can be done simply in a desiccator connected to a dual-stage rotary pump, if the solution is frozen on a watch glass cooled to about − 25 °C prior to evacuation).

Dosage of ^{3}H-DG

Drosophila (weight about 1 mg) needs 20 μCi of injected label in 0.05 μl solution to give an estimated hemolymph concentration of about 1−3 mM DG, a level that is comparable to natural glucose concentrations (Wyatt 1967). Toxicity of this concentration has not been investigated systematically. However, most injected flies still show optomotor responses when running "on the styrofoam ball" (Buchner 1976). If it turns out that proportionally higher concentrations of ^{3}H-DG are required for larger insects, costs of labelled DG may well restrict application of the method to insects weighing less than a few tenths of a gram. Starvation of the animals prior to injection may increase the uptake of ^{3}H-DG into brain tissue by a factor of about 3−5, as measured in rats (Young and Deutsch 1980). Further considerations on dosage are given below.

Stimulation

After application of ³H-DG the animals are subjected to a specific stimulus for a certain amount of time. Our experience so far is limited to visual and mechanosensory stimulation but we do not see any reason why the method should not work for other sensory modalities as well. Clear stimulus-induced labelling was obtained with stimulation periods between 2 and 8 h when deoxyglucose was fed to the animal and between ¾ and 7 h when injected. Low ³H-DG dosage (e.g. in larger insects) may to some extent be compensated by prolonged stimulation periods. In any case, if the various steps of the entire procedure cannot be perfectly controlled, the stimulation regime should be designed to simultaneously incorporate the experimental situation and the control stimuli. Whenever topographic representations of receptor inputs are known to exist (e.g. visual fields, odour compositions, sound frequencies, etc.), restricting stimulation to part of the represented range allows an immediate comparison between stimulated and unstimulated brain regions. By exploiting the bilateral symmetry of the sensory neuropils, additional controls can be incorporated in a single experiment with one side receiving the control stimuli and the other side, the test stimuli.

Fixation, Dehydration and Embedding

Freeze fixation in melting nitrogen results in satisfactory structural preservation in areas of *Drosophila* and *Musca* nervous system at the cellular level. Melting nitrogen is easily prepared by cooling liquid nitrogen in a desiccator connected to a rotary pump. Propane, which according to the literature may freeze biological specimens even faster (Rebhun 1972), has not yet been tried in our laboratory. The frozen preparations are freeze-dried in a simple device at $-70\,°C$ for 6 days (details in Buchner and Buchner 1980). Fixation of the dry tissue for $1-2$ h in OsO_4 *vapour* (BIOHAZARD: use stringent safety precautions) provides contrast for light and electron microscopy. After the dry tissue has been embedded under mild vacuum in EPON (Shell) $(A:B = 3:7,$ Luft 1961), $2-3\,\mu m$ sections are routinely cut on a dry glass knife. This procedure ensures that after freezing the tissue is never exposed to aqueous solutions. Contact with nonpolar fluids occurs only during embedding. As sugars are insoluble in EPON, no diffusion of radioactive material is expected. A small amount of radioactive material in the EPON adjacent to tissue, e.g. in air sacs, may indicate that the hygroscopic EPON components have taken up some moisture. Fresh EPON components may reduce autoradiographic background.

Some alternative procedures have been tried with *Drosophila*. Freeze substitution instead of freeze drying preserves the gross distribution of label. The transition between regions of low and high radioactivity appears, however, to be somewhat more diffuse than in freeze-dried preparations. This indicates that some local diffusion during ice dissolution may have occurred (Nicod 1983). High-resolution light microscope deoxyglucose autoradiographs have been obtained by freeze substitution of molluscan ganglia (Sejnowski et al. 1980). These authors also describe a method of dehydration of the unfixed tissue in anhydrous acetone at + 4 °C. When flies, however, are dehydrated above 0 °C, autoradiographs show an entirely different distribution of label and do not reflect stimulus-induced nervous activity (Buchner and Buchner 1980; Buchner, unpublished results).

Autoradiography

High-resolution autoradiographs can be obtained only if the section is in intimate contact with the autoradiographic film. Since aqueous solutions displace radioactive material, even in EPON-embedded sections (Buchner and Buchner 1981), contact must be established after the film has been thoroughly dried. In standard contact autoradiography the dry sections are positioned on a microscope slide, and a second slide carrying the dry film is firmly pressed down on the sections. Two problems may arise with this procedure. Minute pieces of glass from the slides or debris from sectioning, located between the two slides, may prevent close contact. Furthermore, when the slides are separated at the end of exposure, the sections may stick to the film and get lost during development, fixation and rinsing. This is avoided using the modified method of exposure illustrated in Fig. 2. Flexible, clear sheets of silicone are prepared by mixing 15 parts SilGel 2001 and 1 part hardener 2001 (accelerator) (Wacker, Munich, FRG). The mixture is spread between two clean glass plates separated by 0.5 mm and allowed to polymerize at 70 °C for 12 h. From these sheets two silicone pads are cut (17 × 5 mm) and placed on a microscope slide. The slide with the pads is wrapped with transparent foil (as used in the kitchen) which has been dipped in a 1:9 mixture of Best-Test adhesive and Bestine solvent (Union Rubber, Trenton, N.Y., USA). The sections are placed on the foil and gently spread out. The foil prevents chemography from the silicone and neutralizes forces tangential to the section. Such forces are generated when the slide

Fig. 2. "Sandwich" for contact autoradiography designed to provide intimate contact between section and autoradiographic film (see text). *C* clamp; *F* transparent foil; *MS* microscope slide; *S* section; *SF* stripping film; *SP* silicone pad

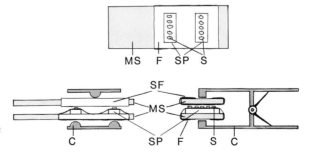

carrying the stripping film (AR-10, Kodak) is pressed firmly onto the sections by the clamp (slightly modified dark-room clamp). The elastic silicone will press all parts of the sections uniformly onto the film surface, and the adhesive will prevent the sections from sticking to the film when the two slides are separated at the end of exposure. Preparation of stripping films and photographic processing of the autoradiographs follow essentially the schedules given by Rogers (1973) (development in D 19 (Kodak), 5 min; stop-bath in distilled water, 1 min; fixation in standard fixer, 12 min; rinsing in two changes distilled water, 5 min each).

Exposure time for *Drosophila* nervous tissue was 10 days for 20 μCi of injected label and 4 days for 50 μCi. There is an intricate relationship determining autoradiographic resolution as a function of the number of disintegrations per unit area (or the product of dosage × time) and latent image fading. A high dose with short exposure is preferable as long as no adverse physiological effects of the compound (deoxyglucose) or the isotope (tritium) are encountered. Due to latent image fading it is not possible to make up for low doses by very long exposures. Thus a compromise has to be found. Spatial resolution of the autoradiographs, however, depends on the density of silver grains in the developed film. Consider, for example, a structure of $2 \times 2\,\mu$m area with 25% higher concentration of label than its homogeneous surround in the section. This structure will be detected as significantly more strongly labelled only if the 25% increase corresponds to at least 2.5 standard deviations of the grain density of the surround. The statistical fluctuations inherent in radioactive decay are described by Poissons's distribution, i.e. repeated counts of n disintegrations will fluctuate with a standard deviation of $\sqrt{n}$. In the above example one calculates that $0.25\,n > 2.5\,\sqrt{n}$ or $\sqrt{n} > 10$ or $n > 100$. If this calculation is used as an approximation to grain statistics in the film, at least 100 grains over the $2 \times 2\,\mu$m structure would be required to detect its increased label. With a grain size of $0.2\,\mu$m, this means that the film should be saturated with ionizing radiation. As the difference in concentration of label can only be detected if the film is not saturated, again a compromise has to be found. In our experience fine details can be optimally resolved in autoradiographs when the grain density is high enough to absorb 90%–95% of the incident light by microphotometric measurement.

Results

Using the described procedure autoradiographs of complete series of semithin sections through *Drosophila* heads have been prepared to localize movement-specific nervous activity. One of the stimulation paradigms employed is shown in Fig. 3a. The autoradiograph (Fig. 3b) of a frontal section displays enhanced labelling in that part of the right medulla whose visual field was stimulated by movement. A similar sector in the flicker-stimulated medulla cannot be recognized. Labelling of the central commissure is probably due to mechanical stimulation of the antennae, which were intact in this experiment (cf. Buchner and Buchner 1980). However, not all the displayed radioactivity arises from these two stimuli. In a control animal the antennae were glued to the head capsule and the fly was kept in the dark after injection. The autoradiograph of a section cutting approximately through the same plane (Fig. 3c) shows no enhanced labelling in the medulla and displays considerably less contrast between the various layers of the optic ganglia. Some activity in the suboesophageal ganglion is present in both preparations. An even more homogeneous distribution of label is obtained when the fly is inactivated by low temperature at 9 °C (horizontal section in Fig. 5b), though some regions of increased glucose uptake are still recognizable. When nonmetabolizable tritiated 3-O-methylglucose is injected, the label is homogeneously distributed throughout the neuropil. A slightly higher uptake is recognizable in the zones surrounding the neuropil where the perikarya are located (cellular cortex) (Fig. 4a). Even this latter difference is absent when tritiated glucose is offered (Fig. 4b). While glucose and deoxyglucose appear to be selectively incorporated in adipose tissue this does not seem to be true for 3-O-methylglucose. With a slightly modified stimulus — the flicker is replaced by stripes moving with sinusoidally modulated speed — habituation of movement-specific activity was tested. The horizontal section in Fig. 5a shows no detectable difference between left and right hemispheres. This means that habituation under conditions of constant speed compared to variable speed does not seem to play a major role for metabolic acitivty of the labelled tissue. The autoradiograph illustrates, however, the selective labelling of the stimulated parts of the medulla and the lobula—lobula plate complex. Recently the technique has also been applied to larger flies (*Musca domestica*). Figure 6 shows an autoradiograph of a horizontal section through the head of a housefly that had been stimulated in a rotating striped drum. The most striking difference to the *Drosophila* autoradiographs is the clear resolution of single labelled nerve axons in the left hemisphere which was stimulated by front-to-back movement ipsilaterally and back-to-front movement contralaterally. These cells which display directionally selective movement-specific labelling were identified by computer-aided reconstruction from complete series of autoradiographs (see Chap. 9).

The results obtained with the deoxyglucose technique so far may be summarized as follows: photoreceptor cells do not accumulate deoxyglucose or its derivatives upon stimulation. Only weak differences in radioactivity are detectable between stimulated and unstimulated parts of the lamina. In the medulla five layers can be identified that show little response to homogeneous flicker but are, however, selectively labelled by movement. Directional preference has so far not been detected in the medulla. In the lobula three layers of retinotopically organized movement-specific activity are found (Fig. 5a). The fronto-caudal position of a similar layer in the lobula plate seems to depend on the direction of movement (Buchner and Buchner

1983). Except for the cell shown in Fig. 6 movement-specific activity proximal to the optic ganglia has not yet been identified unequivocally by the deoxyglucose method.

These results supplement and corroborate electrophysiological data from larger flies (reviewed by Hengstenberg and Hengstenberg 1980; Hausen 1981; Chap. 9, this vol.). The specific advantage of the deoxyglucose technique is exploited when applied to very small insects, such as *Drosophila* and its neurological mutants, whose brains are practically inaccessible to electrophysiological techniques. So far three visual mutants have been investigated (Nicod 1983): outer rhabdomeres absent ora[JK84], optomotor blind omb[H31] and small optic lobe sol[KS58]. The results obtained so far support the functional interpretations inferred from behavioural and neuroanatomical analysis (Heisenberg and Buchner 1977; Heisenberg et al. 1978; Fischbach and Heisenberg 1981).

Extension of the Technique to the Single-Cell Level by Electron Microscope Autoradiography

Clearly the potential of the deoxyglucose technique could be considerably increased if it were possible to associate individual silver grains of the autoradiographs with cellular profiles of identified neurons. As most cellular profiles in synaptic neuropil have a diameter of less than 1 μm, this can in

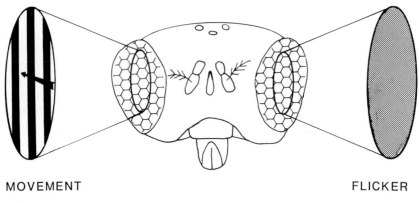

MOVEMENT   FLICKER

a

Fig. 3. a Visual stimulus for comparing nervous activity in movement-stimulated, flicker-stimulated and unstimulated parts of the retina and retinotopically organized optic ganglia. Vertical stripes moving front-to-back at constant speed are presented to the central visual field of the right eye, while the corresponding field of the left eye observes a flickering disc of the same mean luminance. **b** Autoradiograph of a frontal 3-μm section through the head of a *Drosophila* fed with 20 μCi of ³H-deoxyglucose and stimulated for 2 h. Layers and columns of the second optic ganglion (medulla) are recognizable as concentric rings and "beads" of enhanced label, respectively. *Left* right eye. The sector of increased metabolic

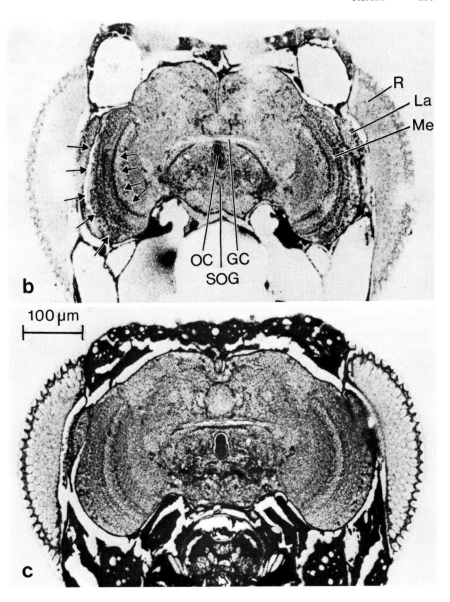

activity in the right medulla (*arrows*) corresponds to the visual field stimulated by movement. This sector therefore maps stimulus-induced movement-specific nervous activity. *GC* grand commissure; *La* lamina; *Me* medulla; *OC* oesophageal canal; *R* retina; *SOG* suboesophageal ganglion. **c** Autoradiograph of a control fly injected with 20 μCi of ³H-deoxyglucose and kept in the dark at room temperature for 3.5 h before shock-freezing. Layers and columns in the medulla are still recognizable but show considerably less autoradiographic contrast than the preparation in **b**. The suboesophageal ganglion exhibits conspicuous metabolic acitivity

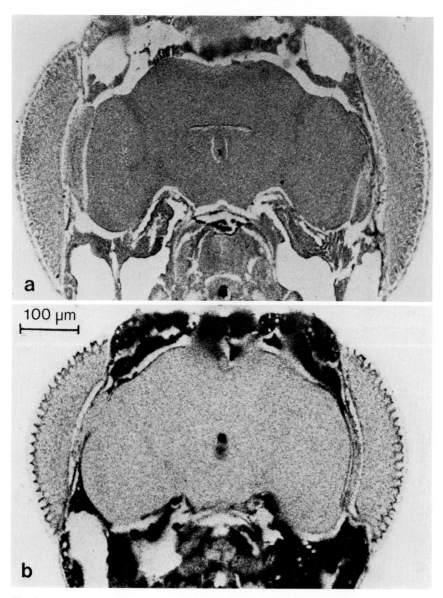

Fig. 4. a Autoradiograph of a fly fed with 40 μCi of ^{3}H-3-O-methylglucose and visually stimulated for 5 h. Stimulus-induced nervous activity cannot be detected. This indicates that phosphorylation of deoxyglucose is essential for metabolic mapping. **b** Autoradiograph of a fly injected with 30 μCi of ^{3}H-glucose and visually stimulated for 5 h. Labelled metabolites of glucose distribute homogeneously throughout the nervous system. Adipose tissue is heavily labelled

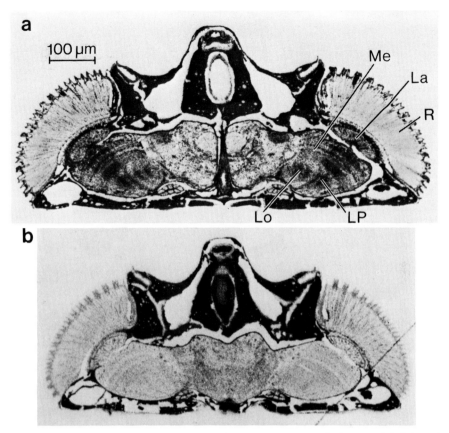

Fig. 5. a Autoradiograph of a horizontal section through the head of *Drosophila* injected with 20 µCi of ³H-deoxyglucose and visually stimulated for 1 h. In this experiment the flicker stimulus (Fig. 3a) was replaced by stripes moving front-to-back at sinusoidally modulated speed. Stimulus-induced nervous activity is mapped in the central sectors of medulla (*Me*), lobula (*Lo*) and lobula plate (*LP*) on either side. *R* retina; *La* lamina. **b** Autoradiograph of a control fly fed with 10 µCi of ³H-deoxyglucose and kept in the dark at a temperature of 9 °C for 5 h prior to shock-freezing

general only be accomplished by electron microscopy (EM). The major difficulties encountered are due to the water solubility of deoxyglucose and its derivatives. It has been mentioned already that "classical" techniques for the preparation of tissue for electron microscopy (aqueous fixation and dehydration at room temperature) do not preserve the distribution of label. We therefore had to determine if structural preservation of the shock-frozen, freeze-dried, osmium vapour-fixed and EPON-embedded tissue was adequate. Furthermore, as ultrathin sections cannot easily be cut on a dry knife, the effect of water contact on the ultrathin sections during sectioning and coating of the sections with photographic emulsion had to be investi-

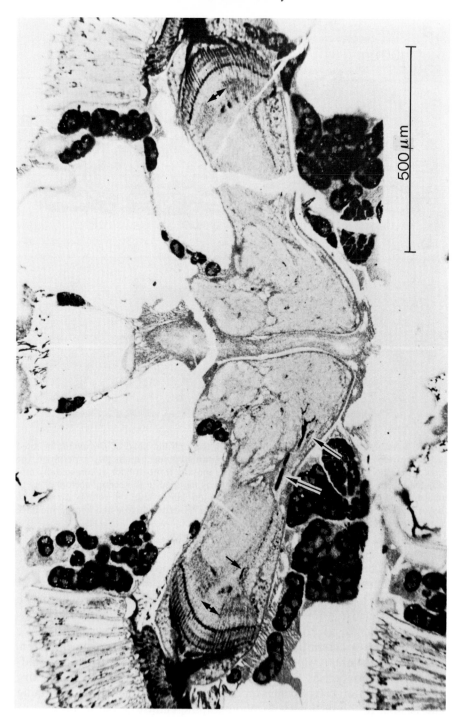

gated. Preliminary results demonstrate adequate structural preservation of cellular membranes. Also, the stimulus-induced distribution is at least partially preserved when the EPON-embedded tissue is exposed to water (Buchner and Buchner 1981). Recently, ultrathin sections of a central fibre tract in the thoracic ganglion were prepared for routine EM-autoradiography (Bouteille 1976) by coating them with a monocrystalline layer of photographic emulsion (Ilford, L4). EM autoradiographs obtained after exposure and development of these sections demonstrate that the deoxyglucose method can provide information on cellular metabolic activity in insect nervous tissue. The dynamic range of this method, as defined by the ratio of grain density over heavily labelled axonal profiles to that over axons showing only background labelling, may well cover an order of magnitude (Buchner and Buchner 1982). These results indicate that the relative levels of metabolic (and possibly electrical) activity can be determined simultaneously for a large number of identifiable neurons by extending the deoxyglucose technique to electron microscopy.

Concluding Remarks

The deoxyglucose method for insects has, so far, been used to qualitatively map functional activity. Quantitative measurements of regional brain glucose consumption as described for vertebrates (Sokoloff et al. 1977) appear difficult if not impossible to achieve in small insects. Neither can the hemolymph concentrations of deoxyglucose required for high resolution autoradiography be considered as tracer amounts nor can glucose and deoxyglucose concentrations in the hemolymph be easily monitored during an experiment. Even for qualitative interpretations of autoradiographs the following points should be considered. Only label densities over regions of identical tissue composition may be compared directly. This can be achieved, e.g. by utilizing the bilateral symmetry of most parts of the insect nervous system or by comparing identical structures in different animals. Glucose consumption, and thus uptake of label, depends not only on electrical activity of a given brain region but also on the surface-to-volume ratio of the cell processes in that region (Schwartz et al. 1979). The difference in labelling, e.g. of various layers in the medulla, even in unstimulated control

---

Fig. 6. High resolution activity staining in *Musca domestica*. The fly was injected with 150 μCi ^{3}H-deoxyglucose and stimulated for 4 h in a rotating striped drum moving anticlockwise. In both eyes a mask occluded the middle retina area and this is reflected by low activity in the inner medulla (*double arrows* left and right). Single cell resolution is achieved to the left. A centrifugal horizontal motion-sensitive neuron (CH-neuron) is labelled on that side which received front-to-back movement stimulation. Its terminal branches, axon and ipsilateral central dendrites are indicated by arrows to the left (*small outer arrow* at lobula plate = terminal, *middle arrow* = axon, *inner arrow* = dendrites)

preparations (Fig. 3c) can, therefore, not be interpreted as reflecting different levels of electrical activity. The darker bands may simply contain a large number of small processes while the lighter bands may be composed of larger-diameter portions of the same or different neurons, all possibly being at the same level of electrical activity. These phenomena impede the detection of small stimulus-induced changes in unpaired structures, such as the central body or central commissures, since considerable variation in labelling occurs from animal to animal due to differences in size, nutritional state, level of arousal and so on. Yet, when combined with appropriately designed stimulation schedules the deoxyglucose method represents a powerful new technique in insect neurobiology. Its potential has barely begun to be exploited.

Acknowledgements. We wish to thank Professor K. G. Götz and Dr. R. Hengstenberg for their encouragement and invaluable advice throughout this work. The excellent technical help of C. Straub and J. Pagel is gratefully acknowledged.

Chapter 12

Strategies for the Identification of Amine- and Peptide-Containing Neurons

Michael E. Adams

Max-Planck-Institut für Verhaltensphysiologie
Seewiesen, F.R.G.

Cynthia A. Bishop and Michael O'Shea

Department of Pharmacological and Physiological Sciences
Chicago, Illinois U.S.A.

Introduction

We have been searching for large, uniquely identifiable neurons which can be used to assess the physiological actions of peptide and amine transmitters on a cellular level. Such identified neurons show consistent properties of form, transmitter substance and pattern of connections. This allows for repeated sampling for precise anatomical, physiological and biochemical analysis. This short paper outlines some staining methods which, when combined with intracellular dye-filling and biochemical and physiological studies, can be used to identify transmitter substances in individually identified, physiologically accessible nerve cells.

Neutral Red: A Nonspecific Stain for Amine-Containing Neurons

Vital staining of monoamine-containing neurons in invertebrates with the dye, neutral red, was introduced by Stuart et al. (1974) and has been used in insects by others (Evans and O'Shea 1978; Dymond and Evans 1979; O'Shea and Adams 1981). The method does not discriminate between different amine substances as do the Falck-Hillarp and glyoxylic acid fluorescent techniques (Klemm 1980; Axelsson et al. 1973; Lindvall and Bjorkland 1974). Neutral red therefore provides a convenient stain for initial screening of the nervous system for a wide variety of putative aminergic neurons. Cells identified in this way can subsequently be subjected to more specific histological and biochemical techniques for transmitter identification. An additional convenient feature of the method is its ability to stain living cells in whole ganglia. It can therefore be used to locate cells for functional studies using electrophysiological recording techniques.

Neutral red is a weakly basic chloride salt of the azin series (Fig. 1). Aqueous solutions are red at acid pH, turning yellow in the pH range 6.8−8.0, making the substance useful as an indicator. In histological appli-

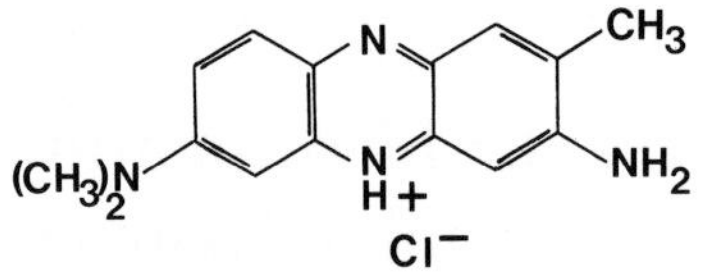

Fig. 1. Structure of the vital stain, neutral red

cations it is a weak nuclear stain and shows specificity for cytoplasmic organelles such as the Golgi apparatus and the Nissl substance. It has been especially useful as a vital stain for Protozoa, blood cells and neurons (Lillie 1969).

Stuart et al. (1974) reported that several weakly basic, three-ring heterocyclic dyes (neutral red, azure A, brilliant cresyl blue, neutral violet, Nile blue A, toluidine blue) specifically stain amine-containing somata in segmental ganglia of the leech. Neutral red proved to be the most useful stain of the compounds tested, combining low toxicity with consistent staining. For staining, live excised ganglia were bathed in saline solutions containing 0.01 mg ml^{-1} neutral red (Chroma Gesellschaft). A 30-min incubation at room temperature was sufficient for consistent and permanent staining of a few specific cells (about seven cells or 3% of the total) in the ganglia. The neutral red-positive neurons consisted of the giant Retzius cells, two types of bilaterally paired somata and a single, possibly unpaired, soma. Falck-Hillarp staining experiments for amine identification showed intense serotonin fluorescence in the Retzius cells. The other neutral red-stained cells showed monoamine fluorescence, suggesting the presence of dopamine. Because the staining could be carried out successfully in the presence of metabolic inhibitors or even on prefixed ganglia, it was concluded that active processes are not involved in the stain's accumulation in aminergic neurons. No explanation is yet available for the specificity of neutral red staining.

Subsequent papers have described neutral red staining in orthopterous insects (Evans and O'Shea 1978; Dymond and Evans 1979). In a detailed study of the DUM (dorsal unpaired median) neurons in the locust metathoracic ganglion, neutral red staining provided an initial clue to the positions, size and aminergic nature of the DUM cells (Evans and O'Shea 1978). The DUM cells appear as clusters of 6−8 large somata situated along the dorsal midline of the ganglion. They send bifurcating axons to the skeletal muscles on both sides of the animals where they modulate myogenicity and synaptic transmission (Evans and O'Shea 1978; O'Shea and Evans 1979). Desheathed ganglia were incubated at room temperature in 0.1 mg ml^{-1} solutions of neutral red for 2−3 h in saline. Alternatively, staining was done at 4 °C overnight. A group of large dorsal median somata was consistently stained in most preparations. One of these neurons (DUMETi, Hoyle et al. 1974) innervates the extensor tibia muscle of the jumping leg. Isolation of the cell body and radioenzymatic assay (Molinoff et al. 1969) showed the presence of about 0.1 pmol octopamine per cell body (Evans and O'Shea 1978).

Neutral red-positive dorsal cell bodies are also found in each segmental ganglion of the cockroach (Dymond and Evans 1979). These cells are probably homologous to the DUM neurons described in locust. Groups of these segmentally repeating somata, collected from the metathoracic and terminal ganglia, showed high levels of octopamine when assayed radiochemically. Staining of specific areas in the cockroach brain (mushroom bodies, optic lobes) was also correlated with high levels of octopamine, although in these regions staining and octopamine presence were not shown to be directly associated with specific neutral red-stained neurons.

The evidence described above shows that neutral red can be used as a convenient and primary probe in the identification of specific amine-containing neurons in insects. It will stain serotonin-, octopamine- and dopamine-containing neurons indiscriminately and must therefore be used in combination with other analytical techniques. Its usefulness, however, in indicating the location of identifiable amine neurons is clear, and mapping of neutral red-positive neurons in insects is a relatively simple and potentially important step in the chemical identification of specific cells.

Neutral Red: An Indicator of Peptidergic Neurons

There is growing evidence in the vertebrates that amine and peptide transmitters may be co-localized in single neurons (Hökfelt et al. 1980a, b). Neutral red may therefore indicate the location of some peptidergic neurons in insects. One example of a neutral red-positive, peptide-containing neuron has been described (O'Shea and Adams 1981). Using a combination of neutral red staining, intracellular dye injection and biochemical analysis, O'Shea and Adams (1981) showed that the LW (lateral white) neuron contains about 0.1 pmol of the bioactive pentapeptide proctolin (Brown 1975, 1977). The intense neutral red staining of this neuron (Fig. 1 a) suggests the co-localization of a monoamine with proctolin in this neuron.

The staining methods used for the LW neuron differ slightly from those described above. Staining was carried out at room temperature on isolated nerve cords from cockroaches (*Periplaneta americana*) and crickets (*Gryllus bimaculatus*) under physiological saline. Ganglia were not desheathed. Nerve cords were stained for 30−60 min in a solution of 0.03 mg ml^{-1} neutral red (Chroma Gesellschaft or Sigma Chemical). The ganglia became more or less uniformly stained, and a destaining period followed in normal saline for 4−6 h. During this time, nonspecific background staining disappeared and several somata appeared bright red against an unstained background.

Figure 2a shows two intensely stained LW somata laterally on the ventral side of a cricket second abdominal ganglion. The anatomy of these neurons was revealed by the precipitation as a sulphide of intracellularly injected cobalt (Pitman et al. 1972) using the modification of Wilson (1981).

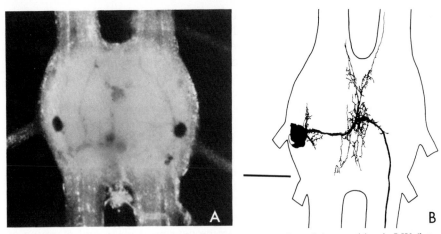

Fig. 2 A, B. Neutral red staining and anatomical reconstruction of the peptidergic LW (lateral white neuron). **A** Ventral view of a living, neutral red-stained abdominal ganglion from the cricket *Gryllus bimaculatus.* A single, bilateral pair of neuron cell bodies have stained intensely. These cell bodies can be penetrated with microelectrodes and dye-injected for reconstruction. **B** A single LW neuron reconstructed from whole-mount preparation following dye injection. The central projections and contralateral descending axons are characteristic of this neutral red staining cell. Bar = 100 μm

Visualization of the cell in whole mount was enhanced by intensification of the cobalt sulphide precipitate using a modification of the Timm's silver intensification method (Bacon and Altman 1977). A reconstructed LW neuron (Fig. 2b) shows central projections concentrated along the middle of the ganglion, a contralateral descending axon and a vacuolated cell body (Adams and O'Shea 1981). This pattern of projections was observed in each segmental pair of LW neurons stained by neutral red. The somata were subsequently isolated, extracted and subjected to biochemical analysis (see below).

Permanent Preparations of Vital Staining with Neutral Red

Whole-mount preparations of the neutral red-stained cells can be fixed and sectioned. Using Karnovsky's formaldehyde−glutaraldehyde mixture (Karnovsky 1965), we have made permanent preparations of vitally stained LW neurons. Alcoholic fixatives, such as Carnoy's, remove the stain and cannot be used. Fixed preparations should be rapidly dehydrated to avoid extraction of the dye: 50%, 70%, 95%, 100% ethanol, 5 min each. Tissue was embedded in soft araldite and sectioned at 40 μm.

Stained LW somata, viewed in section, showed concentration of neutral red in discrete, intracellular granules ranging in size up to about 3 μm in diameter. The cytoplasm surrounding the neutral red granules was devoid of stain. Staining of some large processes could be seen in neuropil, but in general staining was too faint to be of use in tracing connections to cell bodies.

Use of Immunohistochemical Approaches to Neuron Identification

For the purposes of electrophysiological and biochemical studies, it is important to know the precise location of the neuronal cell body. This is because identified insect neurons typically have their somata in characteristic positions; the soma is often the most electrophysiologically accessible part of the cell, and the largest somata can often be individually excised from the ganglion and subjected to biochemical analysis.

Methods have now been developed for whole-mount immunochemical mapping of insect neuronal cell bodies (Bishop and O'Shea 1982, 1983; O'Shea and Bishop 1982). The immunoreactive cell body maps can be used in selecting particular neurons for individual identification. The individual identification of neurons that are immunoreactive to antibodies raised against putative insect neurotransmitters can be achieved by combining intracellular dye injection with immunohistochemistry. Finally, the content of the individual cell body can be checked biochemically for the presence of the transmitter indicated by the immunohistochemistry. This experimental approach to the chemical and morphological characterization of specific neurons can be summarized as follows. It involves three phases: (1) *immunochemical screening* (using whole-mount immunohistochemical mapping), (2) *identification of specific neurons* (requires double-labelling by combined intracellular dye injection and immunohistochemistry), and (3) *biochemical confirmation of immunohistochemistry* (requires identification of cell content by chemical methods independent of immunoreactivity). This strategy is an approach for determining the physiological consequences of neurally released transmitter. The methods employed in this approach to neuronal identification are summarized below.

Immunohistochemical Screening: Whole-Mount Method

We have developed whole-mount methods for mapping serotonin (5-HT) and proctolin (H-Arg-Tyr-Leu-Pro-Thr-OH) immunoreactivity. Antibodies to serotonin can be obtained from Immuno-Nuclear Corp. (Stillwater, MN) and are suitable for both immunochemical staining and radioimmunoassay (RIA). We have raised several specific antisera to proctolin using standard methods. Details of these methods can be found in Bishop et al. (1981).

Proctolin-immunoreactive cells can be visualized in a whole-mount ganglion preparation (see Fig. 3) by introducing several modifications of the peroxidase-antiperoxidase (PAP) procedure (Sternberger 1979) used to stain cells in sectioned material (see Bishop et al. 1981). Three of these were designed to increase the penetration of the antisera in the tissue: (1) use of the detergent Triton X-100 with the antisera and the washes; (2) increasing the incubation times in the second and third antibodies and the washes; and (3) desheathing: removing the neurilemma, a membrane which surrounds the ganglia. Of these, the most critical is desheathing, since staining was almost never achieved without it. The most intense staining of cell bodies was achieved when these modifications were combined with the injection of colchicine, a procedure which also enhances the staining of fibers. Full technical detail of the whole-mount PAP method can be found in Bishop and O'Shea (1982).

The whole-mount procedure allows for the convenient and rapid visualization of cell body location with respect to the positions of other cell bodies and in relation to the whole ganglion. Whole-mount preparations therefore provide a means of cell body mapping without serial-section reconstruction. In addition, the major projections of some axons and processes can be seen. In spite of the advantages offered by whole-mount visualization, it does not reveal certain details of the anatomical relationships between the immunoreactive cells and the internal structure of the ganglion. These features can be visualized only by the reconstruction of serial sections. This can be done by dehydrating and sectioning the whole-mount preparation after immunohistochemical staining. These procedures therefore allow for the combined advantages of both whole-mount and section visualization. The whole-mount method differs from our previous immunohistochemical procedures in which dehydration and sectioning preceded staining (Bishop et al. 1981). The results are for the most part comparable, although some additional cells stain in the whole-mount procedure. This may be due to the loss of proctolin-like antigens from the tissue when dehydration precedes the immunohistochemical reaction.

In our studies of the cockroach central nervous system (Bishop and O'Shea 1982), proctolin immunoreactive neurons can be seen in all ganglia. Mapping, however, is complicated by staining variability in cell number and intensity. A variety of factors, including damage during processing and changes in the position of cell bodies during desheathing, incomplete desheathing, poor infiltration of antibodies, and variation in response to colchicine, may account for the variability. Other factors may be due to intrinsic properties of the cells and may therefore be of biological interest. These include the possibility of cross-reactivity with antigens similar to proctolin and variation in the amount and location of antigen in individual cells. Clearly, great care must be exercised in interpreting the immunological data. In spite of the variability, however, a comparison of several whole-mount preparations indicates that proctolin immunoreactivity in cockroach CNS is confined to a select subset of neurons so that a pattern of staining can be mapped. Mapped cell bodies are identifiable from preparation to preparation by a number of distinguishing characteristics. They can be identified anatomically by their size or the branching pattern of their processes. Cell bodies are also recognizable with respect to their position in the whole ganglion and relative to other staining cells and fiber tracts. Stained cell bodies may occur in bilaterally symmetrical pairs or may be unpaired. They may or may not be clustered with other staining cell bodies.

Although we have been concerned primarily with proctolin staining, we have also applied the techniques in mapping serotonin-immunoreactive cells (Bishop and O'Shea, 1983). In principle, the same methods can also be used to localize cells immunoreactive to other antibodies (against vertebrate neuropeptides, for example). We will discuss the problem associated with the use of antibodies raised against antigens not known to be present in insects below (see Concluding Remarks).

Identification: Immunohistochemistry and Dye Injection

The following methods can be applied in obtaining anatomical evidence for the unique neuronal identity of immunoreactive cell bodies. In general, there are limitations on which individual immunoreactive neurons can be identified in this way. The following criteria established by Bishop and O'Shea (1982) can help in the selection of good candidate neurons for further physiological and biochemical characterization. For immunoreactive neurons to be suitable for this they should be: (1) recognizable from animal to animal in the living ganglion; (2) large enough for single neuron dissection and extraction for biochemical assay (30 μm cell body diameter); (3) accessible to intracellular microelectrode penetration; and (4) easily and reliably distinguished from other immunoreactive neurons (not a member of a cluster of immunoreactive cell bodies, for example).

An example of an immunoreactive cell body that meets these requirements is a giant dorsal bilateral cell in the third thoracic ganglion of the cockroach, *Periplaneta americana* (Fig. 3). In the living metathoracic ganglion, a cell of similar size and position can be seen. It is easily distinguished from neighboring cells by its large size and opaque appearance when the isolated ganglion is illuminated against a dark background. A cell with similar appearance appears in this location consistently in different individuals. We have made a preliminary electrophysiological and anatomical characterization by intracellular recording and by combining dye injection with whole-mount immunohistochemistry (O'Shea and Bishop 1982).

Under visual control, a Lucifer yellow-filled microelectrode can be readily placed in this cell and its anatomy revealed by dye injection and fluorescence microscopy. Figure 3a−d illustrates the central anatomy of this cell in the third thoracic ganglion. That this cell corresponds to the pair of immunoreactive cells shown in Fig. 3a is confirmed by experiments in which a ganglion containing a dye-injected cell (Fig. 3b, c) is processed for whole-mount immunohistochemistry (Fig. 3d). The immunochemical processing reduces the dye fluorescence (Fig. 3d), particularly around the

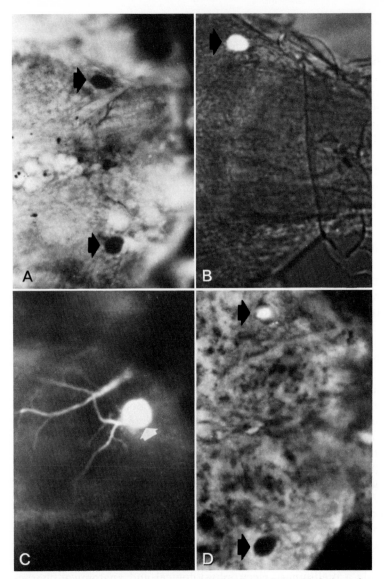

Fig. 3A–D. Dorsal views of the posterior region of the third thoracic ganglion of the cockroach. Whole-mount immunohistochemistry reveals the presence of a single large bilateral pair of neuronal cell bodies (**A**). In the living ganglion (**B**) a single large neuron can be seen in this lateral region, penetrated with a dye-filled microelectrode and injected. The projections of the neuron are then revealed (**C**) by fluorescence microscopy in either the living or the fixed and cleared whole ganglion. A ganglion with a dye-injected neuron can then be prepared for whole-mount immunohistochemistry (**D**). This confirms that the injected and intracellularly characterized neuron is one of the immunoreactive pair shown in **A**. Note that dye is lost primarily from the periphery of the soma during processing for whole-mount immunohistochemistry; cf. **B** and **D**. (After O'Shea and Bishop 1982)

periphery of the cell body, and the presence of dye interferes with the immunoreactivity. It is clear, however, that the injected cell corresponds both to the neuron described above and to the proctolin-immunoreactive neuron on the uninjected side in Fig. 3d. On the injected or double-labelled side, there are no additional immunoreactive cells of appropriate size to correspond with the uninjected side.

Confirmation of Immunohistochemistry: Cell Isolation, Extraction and Assay

Immunohistochemical staining of specific neurons suggests the presence of a substance either identical to or resembling the antigen against which the antiserum was raised. Biochemical analysis of isolated, immunoreactive cell bodies can provide additional evidence for the presence of the substance in question. The procedure will be described with special reference to proctolin-immunoreactive neurons. For serotonin-immunoreactive cells, a sensitive radioenzymatic assay which can be applied to single-cell extracts has been developed and has been described elsewhere (Saavedra et al. 1973).

Using dark-field illumination, the cell bodies of some neurons can be seen. Neurons with particularly granular cytoplasm appear opaque or white when illuminated in this way. Individual neuronal cell bodies can be isolated by removing the ganglionic sheath and dissecting single cell bodies free of the ganglion. Single cell bodies are cleaned of adhering debris from other cells, sucked into a glass capillary ($80\,\mu$m diameter) and transferred to a small test tube containing $50\,\mu$l 2 N acetic acid. The same individually identified neurons from different specimens can be pooled in the same extraction tube. Extraction tubes are sonicated and frozen and thawed after each cell is added. After centrifugation, the extract supernatant is transferred to another tube and evaporated under reduced pressure at $60\,°$C. The dried samples can then be redissolved in either isotonic physiological saline for direct bioassay or in chromatography solvents for HPLC purification (see below).

A highly sensitive and convenient bioassay for proctolin has been described (O'Shea and Adams 1981). The assay has been developed from a specialized bundle of muscle fibres which form part of the main extensor muscle (the extensor tibialis) of the locust (*Schistocerca nitens*) metathoracic leg. These fibers produce a spontaneous myogenic rhythm of contraction and relaxation (Hoyle and O'Shea 1974). The frequency of this rhythm is know to be altered by certain neurotransmitters (Piek and Mantel 1977; Evans and O'Shea 1978; O'Shea and Adams 1981). Octopamine (10^{-6} M), for example, decreases its frequency and proctolin (threshold $10^{-10}-10^{-11}$ M) increases both the rhythm frequency and muscle tone. This assay can be used to test crude and chromatographically purified extracts of single cells. For HPLC, extracts of individually identified proctolin-immunoreactive neurons can be chromatographed on a 10-μm Bondapack C_{18} reverse phase column (Altech). Elution time of proctolin in this system was determined by chromatography of tritium-labelled proctolin (^{3}H-Tyr$_2$-Proctolin). A volatile solvent system should be used to avoid salt-disruption of the bioassay used to monitor proctolin-like bioactivity in dried and saline-suspended HPLC fractions. Chromatographed fractions are bioassayed after evaporating the liquid phase under reduced pressure and redissolving each dried fraction in isotonic physiological saline.

This approach reinforces the conclusion that proctolin-like immunoreactivity is due to the presence of proctolin.

Concluding Remarks

Defining precise functions for neurotransmitters depends on their localization in identified neurons that are amenable to physiological analysis. The staining methods outlined here provide a means for "screening" the nervous system for putative amine- and peptide-containing neurons. Large, easily identifiable neurons can then be chosen for experiments designed to demonstrate release of transmitter substance and action at well-defined post-synaptic targets. Since staining alone cannot provide unequivocal chemical identification, it must be verified by biochemical analysis of staining neurons. We see the power of these staining methods therefore in their ability to locate large neurons that can be further analyzed.

Neutral red, a nonspecific stain for aminergic neurons, provided a lead to the first chemical identification of an insect octopaminergic neuron (Evans and O'Shea 1978) and to the physiological characterization of its modulatory roles at locust skeletal muscle synapses (O'Shea and Evans 1979). This stain also led to the first biochemical and physiological identification of a proctolinergic neuron, the lateral white (LW) neuron (O'Shea and Adams 1981). These examples serve to illustrate the usefulness of these stains as a first approach to the chemical identification of specific neurons.

Our experience with immunohistochemical staining of identified proctolin-containing neurons has shown that certain cells shown to contain proctolin by biochemical methods fail to react in immunohistochemical staining. An example of this is the LW neuron, which does not stain with the proctolin antiserum. Immunohistochemical staining therefore can fail to stain neurally localized antigen in certain types of neurons. This may be related to specific modes of packaging or the amounts of antigen in these nonstaining cells. Conversely, cross-reactivity of the antiserum can lead to nonspecific staining. This may be especially true if staining is obtained using antibodies against compounds that are not known to exist in insects. Thus, for example, although there is much immunochemical evidence for the presence of mammalian peptides in insects and other invertebrates (for review see O'Shea 1982), the immunoreactive compounds are generally unidentified in these studies. Conclusions based on such studies therefore should be made with caution. A recent and notable exception to this, however, has been provided by Duve et al. (1982) who provided biochemical evidence for the presence of a peptide-like vertebrate pancreatic polypeptide in the blowfly.

Acknowledgements. This work was made possible by money provided by the Max-Planck-Gesellschaft, the National Institutes of Health (grants NS-16298 to M.O. and NS-06684 to C.A.B.), the North Atlantic Treaty Organization (fellowship to M.E.A.) and by the Deutsche Forschungsgemeinschaft (grant HU + 35/17 to M.O.). We thank Professor Franz Huber for his valuable criticism, support and encouragement and Dr. Les Williams for his helpful advice, humor and technical assistance.

Immunochemical Identification of Vertebrate-Type Brain-Gut Peptides in Insect Nerve Cells

HANNE DUVE and ALAN THORPE

School of Biological Sciences,
Queen Mary College, University of London
London, England

Introduction

Immunocytochemistry has been used with great success over the past decade in the identification of the sites of the so-called brain-gut peptides of vertebrates. Information resulting from the use of this technique has made it clear that bioactive peptides that were once thought to be strictly gut peptides or hormones, such as cholecystokinin (CCK) and vaso-active intestinal peptide (VIP), are also widespread in the central and peripheral nervous systems. Conversely, neuropeptides such as the enkephalins and substance P have been shown to occur in the gut (cf. Hökfelt et al. 1980a, b). The list of these dual occurrence brain-gut peptides is constantly being extended.

Immunocytochemical localisation of the peptides within specific vertebrate tissues has, in many instances, been followed by their purification and chemical characterisation from tissue extracts. These procedures have necessitated extremely efficient protein and peptide separation techniques such as immuno-affinity adsorption and high pressure liquid chromatography (HPLC), as well as the development of highly sensitive and specific radioimmunoassay monitoring systems.

Interest in the phylogeny of the brain-gut peptides has led to comparative vertebrate studies and also studies of invertebrate phyla. It has become generally accepted that this important group of molecules was present early in evolution and that the basic structure has been strongly conserved.

The present chapter deals with the current state of knowledge of those brain-gut peptides that occur in the nervous system of insects. The immunocytochemical methods used in their localisation are described and discussed, followed by an account of biochemical studies aimed at the chemical characterisation of these peptides in brain extracts.

Immunocytochemistry: Basic Principles

In view of the several reviews on immunocytochemistry (Sternberger 1974; Nakane 1975; Kuhlmann 1977) and more recently Pearse (1980) it is not necessary to give a highly detailed account of the technique here. Instead, we prefer to make reference to the basic principles upon which it is based, to describe the particular methodology that we have found to be satisfactory in studies on the brain of *Calliphora* and finally to discuss some of the more interesting and problematical aspects of the technique as they relate to work on insect nervous tissues.

Theoretical Considerations of the Production of an Antiserum
and its Antibody Content

Essentially, an antiserum is produced by injecting a particular peptide mixed with an adjuvant (usually Freund's) into a mammal such as a rabbit or guinea pig over an immunisation period of 2−3 months. The peptide may be pure i.e. synthetic, or it may be purified from tissue extracts. Depending upon the size of the peptide, it may or may not have to be covalently coupled (conjugated) to a larger protein, such as bovine serum albumin or thyroglobulin, to induce antibody formation. An immune serum produced against a particular antigen or conjugated hapten contains a variety of different antibodies with a range of binding affinities. There will be a series of antibodies directed against the different (perhaps many) antigenic determinants or epitopes on the protein surface and since any one antigenic determinant has a three-dimensional shape, a single determinant may itself induce a heterogeneous population of antibodies. It follows that even for a pure immunising peptide, a heterogeneous series of antibodies is present in the antiserum. If a partially purified peptide or tissue homogenate or extract is used, the antiserum will inevitably contain a complex array of antibodies which could cause serious confusion and error in interpretation if used in immunocytochemistry.

Considerations such as these have led to cautious acceptance of data gained from immunocytochemical studies, most workers preferring to use the suffix "like" when reporting positive cross-reactivity of a particular antiserum with tissue peptides. Thus, in our own insect studies we can only say that particular neurons contain, for example, α-endorphin-like material (Fig. 1). This means that a particular amino acid sequence or sequences of α-endorphin are recognisable within certain proteins of the cells, but it does not mean necessarily that they contain this particular peptide.

Considerable progress has been made in the direction of a more positive immunocytochemical identification of complete protein molecules by the use of a series of region-specific antisera, each containing antibodies directed against a specific antigenic determinant (usually a sequence of 3−8

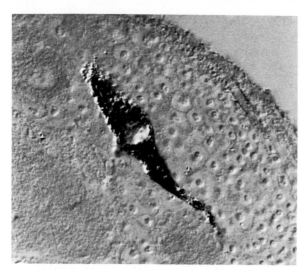

Fig. 1. Cell body of a median neurosecretory cell of *Calliphora erythrocephala* showing α-endorphin-like immunoreactivity. (Antiserum specific for α-endorphin, diluted 1 : 1000 − PAP technique − DAB substrate − Nomarski optics) (1200×)

amino acids). The point is well illustrated by reference to the gastrin/CCK series of peptides which have a common pentapeptide C-terminal sequence (cf. Dockray 1980). Antisera directed against this particular sequence will obviously fail to distinguish between CCK and gastrin. Only when antisera, specific for other regions of both gastrin and CCK, are used, can cells containing one or the other of these two peptides be distinguished (Larsson and Rehfeld 1977a, b). Our own studies on *Calliphora* have so far shown that certain neurons possess peptides with the gastrin/CCK C-terminal sequence (Fig. 3A−C) (Duve and Thorpe 1981; Dockray et al. 1981) but we have not yet been able to identify the chemical nature of the peptides beyond this.

On a purely practical note, antisera containing high avidity antibodies are so highly valued that their storage and subsequent handling require careful consideration. Aliquots of a size suitable for a single series of experiments may be frozen at − 20 °C with or without a small quantity of sodium azide or merthiolate. Repeated freezing and thawing tends to reduce the combining power of the antibodies and should be avoided. Other workers prefer to maintain antisera at 4 °C with a preservative added.

Techniques of Immunocytochemistry

Fluorescence Techniques

The direct fluorescent antibody method (FAB) requires a fluorescent antibody to localise the corresponding antigen. A much more widely used derivation of this technique, however, is the indirect fluorescent antibody method (Coons et al. 1955) in which a second layer is added to form a "sandwich" effect. The antigen in the tissue is combined with an antibody raised in, a rabbit, for example, and this is followed by the application of a fluorescent labelled anti-rabbit IgG produced in, say, a sheep or goat, added as a final layer (Fig. 2). This technique is a sensitive immunocytochemical method (Appendix 1) which has given good results in insect studies (Rémy et al. 1978, 1979; Doerr-Schott et al. 1978; Duve and Thorpe 1979, 1983; Yui et al. 1980).

One advantage of the fluorescent immunocytochemical method for insect studies is that it is possible, at least theoretically, to use two or more fluorescent dyes when investigating more than one tissue antigen, so that different cell types may be visualised simultaneously in the same section. (The fluorescent technique may be combined with an immunoenzyme method to achieve the same effect.) The disadvantages of the method are that the specimens are not permanent and photography of the specimens cannot usually be repeated since fading occurs during the exposure. A further disadvantage is that it is sometimes necessary to use a lower dilution of the primary antiserum (1:20−1:200) than required by other techniques to obtain sufficient quantities of the fluorescing label. For this reason alone, if the antiserum is in short supply, it may be necessary to use a different method.

Immunoenzyme Techniques

The covalent coupling of enzymes to antibodies and the subsequent detection of the particular enzyme in the tissue was a breakthrough in the immunocytochemical method (Nakane and Pierce 1966; Avrameas 1969; Sternberger et al. 1970). The main enzyme used is horseradish peroxidase and the most commonly used substrate, hitherto, has been 3,3'-diaminobenzidine tetrahydrochloride (DAB). This combination results in an intensely stained, golden-brown immunoreaction product within the cells bearing the original antigen. Unfortunately, DAB has been shown to be carcinogenic and its manufacture has been stopped by certain drug firms. Other compounds, however, may be substituted for DAB e.g. 3-amino-9-ethylcarbazole, Hanker-Yates reagent, or 4-chloro-1-naphthol (see Chap. 3, this Vol). The latter compound results in a violet-blue colouration of the immunoreaction product.

The most frequently employed immunoenzyme method is the extremely sensitive, unlabelled antibody enzyme technique in which the three or four layers are as follows: (1) the primary rabbit antiserum, raised against the particular antigen to be detected (2) an antibody to rabbit immunoglobulins (raised in sheep, goat or swine) in excess (3) rabbit immunoglobulins to horseradish peroxidase (4) purified horseradish peroxidase. The stable PAP (peroxidase-anti-peroxidase) complex (Sternberger et al. 1970; Sternberger

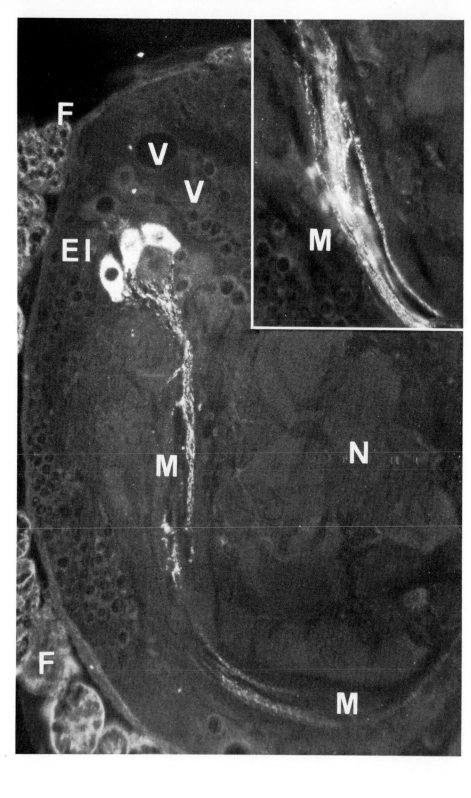

1974) combines stages (3) and (4) and prevents the losses of peroxidase which otherwise occur.

The PAP technique is a widely used immunocytochemical technique for both light and electron microscopy and unlike immunofluorescence, can be used to great effect with high dilution (1:100−1: several thousand) of the primary antiserum. (The method as used in our studies on *Calliphora* is described in Appendix 2.)

The advantages of the PAP method are that it is extremely sensitive, requiring high dilution of the primary antiserum and that the preparations may be either permanent or temporary. Thus, with DAB, sections are taken through xylol and mounted in DPX or Canada balsam. When 4-chloro-1-naphthol is used to visualise the immunoreaction product, the preparation is mounted in glycerine buffer, and may be photographed and subsequently destained and treated with a second antiserum before being restained and rephotographed. In this way, cells with different immunoreactive contents may be studied in the same preparation.

Problems of Specificity

Specificity in immunocytochemistry is an extremely vexed question and one that must be answered in as many ways as possible. An excellent account of the problems and pitfalls in immunocytochemistry, including a discussion on specificity, has been written by Larsson (1980) and prospective practitioners of the art are well advised to begin their studies there. Larsson and Schwartz (1977) have introduced the technique of radioimmunocytochemistry (RICH) in an attempt to "increase immunocytochemical specificity, and also to make immunocytochemistry and radioimmunoassay more easily comparable." Briefly, antisera are reacted in excess with a specific, radioactively labelled antigen to form RICH complexes (bivalent antibodies with labelled antigen on only one of the two binding sites). When the RICH complexes are applied to tissue sections, the free site reacts with tissue-bound antigen and these may later be detected autoradiographically. The only antibodies detected are those able to bind radioactively labelled anti-

---

Fig. 2. Sagittal section through the brain of a sugar-fed blowfly, *Calliphora erythrocephala*, treated with a primary antiserum against α-endorphin (1 : 1000) and a secondary rhodamine-conjugated sheep anti-rabbit antibody (1 : 200). The section passes through the median neurosecretory cell groups (MNC) and the median nerve bundle and shows immunoreactive material fluorescing strongly within certain of the perikarya and axons (240×) Inset. Part of section at higher magnification (470×) showing α-endorphin-like immunoreactive material in the median bundle. *EI* α-endorphin-like immunoreactivity in cell bodies. Note that other cells in MNC remain unstained; *M* α-endorphin-like immunoreactivity from the MNC within axons of the median bundle; *V* vacuoles, posterior to the MNC; *F* fat-body cells; *N* neuropil

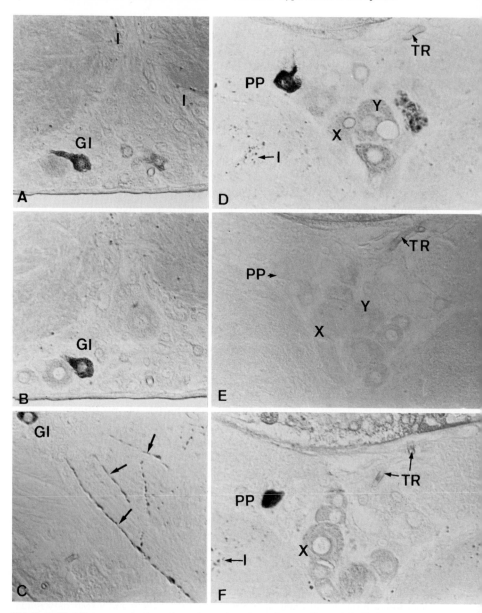

Fig. 3. A, B Adjacent sagittal sections through a thoracic ganglion of *Calliphora* showing gastrin/CCK-like immunoreactive material in cell bodies (*GI*) and in axons (*I*). **C** Transverse section through neuropil of a thoracic ganglion showing gastrin/CCK-like immunoreactive material in axons (*arrows*). (Antiserum specific for tetrapeptide of COOH terminus of gastrin and CCK, diluted 1 : 1000 − PAP technique − Nomarski optics) (500×). **D−F.** Adjacent transverse sections through the MNC region of *Calliphora*. **D** and **F** Sections treated with bovine pancreatic polypeptide (BPP) antiserum (1 : 1000-PAP technique − DAB substrate − Nomarski optics). **E** Intervening section to **D** and **F** treated with anti-

gen, thus allowing a much greater antibody-antigen specificity than is normally possible.

Antibody nonspecificity may result from a variety of different causes. Natural antibodies occur in all sera and may cause nonspecific cross-reactivity. A much more serious difficulty, however, is caused by sera containing multiple antibodies directed against the different antigenic determinants occurring in pure peptides and particularly in mixed immunising antigens. Such determinants may be found in proteins other than the one under investigation. Since positive cross-reactivity could arise from a similarity of only 3—8 amino acids (the antigenic determinant) rather than the complete peptide, the suffix "like" is used to describe immunoreactive findings. For this reason, it is essential to isolate and purify the peptides (using radio-immunoassay as a monitoring system) with the ultimate goal of obtaining the amino acid composition and sequence.

Specificity controls, or absorption controls, are an essential part of any immunocytochemical study. Antisera are reacted with pure or synthetic peptides in excess for 24 h at 4 °C, a procedure which should result in a failure of the antiserum to stain when applied to tissues. Such absorption controls, if positive, suggest at least that the antiserum is able to bind the peptide, but, of course, it could also react with other peptides containing the particular antigenic determinant recognised by the antibodies. For this reason region-specific antibodies have proved useful in determining the existence of more than a short amino acid sequence. The absorption controls when applied to insect tissues require special care in that there are normally only a few immunoreactive neurons in any one particular area. It is usual to treat adjacent tissue sections with (a) normal immune serum and (b) antigen-saturated immune serum. The former should give positive staining; the latter, no staining whatsoever (Fig. 3D—F). A problem sometimes arises in insect tissues where it is difficult to judge whether two sections of the same, single cell are present in adjacent tissue sections. In these instances we have used a subsequent nonspecific stain such as paraldehyde fuchsin (PAF) to ensure that the absorbed-out control section does actually contain the same cell that has stained immuno-positively on the adjacent section (cf. Duve and Thorpe 1980).

Another method that we have adopted is to apply antigen-saturated antiserum to insect nervous tissue sections and then to examine and photograph those areas where the immunoreactive cells are known to exist. The sections are washed thoroughly and the whole process repeated with the reactive im-

---

gen-inactivated BPP antiserum (30 μg BPP per ml diluted antiserum) 24 h at 4 °C. The same *PP* cell is seen in all sections but the lack of immunoreaction product in **E** is obvious. Similarly, transverse sections of PP-immunoreactive axons (*I*) are seen in **D** and **F** but not in **E**. The cut sections of tracheae (*Tr*) and the other nonimmunoreactive cells within the MNC (*X*) and (*Y*) serve for orientation (530×)

mune serum. The sections may then be compared with the previously taken photographs and the immunoreactive cells rephotographed (cf. Duve and Thorpe 1981).

A further point relevant to specificity controls concerns antisera that have been produced as a result of immunisation with an antigen conjugated to a large protein e.g. CCK8 conjugated to bovine serum albumin (BSA). Such antisera require pretreatment with an excess of the conjugating protein to absorb out the corresponding antibodies. This tends to result in sections with a much paler background, presumably because it prevents nonspecific protein-protein interactions. Accordingly, the specific immunoreaction products stand out very clearly (Fig. 4).

Staining controls are used to exclude nonspecific effects due to, for example, autofluorescence or endogenous enzymes, if an immunoenzyme technique is used. The antibodies used in the method, individually or collectively, are eliminated from the procedure and this should always result in disappearance of the staining. Another control frequently used to eliminate the possibility of nonspecific binding of the second antibodies is the blocking test in which the tissue sections are first treated with a dilute solution of the normal serum from the species supplying the second antibody. Finally, a commonly used control (of somewhat doubtful significance) is one in which the primary antiserum is substituted by a pre-immune or non-immune

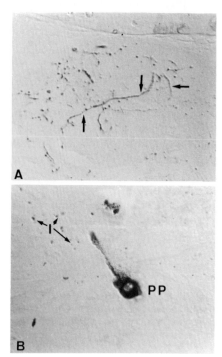

Fig. 4A, B. Transverse section of the brain of *Calliphora erythrocephala* showing **A** BPP-immunoreactive axons within the dorsal part of the protocerebrum (*arrows*). (450×). **B** BPP-immunoreactive cell body (*PP*) in the same region. (*I*) transverse sections of BPP-immunoreactive axons in the neuropil. (Antibody dilution, 1 : 2000 – PAP technique – DAB substrate – Nomarski optics) (450×)

serum. A more useful variation of this test is one which uses a hyperimmune serum directed against an antigen different from the one under investigation.

Finally, when investigating the possible presence of vertebrate-type peptides in invertebrates it is essential to control the study by using, in parallel, vertebrate tissues that are known to contain the peptides.

Brain-Gut Peptides in Insects

Immunocytochemical studies on *Calliphora* (Duve and Thorpe 1979–1983) suggest the presence of insulin-, pancreatic polypeptide (PP)-, gastrin/CCK- and α-endorphin-like peptides in specific neurons of the brain. Work on other species, using both immunofluorescence and immunoperoxidase techniques, has given similar results. Thus, an insulin-like peptide has been located in neurons of the pars intercerebralis of the silkworm *Bombyx mori* with fibres reaching the corpus cardiacum and corpus allatum (Yui et al. 1980). The same study reports, in addition, the presence of both gastrin-like and PP-like immunoreactive neurons. Similarly, El-Salhy et al. (1980) report the presence of at least ten vertebrate-type brain-gut peptides in neurons of the brain of larvae of the dipteran, *Eristalis aeneus* including insulin-, PP- and gastrin/CCK-like materials. Insulin-immunoreactive material has also been located in median neurosecretory cells of the brain and in the storage lobe of the corpus cardiacum of the locust (Orchard and Loughton 1980). PP-like immunoreactive material has been found in the central and visceral nervous systems of the cockroach, *Periplaneta* (Endo et al. 1982). The studies of Remy and co-workers (1977–1981) have, furthermore, shown the presence of α-endorphin-like material in the lepidopterans, *Thaumetopoea pityocampa* and *Bombyx mori*, and a neurophysin-vasopressin-like substance in neurons of both *Locusta migratoria* and the phasmid, *Clitumnus extradentatus*.

Extraction and Purification

Immunocytochemistry has proved to be an extremely useful indication for the presence of particular vertebrate-type peptides within cells, but the necessity for complementary studies on the isolation, purification and characterisation of the immunoreactive peptides is obvious. We have undertaken a research programme aimed at purifying certain of the peptides of *Calliphora* with the specific objective ultimately of obtaining the amino acid composition and sequence. Supplies of the pure peptides are also required for studying their physiological role in insects.

The following sections give a brief account of the methods we have found suitable for purifying the *Calliphora* insulin- and PP-like peptides.

Care and Maintenance of Flies

The number of cells in the brain of *Calliphora* that are immunoreactive against either PP or insulin antisera, has been shown from cytological studies to be small (12—30). It is apparent, therefore, that extremely large amounts of tissue are necessary for the purification of even microgram quantities of peptide.

In a recent study (Duve et al. 1981) we have used ca. 10^6 blowflies from which we obtained 5 kg of heads for extraction. Batches of up to 20,000 pupae can be handled relatively easily in large polythene tanks. The adults are fed on sucrose and water for 5—6 days and then anaesthetised and frozen by means of Cardice. Heads can be collected by shaking the flies, together with powdered Cardice, in large polythene bags and sieving the detached heads.

Homogenisation and Extraction

The large volume occupied by 5 kg of blowfly heads requires a considerable scaling-up of the procedures normally used in the laboratory for extraction of pancreatic insulin or PP, where the yield per gram of tissue is so much higher. We have found that the most satisfactory homogenisation method is to mix the heads in a freezing mixture of ethanol and Cardice and to pass this through an industrial meat mincing machine. Extraction in acidified alcohol (ethanol/phosphoric acid) is carried out for 24 h at 4 °C with re-extraction of the centrifuged residue for a further 4 h. Large volumes of tissue necessitate the use of equally large volumes (up to 70 l) of alcohol and the subsequent evaporation of this down to a suitable volume for gel filtration requires the use of a cyclic still. The evaporation is carried out under vacuum and the temperature does not exceed 30 °C. At all stages in the purification procedure the possibility of contamination from mammalian peptides must be guarded against.

Sephadex Gel Filtration and Ion-Exchange Chromatography

The extracts are filtered on a series of Hyflo beds and desalted either on Sephadex G15 or G25. The relevant peptide-containing fractions are then freeze-dried, reconstituted in an appropriate buffer (1 M acetic acid) and applied to Sephadex G50 for molecular weight fractionation. It is essential to have a monitoring system for the peptides which are being analysed and this may be either bio- or radioimmunoassay. The insulin-like and PP-like peptides of *Calliphora* both cross react in mammalian (bovine) radioimmunoassay systems and show both linear dilution and parallelism with the bovine standard peptides (Fig. 5).

Under normal circumstances, two peptides of MW 6000 (insulin) and 4200 (pancreatic polypeptide) would separate from each other on Sephadex

Fig. 5. Radioimmunoassay dilution curves of *Calliphora* insulin- and PP-like peptides. Following acid alcohol extraction and Sephadex gel filtration of the peptides, samples of insulin- and PP-immunoreactive materials were tested for linearity of dilution and parallelism with the corresponding bovine insulin and PP standards. The linear dilution is an expression of specificity of the immunoreactive material in the assay and the parallelism denotes that the antigenic determinants are the same as those of the standards. ■ Bovine insulin and PP standards; ● *Calliphora* PP-immunoreactive material; ▼ *Calliphora* insulin-immunoreactive material

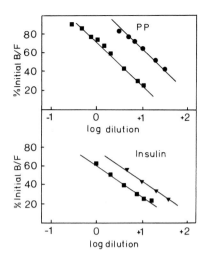

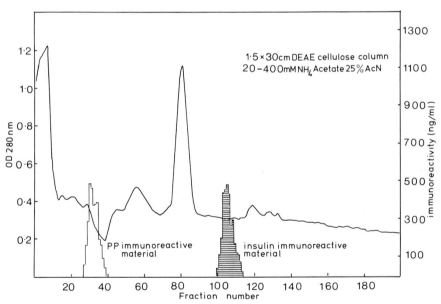

Fig. 6. DE-52 DEAE-cellulose chromatography of *Calliphora* insulin- and PP-immunoreactive peptides. The peptides were extracted from *Calliphora* heads in acid alcohol and gel filtered using Sephadex G15 followed by G50. Freeze-dried immunoreactive fractions from Sephadex G50 were reconstituted in the running buffer and applied to the DEAE column. Column size 1.5×30 cm; gradient 20–400 mM ammonium acetate in the presence of 25% acetonitrile pH 8.5; flow rate 20 ml h^{-1}; 4 ml fractions. Samples (10 μl) immunoassayed for PP and insulin. Reproduced by kind permission of the Biochemical Journal

G50 but PP tends to aggregate so that the two peptides run together. However, application of the G50 immunoreactive fractions to either DE-52 DEAE-cellulose or QAE-Sephadex gives excellent separation of the peptides (Fig. 6) as does polyacrylamide gel electrophoresis (Fig. 7).

High Pressure Liquid Chromatography (HPLC)

HPLC is undoubtedly an extremely valuable technique in the purification of insect neuropeptides. This can be judged from experiments in which insulin-like material or PP-like material from Fig. 6 is applied to an HPLC preparative column such as Ultrasphere ODS. The result is a loss of the majority of contaminating protein (Fig. 8). Transference of this immunoreactive material to an analytical column (e.g. propyl CN) yields pure peptides suitable for amino acid composition and sequence studies (Fig. 9).

Amino Acid Composition Studies

The ultimate aim of all peptide purification studies is the amino acid composition and sequence. We have recently analysed the composition of the PP-like peptide of *Calliphora* and it is clear that there is a close homology with the known mammalian PP species. The peptide is composed of

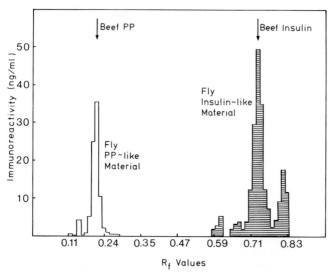

Fig. 7. Polyacrylamide gel electrophoresis of 3 mg of freeze-dried *Calliphora* immunoreactive material from Sephadex G50. (For details see Duve et al. 1981). Reproduced by kind permission of the Biochemical Journal

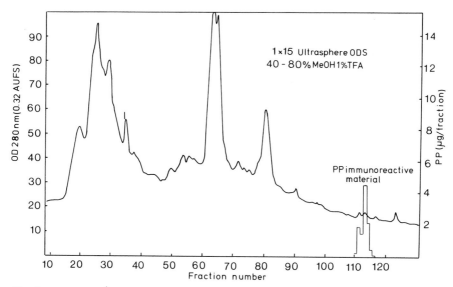

Fig. 8. Preparative HPLC of PP-immunoreactive material from the DEAE column (Fig. 6). Samples (5 μl) were immunoassayed for PP. (For details see Duve et al. 1982). Reproduced by kind permission of the Biochemical Journal

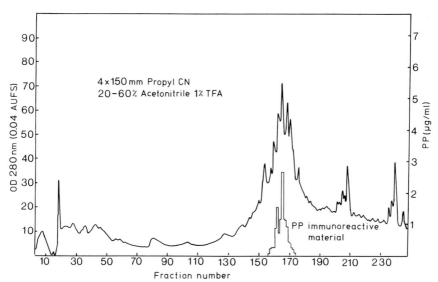

Fig. 9. Analytical HPLC of PP-immunoreactive material from Fig. 8. The main peak was further purified on an analytical Ultrasphere ODS column before the amino acid composition was determined. Reproduced by kind permission of the Biochemical Journal

36 amino acids, it lacks histidine, phenylalanine, lysine, tryptophan and cysteine and it contains the other amino acids in proportions similar to those of the mammalian PP species (Duve et al. 1982). The composition is as follows: Asx (3) Thr (2) Ser (1) Glx (6) Pro (4) Gly (2) Ala (4) Val (1) Met (1) Ile (2) Leu (3) Tyr (4) Arg (3).

Other Biochemical Studies on Insect Brain-Gut Peptides

In addition to the PP- and insulin-like peptides of *Calliphora* we have also partially characterised the gastrin/CCK-like peptide using immunoaffinity adsorption to a C-terminal-specific gastrin/CCK antiserum (Dockray et al. 1981). The data suggest a similarity to C-terminal fragments of the mammalian brain-gut peptide CCK and the related hormone gastrin. Similarly, Kramer et al. (1977) have demonstrated a gastrin-like immunoreactive peptide in extracts of the brain and corpora cardiaca/corpora allata complexes of the lepidopteran *Manduca sexta*. Also in *Manduca*, the presence of insulin-like and glucagon-like peptides has been demonstrated in neuroendocrine tissues (Tager et al. 1976) and in the haemolymph (Kramer et al. 1980). More recently, LeRoith et al. (1981) have reported the presence of an insulin-like peptide in extracts of *Drosophila*.

Conclusions

Immunocytochemical evidence for the presence of a wide range of the vertebrate-type, brain-gut peptides in insect neurons, supported by the confirmation of their presence in tissue extracts, is of considerable interest. The presence of these peptides in other invertebrate neurons including those of coelenterates (Grimmelikhuijzen 1980), annelids (Sundler et al. 1977; LeRoith et al. 1981), molluscs (Van Noorden et al. 1980) and tunicates (Fritsch et al. 1979) suggests that the peptides had their evolutionary origins in nervous tissue. If so, during evolution, certain of the peptide-producing cells must have migrated from the nervous tissue to their secondary sites in the gut. Their new physiologcal role may have become the dominant one e.g. pancreatic insulin as opposed to brain insulin of mammals. The question of the physiological role of peptides such as pancreatic polypeptide in the brain of insects remains to be answered. The insulin-like peptide, however, has been shown to be present in haemolymph in both *Calliphora* (Fig. 10) and *Manduca* (Kramer et al., 1980) and it probably arrives there via the axonal pathway from the median neurosecretory cells to the aorta. It would seem that a truly hormonal role for this peptide in the control of carbohydrate metabolism is a distinct possibility (Duve et al. 1979), whereas for other vertebrate-type brain-gut peptides in insects, neurotransmission or neuromodulation appears to be a more likely function.

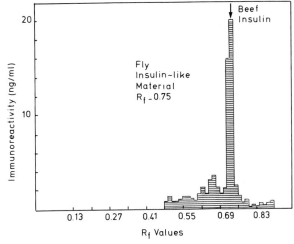

Fig. 10. Polyacrylamide gel electrophoresis of *Calliphora* haemolymph. (The animals were fed for 3 days on sugar and water followed by 1 h of meat, sugar and water). Pooled haemolymph (approx. 300 μl) was applied to a 0.5×22 cm glass column, running gel 10% polyacrylamide, spacer gel 4% polyacrylamide. Gel slices (1.5 mm) were eluted overnight with 125 μl NH_4HCO_3 pH 8.0 with 10 μl 0.25% human serum albumin. 100-μl samples were used for radioimmunoassay of insulin. The position of bovine insulin is *arrowed*

Appendix 1. Immunofluorescence: The Indirect Method

Fixation in aqueous Bouin's fluid for 3−6 h gives good results for most peptides and we have used this method routinely. Other useful fixatives for immunocytochemistry include 4% formaldehyde in physiological saline 24 h at 20 °C. Tissue fixed in this way may require trypsinization to increase the responsiveness to the PAP method. Deparaffinised sections are placed in Tris-saline buffer containing 0.1% trypsin at 37 °C for 5−30 min. If 5% acetic acid is added to the formaldehyde fixative, trypsinization is not required. Alternative fixatives including parabenzoquinone, and diethylpyrocarbonate in the vapour phase, may be used where antigens are denatured by the aldehyde fixatives.

For Bouin-fixed material, paraffin wax sections are taken through xylol, absolute alcohol and petroleum ether and are allowed to air dry. Antiserum is applied for 1 h or longer dependent upon the dilution used. The second antibody (1:50−1:200) is a fluorescent IgG raised against the IgG of the species used in the production of the primary antibody e. g. goat anti-rabbit or rabbit anti-guinea pig. The most commonly used fluorescent labels are fluorescein isothiocyanate (FITC) or rhodamine. Dilution of antisera is normally by phosphate buffered saline (PBS) with or without 1% bovine or human serum albumin to reduce nonspecific background staining. After 1 h the sections are washed thoroughly in PBS, mounted in PBS-buffered glycerine and examined with a UV microscope.

Appendix 2. Immunoperoxidase: The PAP Method

The fixation is as described for immunofluorescence. Wax sections are deparaffinised in xylene and treated with a solution of 0.03%−0.5% H_2O_2 in methanol to block any endogenous peroxidase activity. They are then washed in Tris-HCl buffer pH 7.2 containing 0.5% human or bovine albumin (to reduce nonspecific staining) for 0.5−1 h followed by Tris-HCl buffer alone for 15−30 min. (The "blocking" of nonspecific tissue-binding sites may also be achieved by the use of normal serum (1%) from the species in which the second antibody was raised.) The primary antiserum (raised in rabbit) is then applied, usually overnight at 4 °C, at a concentration of between 1 : 500 and 1 : 5000 or even higher. The use of a sealed moist chamber is essential. It is well worth trying much shorter application times for the primary antisera; we have found as little as 1 h can give perfectly satisfactory results. Sections are washed thoroughly (3 × 15 min) in Tris-HCl buffer pH 7.2 and then treated with antibody against rabbit IgG raised in swine (1:20) for 1 h at room temperature. Sections are washed once more in Tris-HCl buffer pH 7.2 (3 × 15 min) after which the PAP is applied at a concentration of 1 : 50 for 30 min. After final washing in Tris-HCl buffers pH 7.2 (2 × 15 min) and pH 7.6 (15 min) the sections are developed using one of the following; (a) 0.05% 4-Cl-l-naphthol in Tris-HCl buffer pH 7.6 containing 0.005% H_2O_2 (b) 0.05% 3,3′-diaminobenzidine tetrahydrochloride (DAB) in 0.01% H_2O_2 in buffer (c) 3-amino-9-ethylcarbazole 10 mg, dimethylsulphoxide 6 ml, 0.02 M acetate buffer (pH 5.0−5.2) 50 ml, 0.3% H_2O_2 4 ml, giving a final pH of 5.5 (d) Hanker-Yates reagent (PDP) 7.5−15 mg, Tris-HCl buffer pH 7.2 10 ml, 1% H_2O_2 0.1 ml. Development times are between 3 and 20 min, after which the sections are washed in distilled water. Counterstaining at this stage, with a weak solution of haematoxylin may be found useful. For the DAB and PDP development, sections are dehydrated, cleared in xylene and mounted in DPX or Canada balsam, whereas for 4-Cl-l-naphthol or 3-amino-9-ethyl-carbazole an aqueous mountant such as glycerine buffer pH 2.3 is used.

Acknowledgements. We gratefully acknowledge the collaboration of our colleague Dr. N. R. Lazarus, Wellcome Foundation, in the extraction and purification of the insulin- and PP-like peptides of *Calliphora* and for kind permission to use our joint data in this article. We also wish to thank Dr. P. J. Lowry, St. Bartholomew's Hospital, London for permission to use the results of collaborative biochemical studies.

We also express sincere thanks to Prof. G. J. Dockray, Liverpool University, Prof. J. F. Rehfeld, University of Copenhagen, Dr. R. E. Chance, Lilly Research Laboratories, and Dr. P. J. Lowry, for gifts of antisera used in the immunocytochemical studies.

We are pleased to acknowledge support of the Science and Engineering Research Council of Great Britain in the form of a grant to A. T. (GR/B88136)

Chapter 14

Immunocytochemical Techniques for the Identification of Peptidergic Neurons

MANFRED ECKERT and JOACHIM UDE

Wissenschaftsbereich Tierphysiologie der Sektion Biologie
Friedrich-Schiller Universität
Jena, G.D.R.

Introduction and Survey

Immunocytochemistry overcomes disadvantages of unspecificity common to most classical histochemical techniques used for the identification of neuronal pathways. However, it cannot simply replace these techniques. The problem is to produce highly specific antisera against various substances of the nervous tissue. Thus, progress in neuroimmunochemistry mainly depends upon the chemical isolation of many different types of neuronal antigens. Especially in vertebrates (mostly over the past 10 years), it has been shown that immunocytochemistry is a powerful tool in the study of the nervous system (Livett 1978; Setalo and Flerko 1978; Hökfelt et al. 1980 b). Immunohistochemical investigations have led to the discovery of previously unknown neurotransmitter pathways, the distribution of neuropeptides in the central and peripheral nervous system, and the coexistence of classical neurotransmitters and neuropeptides within one neuron. Knowledge of the distribution of neuropeptides in the nervous system of vertebrates has been greatly increased. In evolutionary terms, vertebrate neuropeptides are obviously very ancient functional molecules, since they act as "neuroregulators" throughout the whole animal kingdom. Compounds with a high resemblance to vertebrate neuropeptides have also been identified immunohistochemically in the lowest invertebrates, for example in Hydra (Taban and Cathieni 1979; Grimmelikhuijzen et al. 1981 a, b). In insects numerous vertebrate-like peptides were localized immunohistochemically in the brain (somatostatin, pancreatic polypeptide, gastrin, glucagon, secretin, substance P, enkephalin, β-endorphin) or in the ventral nerve cord (vasopressin, neurophysin II, α-endorphin, pancreatic polypeptide, gastrin, cholecystokinin, somatostatin, secretin, enkephalin) cross-reacting with well-characterized antisera against vertebrate neuropeptides (Doerr-Schott et al. 1978; Duve and Thorpe 1979, 1981; Yui et al. 1980; El-Salhy et al. 1980; Remy and Dubois 1981; El-Salhy 1981; Strambi et al. 1979; Remy et al. 1977, 1979; Remy and Girardie 1980). Structural homologies between insect and vertebrate peptides cannot be demonstrated by immunochemistry alone. The combination of radioimmunoassay, chemical, and physiological methods is necessary to characterize these compounds in insects. At present such investigations exist only for insulin-like, glucagon-like, gastrin-like, and enkephalin-like peptides (Tager et al. 1975; Tager and Kramer 1980; Kramer et al. 1977a, 1980; Kramer 1980; Gros et al. 1978; Moreau et al. 1981; Duve 1978; Duve et al. 1979). On the other hand, an antiserum to the molluscan cardio-excitatory tetrapeptide FMRF-amide stains neurons in Hydra (Grimmelikhuijzen et al. 1982), molluscs, insects, and vertebrates (Boer et al. 1980; Weber et al. 1981; Cottrell et al. 1983).

Still very little is known about the distribution of immunocytochemically characterized insect neuropeptides or neurosecretions. From the two isolated and synthesized insect peptides, proctolin and the adipokinetic hormone (Brown and Starrat 1975; Starrat and Brown 1975; Stone et al. 1978), only proctolin was localized in nerve cells, fibres, and terminals in the ventral nerve cord and in the hindgut musculature of the cockroach *Periplaneta americana* (Eckert et al. 1981; Bishop et al. 1981). These findings, and physiological results (Adams and O'Shea 1983), lead to the suggestion that proctolin acts not only as a neurotransmitter at the hindgut musculature, but also as a neurotransmitter or neuromodulator in the central nervous system itself. Proctolin is commonly distributed in insects (Brown 1977), and we have found proctolin-like immunoreactivity in the nervous system of many different insects and a crayfish, e.g. *Locusta migratoria, Dixippus morosus, Calliphora erythrocephala,* and *Orconectes limosus* (Figs. 1−4, 11, 12). Earlier investigations were concerned with the characterization of different neurosecretions of the insect brain-retrocerebral complex system (Eckert 1973, 1977, 1981a, b; Eckert and Gersch 1978; Friedel et al. 1980). Antisera were generated after immunization of rabbits with purified neurosecretion fractions (MW 60,000−6,000) from several thousand retrocerebral complexes of the cockroach *Periplaneta americana* (Eckert 1973, 1977). Eight neurosecretory cell types (Figs. 7−9, 13−18) could be distinguished in the cockroach brain, using polyspecific or monospecific anti-neurosecretion sera. It was demonstrated that the paraldehyde fuchsin-positive cells of the pars intercerebralis do not represent a uniform cell population, previously shown by numerous histochemical investigations (review: Panov 1980). A monospecific anti-neurosecretion serum stains only a part of the paraldehyde fuchsin-positive cells (R-cells) (Fig. 15).

Antibodies against neurosecretions of the cockroach *Periplaneta americana* cross-react with neurosecretory substances in the brain of different insects (Fig. 10). The pathways of the neurosecretory cells of the optic lobes, the lateral neurosecretory cells, and the neurosecretory cells of the pars intercerebralis have been reconstructed by using different neurosecretion-specific antibodies in combination with unilateral corpus cardiacum ablation and cobalt chloride iontophoresis techniques, (Eckert 1981b). The neurosecretions from the cells of the optic lobes (Fig. 16) and R-cells (Fig. 15) of

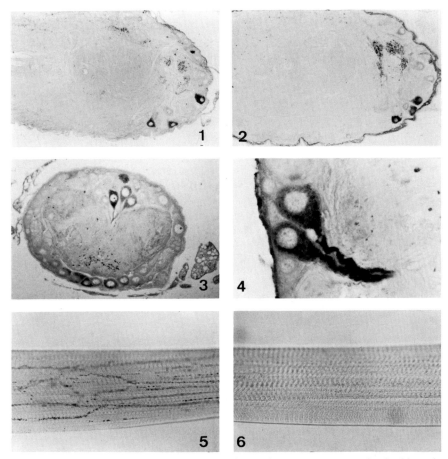

Fig. 1. Posterior region of terminal ganglion of *Periplaneta americana* stained with anti-proctolin antibodies against proctolin-glutaraldehyde-thyroglobulin conjugate. PAP method. Picric acid, glutaraldehyde, acetic acid fixation. × 100

Fig. 2. Same region as Fig. 1, but stained with antiproctolin antibodies against proctolin-carbodiimide-BSA conjugate. An identical pattern of cells and fibre profiles as in Fig. 1 can be observed. × 100

Fig. 3. Cells with proctolin-like immunoreactivity in terminal ganglion of *Locusta migratoria.* PAP method. × 110

Fig. 4. Proctolin-positive cells in terminal ganglion of *Periplaneta americana.* Perikarya show only weak immunoreactivity. However, axons exhibit intense immunocytochemical staining. PAP method. × 440

Fig. 5. Whole-mount preparation of a longitudinal muscle band of rectum of *Periplaneta americana* stained with purified antiproctolin antibodies. PAP method. Formaldehyde fixation. × 150

Fig. 6. As Fig. 5. Control staining with antiproctolin antibodies absorbed with proctolin-ovalbumin conjugate. × 150

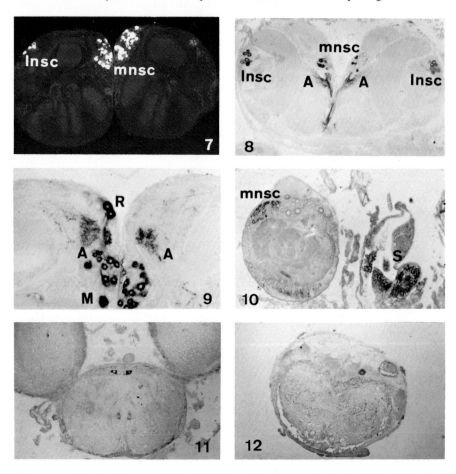

Fig. 7. Neurosecretory cells in brain frontal sections of *Periplaneta americana* visualized by polyspecific antineurosecretion serum by means of indirect immunofluorescence method. *lnsc* lateral neurosecretory cells; *mnsc* median neurosecretory cells. × 45

Fig. 8. As Fig. 7. Immunocytochemical staining performed by indirect immunoperoxidase technique using one-step conjugate. *A* arborization area of ncc I. × 45

Fig. 9. As Fig. 7, using two-step conjugate. R- and M-cells are distinguishable. × 100

Fig. 10. Neurosecretory elements in brain-retrocerebral complex of *Locusta migratoria* demonstrated with antineurosecretion serum against cockroach neurosecretions by means of indirect immunoperoxidase technique. *S* storage part of corpora cardiaca. × 45

Fig. 11. Proctolin-like immunoreactivity in cells of suboesophageal ganglion of last larval instar of *Musca domestica.* PAP method. Picric acid, glutaraldehyde, acetic acid fixation. × 105

Fig. 12. Neuron with proctolin-like immunoreactivity in terminal ganglion of crayfish *Orconectes limosus.* × 50

the pars intercerebralis are stored in different regions within the corpora cardiaca (Figs. 17, 18). The relationship between the neurosecretory cells of the optic lobes and the corpora cardiaca was demonstrated after unilateral removal of the corpus cardiacum-corpus allatum complex. Four days after the operation, neurosecretions accumulated only in fibres of the operated side (Fig. 14), indicating that there is a direct connection between the neurosecretory cells of the optic lobes and the corpus cardiacum (Figs. 14, 16, 18). This system is selectively stainable using a monospecific antiserum (Eckert 1981 b). The immunoreactive fibres originate from more than 100 small neurosecretory cells in the region between the lamina and the medulla of the optic lobe (Fig. 16). These immunoreactive neurosecretory cells are probably identical to the neurosecretory cells described by Beattie (1971).

Drastic accumulation of neurosecretory material in the ncc (nervus corpus cardiacum) I and ncc II appears in preparations after staining with a polyspecific anti-neurosecretion serum (Fig. 14) and intensive synaptic contacts can be observed between the arborization area of the ncc I and ncc II (Figs. 8, 9). The double staining of brain slices using both $CoCl_2$-iontophoresis and immunofluorescence techniques (Fig. 20) has enabled the identification of neurosecretory neurons projecting to the cardiaca nerves. After cephalic $CoCl_2$-iontophoresis via the nervus corporis allati I, numerous cobalt-filled cells were observed in the contralateral pars intercerebralis and in the ipsilateral pars lateralis. The immunoreactive neurons in the pars intercerebralis lacked cobalt precipitate. However, the immunoreactive lateral neurosecretory cells were simultaneously cobalt-stained. We presume that the neurosecretory cells of the pars intercerebralis terminate in the corpora cardiaca. The fibres of the lateral neurosecretory cells, however, pass by the corpus cardiacum. This is in agreement with the morphological and physiological findings of Gilbert et al. (1981) in *Manduca sexta*. The source of the prothoracicotropic hormone is probably one lateral neurosecretory cell in each hemisphere. These cells terminate in the acellular sheath surrounding the corpus allatum. Using immunological, chemical, and physiological methods, Friedel et al. (1980) characterized an A-cell neurosecretion from the brain of *Locusta migratoria*. This protein (MW about 11,000) was localized immunohistochemically in the A-cells of the brain and in the storage lobes of the corpora cardiaca.

Preparation of Antigens

Progress in vertebrate neuropeptide research was much aided by purified or synthetic peptide antigens which could be used to produce antibodies: but this is not the case in insects. It is usually necessary to isolate and purify the antigens directly from nervous tissue. Therefore, the application of immunocytochemistry as a method for neuroanatomical research in insects is still largely restricted by the availability of antibodies against nervous system specific antigens. However, many separation techniques have now been developed allowing the isolation of substances in microquantities sufficient for immunization of rabbits (Eckert 1977; Friedel et al. 1980). A very simple, quick, and elegant method is the separation of small amounts of antigens by means of polyacrylamide gel electrophoresis according to Hartmann and Udenfriend (1969), Stumph et al. (1974), and Lompré et al. (1979). The antigen bands can be cut from the gel and can be used for immunization directly after homogenization (Eckert 1977). Peptides and proteins with a molecular weight over 6,000–10,000 are antigens which induce antibody generation after injection in mammals. Smaller molecules, so-called haptens, are not immunogenic, but can be rendered immunogenic by conjugation to carrier proteins. Also nonpeptidergic haptens, e. g. cAMP (Steiner et al. 1970; Richman et al. 1980) or serotonin (Peskar and Spector 1973; Steinbusch et al. 1978), can be coupled to carrier proteins for immunization. Serotonin has been localized by application of these antibodies in whole-mount preparations of the nervous system of the cockroach, *Periplaneta americana* (Bishop and O'Shea 1982), and in *Rhodnius* (see Chapter 16, this Vol.). The most frequently used carrier proteins are bovine serum albumin (BSA), thyroglobulin, ovalbumin, hemocyanin, and edestin, the latter being a plant protein. Experience has shown that thyroglobulin is a particularly suitable carrier for small peptides (Skowsky and Fisher 1972).

Various methods for the cross-linking of haptens to carrier proteins have been developed (Avrameas et al. 1978), but it is impossible to describe all these techniques in detail here. When there is a free carboxy group in the hapten, the coupling with water soluble 1-ethyl-3(3-dimethylaminopropyl)-carbodiimide is the method of choice (Sofroniew et al. 1978; Stocker et al. 1979). Glutardialdehyde is another bifunctional reagent which reacts preferably with primary amino groups and/or the ε-amino group of lysine present in the haptens. Glutardialdehyde is a very useful agent for the cross-linking of peptides to proteins (Dubois 1976; Remy et al. 1979). TRH (no free amino or carboxy group) and other peptides can be successfully attached to carrier proteins by the bifunctional reagent 1,5-difluoro-2,4-dinitrobenzene (Visser et al. 1977; Stocker et al. 1979). Other hapten conjugates have been prepared using bis-diazotized benzidine (BDB). The coupling is effected via α- and ε-amino, imidazole, phenolic, and a variety of other groups of amino acids and proteins (Bassiri and Utiger 1972).

The specificity of the generated antibodies is influenced by the type of coupling procedure used (Pique et al. 1978; Larsson and Rehfeld 1979; Rehfeld 1980). This was impressively demonstrated for the luteinizing hormone-releasing hormone (LH-RH) immunocytochemistry in the mouse brain. For generating anti-LH-RH antisera Hoffman et al. (1978) used immunogens obtained using four different conjugation procedures {glutaraldehyde, 1-ethyl-3(3-dimethylaminopropyl)carbodiimide hydrochloride, 1-cyclohexyl-3[2morphonyl-(4)-ethyl]carbodiimide metho-p-toluene sulphonate, bis-diazotized benzidine}. Antisera generated using glutaraldehyde con-

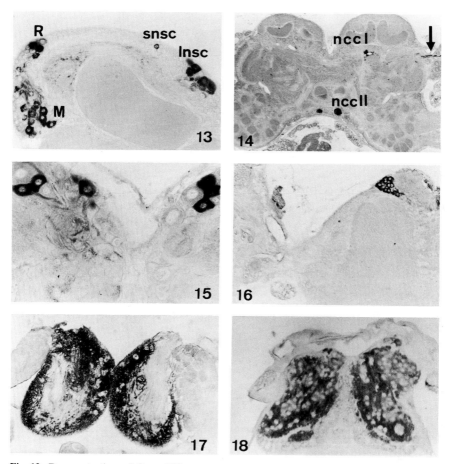

Fig. 13. Demonstration of four different neurosecretory cell types by polyspecific anti-neurosecretion serum in cockroach brain. Transverse section. *M* M-cells; *R* R-cells; *lnsc* lateral neurosecretory cells; *snsc* subocellar neurosecretory cells. × 100

Fig. 14. Immunocytochemical staining of cockroach brain with antineurosecretion serum on 4th day after ablation of right corpus cardiacum. Note accumulation in ncc I and ncc II as well as in fibres running from neurosecretory cells of optic lobes (*arrow*) only on ablated side. Indirect immunoperoxidase technique with two-step conjugate. × 40

Fig. 15. Localization of PAF-positive R-cells of pars intercerebralis (*Periplaneta americana*) with monospecific antineurosecretion serum. Indirect immunoperoxidase technique. × 240

Fig. 16. Cluster of neurosecretory cells (PAF-negative) in optic lobes of *Periplaneta americana* identified with monospecific antineurosecretion serum. Indirect immunoperoxidase technique. × 100

Fig. 17. Selective staining of R-cell secretion in lateral region of corpora cardiaca. × 425

Fig. 18. Selective staining of secretion of neurosecretory cells from optic lobes in central region of corpora cardiaca. × 425

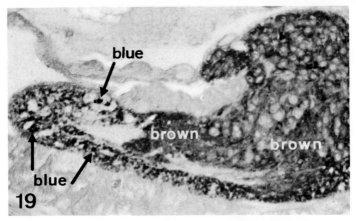

Fig. 19. Simultaneous localization of optic lobe neurosecretory cell neurosecretion (*brown*) and R-cell neurosecretion (*blue*) in corpora cardiaca of *Periplaneta americana* using an immunocytochemical double staining method (Eckert 1977). × 850

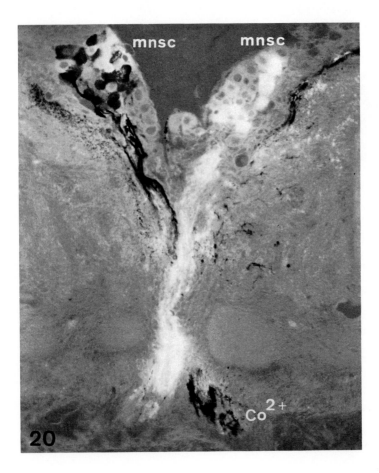

jugates show neurons in field 1 of the brain, but antisera generated using carbodiimide conjugates stain perikarya in field 2. Antisera against the tyrosyl conjugate (cross-linking with bis-diazotized benzidine) react neither with perikarya in field 1 nor field 2. However, all these antisera show the free LH-RH in nerve terminals suggesting that two different populations of LH-RH neurons may exist in the mouse brain.

Another way to generate antibodies against low molecular weight haptens (e.g. cAMP, ecdysone, biotin) is via succinylation followed by conjugation with carbodiimide to a carrier protein (Steiner et al. 1970; Horn et al. 1976; Hsu et al. 1981).

Production and Isolation of Antibodies

Immunization of Rabbits

Rabbits are the animals of choice for immunization because only small amounts of antigen are needed to produce good antibody titre. We use females with a body weight of 3 kg or heavier. Different procedures are used for the bleeding of the animals. Techniques are described in detail in textbooks of immunology, e.g. Campbell et al. (1964). Rabbits may give up to 50 ml of blood per kg body weight by means of heart puncture or by bleeding after cutting of the carotid arteries of deep ether anaesthetized animals.

Many immunization schedules are described (Peetoom 1967), but the best method for the preparation of antibodies against insect nervous tissue antigens is immunization with complete Freund's adjuvant. It consists of 1.5 ml Arlacel (Serva), 8.5 ml Bayol F (Serva) and 5−10 mg dried, heat-killed mycobacteria (Difco). We use, with success, lyophilized heat-killed BCG vaccinia instead of mycobacteria. Rabbits were injected with 0.6 ml of complete Freund's adjuvant (water in oil emulsion: 0.3 ml antigen solution and 0.3 ml complete Freund's adjuvant). The antigen mixture is prepared with a glass Potter homogenizer by adding the antigen solution drop-by-drop to the adjuvant. The emulsion must be stable for several hours. The antigen mixture should be injected subcutaneously at different points of the ventral region of the body and a small amount (ca. 0.1 ml) in the musculature of the hindlegs to stimulate more lymphatic regions. Animals receive 3−5 booster injections at 21-day intervals, and are killed for bleeding 10 days after the last booster injection. After coagulation (overnight at 4 °C) the antiserum is filtered using a glass bacteria filter and stored at − 20 °C in

---

Fig. 20. Combination of cobalt chloride iontophoresis and immunofluorescence technique. Cells matching the right nervus corporis allati I are cobalt-filled on left side of pars intercerebralis. Neurosecretory cells stained by polyspecific antineurosecretion serum are negative. It demonstrates that these cells probably terminate in the corpora cardiaca. × 220

1.0 ml portions. Merthiolate (0.01%) or sodium azide (0.02%) is added to working solutions to suppress bacterial growth. For enzyme-immunocyto-chemistry sodium azide is an unsuitable reagent, because it blocks the enzymatic activity of horseradish peroxidase. The amounts of soluble antigen per injection range from some hundreds of μg to a few mg, depending upon the immunogenicity and antigenicity of the used antigens.

Testing the Antiserum

Although immunodiffusion methods (Ouchterlony technique and immunoelectrophoresis) are the best techniques to estimate the quality of the antisera, they are restricted to precipitable high molecular weight antigens. The quality of antisera against low molecular weight haptens must be investigated by means of the radioimmunoassay (RIA) or enzyme-immunoassay (EIA). Immunodiffusion methods are successful when soluble hapten-protein conjugates are available, as has been demonstrated for proctolin. Antibodies were generated in rabbits with thyroglobulin-glutaraldehyde-proctolin conjugates (Eckert et al. 1981). Immunoelectrophoresis was performed with different protein-proctolin conjugates to decide whether antiproctolin antibodies were present in the antiserum (Fig. 21). Antiproctolin serum precipitated thyroglobulin-proctolin, ovalbumin-proctolin, and BSA-proctolin,

Fig. 21. Immunoelectrophoretic control of antiproctolin serum generated against proctolin-glutaraldehyde-thyroglobulin PGT. Antiserum precipitated (*PGT*), proctolin-glutaraldehyde-BSA (*PGBSA*), bovine serum albumin (*BSA*), and thyroglobulin (*T*). It does not react with *BSA* nor with T after preabsorption of antiserum with T. Therefore, precipitation with PGBSA must be caused by specific antiproctolin antibodies

but did not react with ovalbumin or BSA, demonstrating the presence of specific antiproctolin antibodies in the antiserum. The relative estimation of the antibody titre is easily made by the Ouchterlony technique with dilution series of the antiserum. In this technique six wells containing different antiserum dilutions $(1:1-1:32)$ are arranged in agar plates around a central well containing antigen solution $(0.5-1.0$ mg ml$^{-1})$. High antibody titres are indicated by precipitation using dilutions higher than $1:4$.

Absorption of the Antisera Before Use for Immunocytochemistry

Antigens for immunization are mostly prepared by means of fractionation of nervous tissue extracts contaminated with fat body or hemolymph antigens. Therefore, the antisera need to be absorbed to eliminate the tissue-unspecific antibodies. For example, the polyspecific antineurosecretion sera obtained after immunization with Sephadex G-200 fractions of corpora cardiaca extract (Eckert 1973, 1977), were absorbed with fat body-hemolymph-acetone powder (100 mg ml^{-1} undiluted antiserum) at 4 °C overnight under agitation. The immunoaggregates formed were removed by high-speed centrifugation in the cold. Figure 22 exhibits the immunoelectrophoretic patterns of corpora cardiaca-extract antigens (1000 glands ml^{-1}) with unabsorbed and absorbed antineurosecretion serum. Antibodies against hemolymph antigens are removed. The absorbed antiserum selectively stains neurosecretory cells in the brain of *Periplaneta americana* (Fig. 13).

Antisera generated against protein-hapten conjugates contain a mixture of antibodies specific to the carrier protein and the appropriate hapten. The carrier-specific antibodies can be eliminated by absorption with the carrier protein. Another important problem is the cross-reactivity of the hapten-

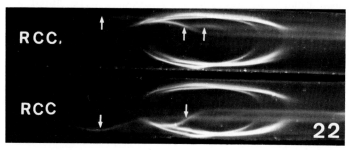

Fig. 22. Immunoelectrophoretic pattern of retrocerebral complex extract (*RCC*) (1000 glands ml^{-1}) with unabsorbed (*above* and *below*) and absorbed (100 mg hemolymph-fat body powder ml^{-1} antiserum) antineurosecretion serum (*middle*). Antiserum was generated in rabbits with a selected Sephadex G-200 fraction of RCC-extract. Antibodies against hemolymph and fat body antigens (*arrows*), causing unspecific immunohistochemical staining, were removed from antiserum

specific antibodies with other closely related haptens. This question is discussed in detail for neuropeptides by Vandesande (1979) and Zimmermann et al. (1980). In this case, the application of solid phase-affinity purification of the antiserum is a better method of resolution than preabsorption.

Isolation of Hapten-Specific Antibodies

Antiproctolin antibodies were isolated in our laboratory by means of affinity chromatography. The affinity gel was prepared by the Pharmacia method: 8 mg proctolin-ovalbumin was cross-linked with 1 g cyanogenbromide-activated Sepharose-4B (Pharmacia) and packed in a small column (6×140 mm). Proctolin-ovalbumin conjugates were produced according to the method of Dubois (1976). The polymerization was started under gentle stirring by dropwise addition of 2.0 ml of a 0.5% glutaraldehyde solution in 0.1 M phosphate buffer, pH 6.8, to 2 ml 0.1 M phosphate buffer, pH 6.8, containing 2.4 mg proctolin and 20 mg ovalbumin. This solution was agitated for 3 h at room temperature. The reaction was stopped by addition of sodium metabisulfite (20 mg in 0.2 ml water). The solution was dialyzed at 4 °C against water (3 to 5 times 1000 ml) containing 0.01% sodium azide. After dialysis the volume was made up to 10 ml and divided into 0.5-ml portions. These were lyophilized and stored at −20 °C until use.

The elution of antiproctolin antibodies from the affinity gel was performed according to Folkersen et al. (1978). The column was washed with 100−150 ml PBS-0.5 M NaCl (0.01 M phosphate buffer, pH 7.2, containing 0.5 M NaCl) before use, after which 2−3 ml of the antiserum was added to the top of the affinity gel. The column was then perfused with the antiserum for 120 min. The effluent was recycled and incubated in the gel for 180 min to saturate the affinity gel; the gel was then washed with 150 ml PBS-0.1 M NaCl. Finally the antiproctolin antibodies were eluted with 3 M $MgCl_2$-0.6 M NaCl (61 g $MgCl_2 \cdot 6\, H_2O$, 3.5 g NaCl to 100 ml water), pH 4.8. Four-ml samples were collected and the antibody concentration was measured at 280 nm ($E^{1\%}_{280\,nm} = 13.4$). The antibody-containing samples were dialyzed against two changes of 1 l 0.15 M NaCl and 1 l PBS and afterwards concentrated by vacuum dialysis in collodium tubes (Sartorius-Göttingen, Germany) to a volume of 1 ml and stored at −20 °C until use for immunocytochemistry (Figs. 1−6).

Methods for Antibody Isolation

The immunoglobulin fraction or specific anti-immunoglobulin-G (IgG) antibodies must be isolated for the preparation of immunoglobulin conjugates for use in indirect immunofluorescence or immunoenzyme techniques. Hyperimmune sera or conjugated immunoglobulin to rabbit IgG can be synthesized or purchased commercially.

A goat was immunized subcutaneously at various points on the back with an emulsion consisting of 5 mg rabbit IgG (prepared by DEAE-Sephadex A-50 chromatography, see below, in 1.0 ml 0.85% NaCl and 2 ml complete Freund's adjuvant). The goat received 4 to 6 booster injections at intervals of 21 days (booster injection 1−3:7.5 mg rabbit IgG in 1.5 ml 0.85% NaCl and 1.5 ml complete Freund's adjuvant; booster injection 4−6: in the same way, but with incomplete Freund's adjuvant). The goat was bled from the jugular vein 10 days after the 4th, 5th, and 6th booster injections. The titres of antisera varied from 1:16 to 1:32 (determined by the Ouchterlony technique − see above). These antirabbit-IgG antisera have proved useful for conjugation techniques described below.

Ammonium Sulfate Precipitation

One volume of precooled (4 °C) saturated [560 g $(NH_4)_2SO_4$ per 1000 ml] aqueous ammonium sulfate was slowly added, under gentle stirring, to one volume of 1:2 diluted (0.85% NaCl) antirabbit-IgG serum, keeping at 0 °C in an ice bath. After standing for 30 min at 4 °C, the mixture was centrifuged for 15 min at 5000 g at 0 °C. The supernatant was discarded. The pellet was washed twice with 50% saturated ammonium sulfate solution and dissolved in a small volume of distilled water. This solution was dialyzed for 24 h against a large volume of PBS (5 changes of PBS) at 4 °C and adjusted to the desired volume. BSA may be used as a reference for the determination of the protein concentration. Alternatively the solution may be measured at 280 nm ($E_{280 \text{ nm}}^{1\%} = 6.5$). IgG preparations of higher purity can be obtained after precipitation with lower concentrations of ammonium sulfate solutions (Clark et al. 1972; Herbert 1976), but the yield of IgG is much lower.

DEAE-Sephadex A-50 Chromatography

Immunoelectrophoretic pure rabbit IgG or antirabbit IgG from goat serum can be easily isolated by means of DEAE-Sephadex A-50 chromatography according to Dedmon et al. (1965). DEAE-Sephadex A-50 is equilibrated with 0.02 M phosphate buffer, pH 7.6, and packed in a column. Eighty ml gel volume is sufficient for purification of 10 ml serum. After penetration of the serum in the gel, the gel is incubated for 30 min, then the gel is eluted with the equilibration buffer in a relatively large volume. The efficiency is controlled by measurement at $E_{280 \text{ nm}}^{1\%} = 13.5$. The IgG is precipitated by addition of crystalline ammonium sulfate to 50% saturation [280 g $(NH_4)_2SO_4$ to 1000 ml water gives 50% saturation]. After 30 min in the refrigerator, the mixture is centrifuged at 4 °C and after that treated as described before.

Immunoadsorption Techniques

Specific antibodies can be isolated from whole immune sera by means of different immunoadsorption techniques.

Cross-Linked IgG-Immunoadsorbent. For example, IgG and BSA should be cross-linked to isolate antirabbit IgG antibodies to the method of Avrameas and Ternynck (1969). One hundred mg rabbit IgG (e. g. isolated by means of DEAE-Sephadex A-50) and 400 mg BSA were dissolved in 5 ml 0.9% NaCl and mixed with 5 ml of 0.4 M acetate buffer, pH 5.0. The mixture was stirred and 2 ml of a 2.5% aqueous glutaraldehyde solution is added drop by drop. The gel was then allowed to stand for 3 h at room temperature. Small portions of the unsolubilized protein were homogenized in PBS with a glass Potter homogenizer. This suspension was centrifuged for 15 min at 3000 g and 4 °C. The immunoadsorbent was resuspended and the whole procedure was repeated 2 or 3 times; the gel was then suspended overnight in 0.2 M ethanolamine-HCl buffer, pH 7.4, to block the remaining free aldehyde groups. Before use, the immunoadsorbent was washed twice with 0.1 M glycine-HCl buffer, pH 2.8, neutralized in 1.0 M K_2HPO_4 and washed with PBS until the supernatant had an optical density of less than 0.04 at 280 nm.

For isolation of antirabbit IgG, adequate amounts of the suspended immunoadsorbent were incubated for 60 min at room temperature with antirabbit IgG serum. This mixture was centrifuged, the pellet resuspended and washed with PBS until the supernatant did not show extinction at 280 nm. Antibodies were isolated from the immunoadsorbent by three incubations in 0.1 M glycine-HCl buffer, pH 2.8 (in each case 15 min). The supernatants were immediately neutralized by addition of 1 M K_2HPO_4, dialyzed against PBS and concentrated by vacuum dialysis. The elution can also be performed with 3 M NaSCN in 0.1 M phosphate buffer, pH 7.4 or 3 M $MgCl_2$−0.6 M NaCl, pH 4.8.

Sepharose 4B-IgG Gel

The technique was modified from that described by Hazlett (1977) and Kuhlmann (1977). An amount of CNBr-activated Sepharose 4B (Pharmacia), calculated to swell to a volume equal to that of the serum, was washed and re-swollen for 15 min at 4 °C in 200 ml 0.001 M HCl per gram of gel. Then the gel was quickly equilibrated with the coupling buffer (0.2 M citrate buffer, pH 6.5, containing 0.5 M NaCl) at 4 °C until the pH reached 6.5 (use a glass filter funnel). The IgG to be coupled was added immediately to the gel in test tubes to give a final concentration of 6.0 mg IgG per ml packed Sepharose. Coupling followed during the next 36 h at 4 °C and the test tubes were rotated end-over-end. Non-bound proteins were washed off with 20 vol of coupling buffer. Unreacted groups of the gel were saturated with 0.2 M ethanolamine-HCl buffer, pH 7.4. Finally, the immunogel was washed once more with 20 vol of coupling buffer and packed into a column. The gel in the column was washed with an excess of PBS-0.5 M NaCl. After charging the column with the antiserum the gel was washed with PBS-0.5 M NaCl followed by PBS until no extinction was measured at 280 nm. The elution of anti-IgG antibodies was done by treatment of the immunogel with 3 M NaSCN in 0.1 M phosphate buffer, pH 7.4, or 3 M $MgCl_2$-0.6M NaCl, pH 4.8. The eluates were desalted by column chromatography on Sephadex G-25 followed by dialysis of the desalted immunoglobulin solution against PBS. If required the solution can be concentrated by vacuum dialysis.

Immunocytochemical Techniques

The principle of immunocytochemistry is that the antibody reaction with a fixed tissue antigen can be made visible by means of an antibody-bound marker molecule. The pioneer work of Coons (1941), who introduced fluorescent-labelled antibodies, stimulated the development of many other immunocytochemical staining methods. For light microscopy, immunofluorescence (Coons 1956) and immunoenzyme techniques (Sternberger 1974) are of significance. It is not possible to describe all these techniques here but there are textbooks and monographs on their application (Wagner 1967; Nairn 1976; Sternberger 1979).

The labels must be cross-linked with the antibody molecule by means of a bifunctional group or bifunctional reagent. The principle is illustrated in Fig. 23. The most important fluorescent stains are fluorescein isothiocyanate (FITC) and lissamine rhodamine B 200 sulfonylchloride (RB 200 SC). The enzymatic label of choice is horseradish-peroxidase (HRP).

An alternative immuno-enzyme technique (the unlabelled antibody enzyme method) was developed by Sternberger et al. (1970) without covalent binding of HRP to the immunoglobulin. The attachment of HRP is realized by use of an antiperoxidase antibody (antiperoxidase). A soluble peroxidase-antiperoxidase (PAP) complex is formed in the presence of antiperoxidase and an excess of peroxidase at low pH:

$$3\,HRP + 2\,IgG \xrightarrow{pH\,2.3} 3\,HRP - 2\,IgG \quad (PAP\ complex)$$

After immediately neutralizing the solution, the PAP complex can be maintained in solution with small amounts of free peroxidase. The advantages of the immunoenzyme techniques over the immunofluorescence techniques are:

1. Specialized UV microscopic equipment is not required.
2. Preparations stained with 3,3-diaminobenzidine-4 HCl (DAB) are permanent and can be mounted in Canada balsam, or other mounting medium.

ISOTHYOCYANATES:

$$R-N=C + NH-Ig \longrightarrow R-N-C-N-Ig$$
$$\quad\quad \| \quad | \quad\quad\quad\quad | \quad \| \quad |$$
$$\quad\quad S \quad H \quad\quad\quad\quad H \quad S \quad H$$

SULPHONYLCHLORIDES:

$$R-SO_2Cl + NH_2-Ig \longrightarrow R-SO_2-NH-Ig$$

GLUTARALDEHYDE   Avrameas (1969):

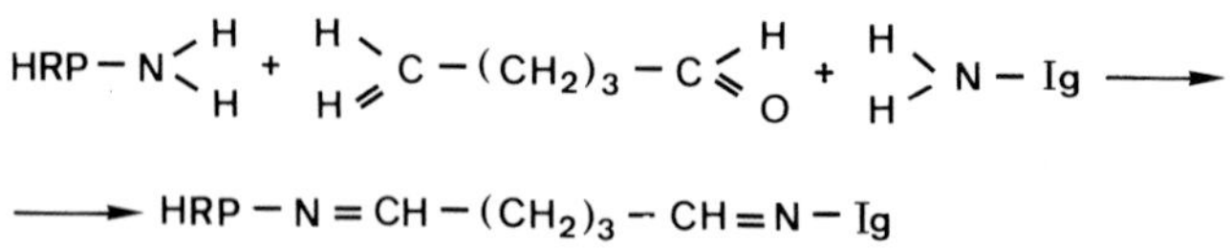

$$HRP-N\begin{subarray}{c}H\\H\end{subarray} + \begin{subarray}{c}H\\H\end{subarray}C-(CH_2)_3-C\begin{subarray}{c}H\\O\end{subarray} + \begin{subarray}{c}H\\H\end{subarray}N-Ig \longrightarrow$$

$$\longrightarrow HRP-N=CH-(CH_2)_3-CH=N-Ig$$

4.4-DIFLUORO-3.3-DINITROPHENYLSULFONE (FNPS)   Nakane
and Pierce (1966):

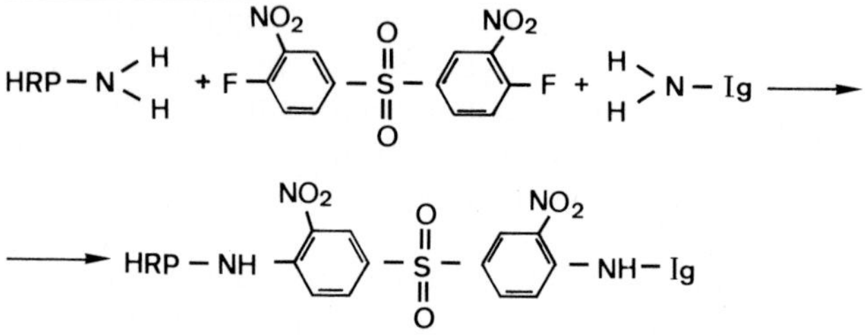

SODIUM PERIODATE   Nakane and Kawaoi (1974):

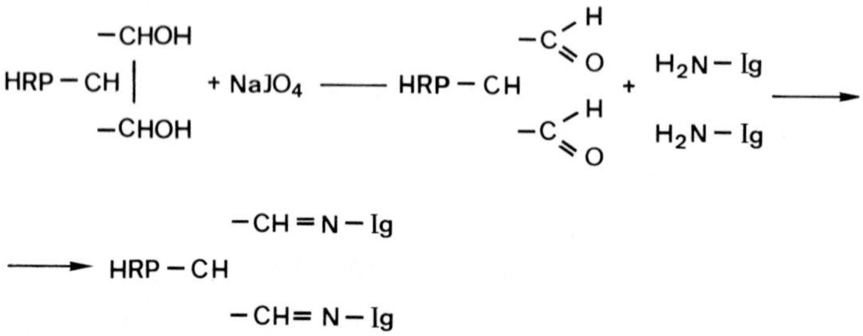

$$\longrightarrow HRP-CH\begin{subarray}{c}-CH=N-Ig\\-CH=N-Ig\end{subarray}$$

CH — CARBOHYDRATE MOIETY OF HRP

PEROXIDASE - ANTIPEROXIDASE (PAP)   Sternberger et al (1970):

$$3HRP + 2IgG \xrightarrow{pH 2.3} 3HRP - 2IgG \text{ (PAP complex)}$$

Fig. 23. Principles of various conjugation methods of antibodies with marker molecules

3. Immunoperoxidase methods are applicable to electron microscopic examination.
4. The PAP technique is 100−1000-fold more sensitive than other immunofluorescence and immunoperoxidase techniques.

The high staining contrast of the preparations is an advantage of the immunofluorescence techniques. This is an important fact for the localization of fine nerve fibres or terminals. Immunofluorescence is a powerful method in combination with immunoenzyme techniques for the simultaneous identification of different antigens or $CoCl_2$-iontophoresis or retrograde HRP-filling techniques. Immunofluorescence conjugates, immunoenzyme conjugates and the PAP complex are commercially available, but the costs are relatively high. For laboratory use it is easy to prepare the immunoglobulin conjugates.

Production of Fluorescence- and Enzyme-Labelled Antibodies

The immunocytochemical staining of tissue antigens can be performed by (1) the direct method, (2) the indirect (sandwich) method or (3) the PAP method (see below). Depending upon the technique used, either the primary antibody directed against the antigen to be demonstrated must be directly labelled or the first antibody remains unlabelled and the second antibody, which is directed against the immunoglobulin of the species in which the first antibody is generated, must be conjugated. The application of the PAP method requires the preparation of the PAP complex.

Conjugation of Antibodies with Fluorochromes

The first step in this process is the isolation of the IgG fraction of the antigen-directed antiserum or of the antirabbit-IgG-serum by means of 50% ammonium sulfate precipitation (see above). The protein concentration is calculated by reading the optical density $E_{280 \text{ nm}}^{1\%, 1 \text{ cm}} = 6.5$ (about 50 mg protein ml^{-1} antiserum). A very rapid and simple method is the labelling of the IgG fraction by means of fluorescein isothiocyanate 10% adsorbat on silica gel (Serva-Heidelberg, Germany) or lissamine rhodamine B 200 sulphonylchloride 5% adsorbat on silica gel (Serva-Heidelberg, Germany) according to Serva instructions. 150 mg protein in 3 ml 0.85% NaCl (IgG fraction) was mixed with 6 ml of 0.1 M $NaHCO_3/Na_2CO_3$ buffer (pH 9.0) and 150 mg FITC adsorbat was added. The mixture was allowed to react for 30 min at room temperature. The solution was then decanted and the unbound FITC was removed by gel filtration on Sephadex G-25 equilibrated with PBS. The solution was concentrated to a volume of 10 ml by means of vacuum dialysis, divided into doses of 0.5 ml and stored at −20 °C. This stock solution was absorbed with 100 mg hemolymph-fat body acetone powder overnight at 4 °C, centrifuged and diluted 1 : 10 for tissue staining prior to use (Fig. 7). RB 200 SC conjugates were prepared in a similar manner.

Conjugation of Antibodies with HRP

Peroxidase-IgG conjugates prepared by different conjugation methods exhibit similar properties for immunohistological staining (Eckert 1980). The quality of conjugates is mainly influenced by the grade of purity of the peroxidase and used IgG fraction. (Avrameas et al. 1978; Kuhlmann 1977; Boorsma and Kalsbeek 1975; Boorsma and Streefkerk 1979). The sensitivity of the immunoenzyme technique performed with conjugates containing covalently bound peroxidase matches the sensitivity of immunofluorescence techniques. Good results were obtained with peroxidase preparations having an RZ > 2.0. Commercially available HRP may require further purification (Wagner 1975; Slemmon et al. 1980; Porstman et al. 1981). The RZ is determined as followed: $RZ = \frac{E\,403\ nm}{E\,280\ nm}$. The concentration of HRP is measured at $E_{280\ nm}^{1\%,\ 1\ cm} = 22$.

For immunocytochemical investigations, glutaraldehyde (GA) conjugates are more suitable than conjugates prepared with periodate (Boorsma and Streefkerk 1979). GA conjugates can be prepared by a one-step or a two-step procedure (Avrameas 1969). Conjugates resulting from the two-step method exhibit a molar ratio of antibody to peroxidase of 1:1. A heterogeneous population of polymerization products with high molecular weight is obtained using the two-step procedure (Boorsma 1977; Kuhlmann 1977; Avrameas et al. 1978).

For routine preparations the IgG fraction obtained by 50% ammonium sulfate precipitation can be used for the conjugation with HRP. The advantage of one-step and two-step conjugates compared with fluorochrome conjugates is that there is less background in the tissue preparations (Fig. 8). The two-step procedure is the method of choice for the preparation of low molecular weight conjugates, e. g. Fab conjugates for incubation of tissue blocks or whole cells.

One-Step Method. 12 mg HRP is added to a solution of 5 mg protein (IgG ammonium sulfate fraction or purified IgG) in 1 ml 0.1 M phosphate buffer, pH 6.8 and 0.05 ml of a 1% aqueous glutaraldehyde solution is added dropwise to this solution while stirring gently. The reaction mixture is kept for 2 h at room temperature. A dialysis bag is filled with the mixture and the mixture is concentrated by means of polyvinylpyrrolidone 350 (Serva) or vacuum dialysis. The unbound glutaraldehyde is removed from the sample by Sephadex G-25 filtration in PBS. Then 0.01% merthiolate is added and small portions are frozen and stored at − 20 °C.

Two-Step Method. The first step consists of the reaction of glutaraldehyde in excess (approx. 10-fold molar excess) with one or two free amino groups of peroxidase. The activated peroxidase reacts in the second step with immunoglobulin.

For the first step, 15 mg HRP dissolved in 0.2 ml 0.1 M phosphate buffer, pH 6.8, containing 1.25% glutaraldehyde was incubated for 18 h at room temperature. After that the solution was filtered through a Sephadex G-25 column equilibrated with 0.15 M NaCl to remove unreacted glutaraldehyde. The brown-coloured HRP-containing fractions were concentrated to 1 ml by vacuum dialysis. One ml of 0.15 M NaCl containing 5 mg antirabbit IgG was mixed with this solution, followed by the addition of 0.1 ml carbonate-bicarbonate buffer, adjusted to pH 9.5, and kept at 4 °C for 24 h. Finally, free aldehyde groups were blocked by adding of 0.1 ml 0.2 M lysine. Two hours later the solution was dialyzed against PBS overnight at 4 °C. The conjugate was further purified by ammonium sulfate precipitation. An equal volume of precooled (4 °C) saturated ammonium sulfate was added dropwise to the conjugate solution under gentle stirring. After standing for 30 min at 4 °C, the conjugate solution was centrifuged at 4 °C and the supernatant was discarded. The pellet was washed twice with 50% aqueous ammonium sulfate solution. Afterwards the pellet was dissolved in a small volume of water. The conjugate was dialyzed over 24 h with PBS, containing 0.01% merthiolate, at 4 °C. The conjugate was stored at −20 °C in small portions.

Preparation of PAP Complex

PAP complex can be prepared according to the original method of Sternberger et al. (1970) and Sternberger (1974, 1979). In our laboratory a modified method is used to determine the equivalence zone of the antiperoxidase serum generated in rabbits (immunization of rabbits with peroxidase (RZ = 3.0, e. g. Sigma, type VI) and complete Freund's adjuvant). The determination of the optimal antibody ratio was performed by gel diffusion (Nowotny 1969). The arrangement of gel pattern is illustrated by Fig. 24. First, serial double dilution of antiperoxidase serum (1:1 up to 1:64) are prepared. Equal amounts of 0.125 mg HRP per ml are added to each diluted antiserum sample and the samples are incubated at 37 °C for 1 h. Second, in the meantime, a microslide is covered with 2.0 ml 1% agarose in PBS. (The agarose is solubilized by using a boiling water bath). The agar plate is cut according to the pattern in Fig. 24. The upper channel is filled with antiperoxidase serum (undiluted) and the lower channel with antigen solution (0.125 mg HRP ml⁻¹ PBS). After the 1 h incubation of the antiperoxidase/HRP mixtures the middle wells are filled with these samples, then the plate is incubated in a moist chamber at room temperature until precipitation is visible. The equivalence point is indicated when no precipitation lines appear on the side of the antigen or antibody excess (Fig. 24).

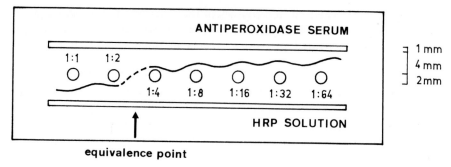

Fig. 24. Scheme of gel diffusion plate to determine equivalence point

The PAP preparation should be started by precipitation of the anti-peroxidase with about 1.5 times HRP equivalence proportion. For the calculation of the amount of antiserum, which is required for precipitation of 0.125 mg HRP ml^{-1} we use that antiserum dilution, which is characterized by the first precipitation line on the side of the antiperoxidase channel, e.g. 1:4 (Fig. 24).

For rapid preparation of PAP complexes the original technique was modified by Mason and Sammons (1978) and Bosman and Cramer-Knij-nenburg (1980).

Fixation of Tissue

Localization of antigens in tissue sections requires the immobilization of the antigen and a fixation procedure that does not damage the antigenicity too extensively. Therefore, the optimal immunohistochemical conditions must frequently be a compromise between a good preservation of tissue structure and the maintenance of the immunoreactivity for the antigen (Pearse and Polak 1975; Bishop et al. 1978; Konttinen and Reitamo 1979; Clayton et al. 1981; Berod et al. 1981; Ørstavik et al. 1981; Treilhou-Lahille et al. 1981). There are no universal fixation mixtures. Formaldehyde fixation perhaps realizes this demand best. Both small peptides as well as larger proteins seem to remain unaffected by formaldehyde fixation (Berod et al. 1981). The source of formaldehyde must be depolymerized paraformaldehyde (see below), because commercially available formaldehyde solutions contain methanol and stabilizers diminishing the immunoreactivity of tissue antigens.

Besides formaldehyde solutions picric acid-containing fixation mixtures have frequently been used, e.g. (1) Bouin: 15 ml aqueous saturated picric acid, 5 ml formalin, 1 ml glacial acetic acid. (2) Bouin-Hollande without acetic acid: 80 ml picric acid solution containing copper ions [dissolve 2.5 g ground neutral copper acetate in 100 ml water; picric acid (4.5 g) is subsequently added and the solution is filtered and stored], 10 ml formalin and 10 ml saturated aqueous solution of $HgCl_2$. When immunoperoxidase technique is to be applied, heavy metal ions must be removed. After treatment of the sections with 70% alcohol, they are incubated in iodated alcohol for 3 min, 3.5% sodium bisulfite for 3 min, then running water, distilled water, and finally buffer solution. (3) Formaldehyde-picric acid (Stefani et al. 1967): the fixative contains 2% formaldehyde and 15% saturated picric acid in 0.1 M phosphate buffer, pH 7.3. 20 g of paraformaldehyde is added to 150 ml of double-filtered saturated aqueous picric acid. The mixture is heated to 60 °C and several drops of 2.5% NaOH are added to depolymerize the paraformaldehyde. Finally the solution is cooled, filtered, and the volume adjusted to 1 l with 0.1 M phosphate buffer, pH 7.3. If necessary, the pH is adjusted to 7.3. The final osmolarity is 900 mOsm. The mixture is stable and can be stored at room temperature for 12 months. (4) Glutaraldehyde-acetic acid (Boer et al. 1979): 15 ml of aqueous saturated picric acid, 5 ml glutaraldehyde (25%), 0.1 ml glacial acetic acid.

In insect immunohistochemistry Bouin's fluid was an excellent fixative for the localization of high molecular weight neurosecretions (Figs. 7–10, 13–18), but not so for proctolin. The demonstration of proctolin in nerve cells, fibres, or terminals depends upon the fixation mixture used (Fig. 25). The number of nervous elements with proctolin-like immunoreactivity was increased with increasing aldehyde concentration. Proctolin-immunoreactive cells and good tissue preservation were observed after glutaraldehyde-picric acid-acetic acid fixation. The clarity of cell and fibre patterns which depend on the kind of fixation raises the question whether the same or different

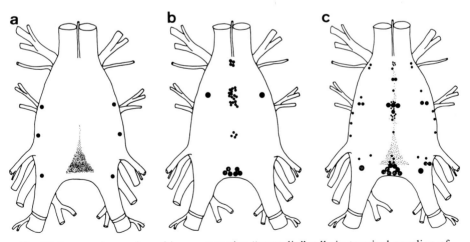

Fig. 25. Increase in number of immunoreactive "proctolin"-cells in terminal ganglion of *Periplaneta americana* depending upon fixation mixtures used. *Left to right* Bouin, paraformaldehyde, picric acid-glutaraldehyde-acetic acid

proctolin molecular forms exist in different neurons. In some cases perikarya of certain cells stain very weakly with antiproctolin antibodies whereas their axons are intensively stained (Fig. 4). Possibly there is a precursor molecule for a proctolin-like substance. For immunostaining of whole-mount preparations the formaldehyde-picric acid fixation gives the best results (Figs. 5, 6).

Sometimes, loss of the immunoreactivity of protein antigens in tissue sections fixed with formaldehyde, glutaraldehyde, or Bouin can be recovered by treatment with proteolytic enzymes (Curran and Gregory 1977; Denk et al. 1977; Reading 1977; Brozman 1978; Mepham et al. 1979; Qualman and Keren 1979; Suganuma et al. 1980; Hautzer et al. 1980).

Immunocytochemical Staining Methods

The principle of the three most commonly used techniques is illustrated in Fig. 26. The direct and indirect (sandwich) method can be performed equally well with the immunofluorescence or immunoperoxidase technique.

Direct Method

The antiserum directed against the antigen to be demonstrated is directly labelled with fluorochrome or peroxidase (Fig. 26). The method is less sensitive than multiple-step methods. The method is suitable for simultaneous identification of different antigens, e. g. by antibodies labelled with different fluorochromes or in a combination of peroxidase-labelled and FITC-labelled antibodies.

Indirect (Sandwich) Method

The primary antigen-specific antibody generated in species A is used unlabelled. A second antibody produced in species B, which is directed against the immunoglobulin of species A, is conjugated to FITC or peroxidase (Fig. 26). In this way the primary antibody, which reacts with the tissue antigen, acts as antigen for the labelled second antibody from species B. The advantages over the direct method are increased sensitivity and the smaller amount of conjugates required. By means of the labelled anti-IgG different primary antibodies can be stained.

Procedure for Staining Paraffin Sections. (1) Removal of paraffin with xylene, hydration through a graded series of ethanol. (2) Wash in cold PBS (3×5 min) containing $2-3$ drops Tween 80 per 100 ml (PBS-Tween). (3) Incubate sections with diluted antiserum overnight at 4 °C in a wet chamber.

Fig. 26. Most commonly used immunocytochemical techniques for light microscopy

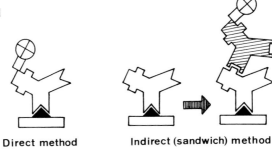

Direct method   Indirect (sandwich) method

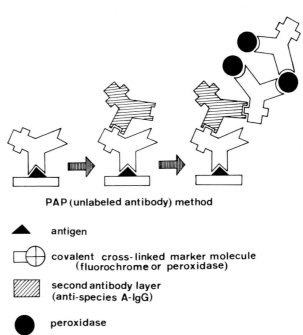

PAP (unlabeled antibody) method

▲   antigen

⊕   covalent cross-linked marker molecule
     (fluorochrome or peroxidase)

▨   second antibody layer
     (anti-species A-IgG)

●   peroxidase

(4) Wash in cold PBS-Tween (3 × 5 min). (5) Incubate with FITC-labelled or peroxidase-labelled second layer anti-IgG (dilution 1:10 to 1:40) for 60 min at room temperature. FITC-labelled anti-IgG was previously absorbed with hemolymph-fat body-acetone powder (see above). Unspecific staining with FITC-conjugate can be abolished by adding Triton X-100 to a final concentration of 0.1%−0.5% to the conjugate solution. (6) Wash in PBS-Tween (3 × 5 min). (7a) FITC-stained sections are now mounted in buffered glycerol (90 vol of bidistilled glycerol, 10 vol of veronal sodium acetate buffer, pH 8.5, I = 1.0:9.75 g veronal sodium, 6.47 g sodium acetate, 60 ml 0.1 N HCl, water to 1000 ml). The mounted sections can be stored for a long time

at $-20\,°C$. (7b) If peroxidase conjugates are used, incubate in freshly prepared 3,3'-diaminobenzidine-4 HCl (DAB). *Handle carefully* as DAB is a
potential carcinogen. Five mg DAB is dissolved in 10 ml of 0.05 M Tris-HCl
buffer, pH 7.6 and H_2O_2 is added to the final concentration of 0.001% (to
0.015%). Filtering of freshly prepared DAB solution before use can avoid
unspecific background staining. Instead of DAB other chromogens can also
be applied in peroxidase immunohistochemistry (Morell et al. 1981; Tubbs
and Sheibani 1981). (8) Wash with water. (9) Incubate for some minutes in
1% OsO_4. (10) Wash and dehydrate through a graded series of ethanols, xylene and mount in Canada balsam.

Staining according to the direct method is performed without steps 5 and
6. Counterstains for immunofluorescence (Schenk and Churukian 1974; Micheel 1978) and immunoenzyme techniques (Ormanns and Pfeifer 1981) can
be used.

PAP (Unlabelled Antibody) Method

The first antigen-specific antibody (from species A) and second anti-IgG
antibody (from species B) are unlabelled, and the PAP complex is used as
the third layer (Fig. 26). The second antibody acts as a bridge between the
primary antibody and the PAP complex consisting of two antiperoxidase
molecules generated in species A and three HRP molecules.

Procedure for Staining Paraffin Sections. Steps (1) and (2) as described for
the direct and indirect method, before. (3) Incubate sections with 1:3 diluted normal serum from species B for 15 min to saturate unspecific IgG-bind-
ing sites. (4) Wash in cold PBS (1 × 5 min). (5) Incubate sections in optimal
diluted primary antiserum (general 1:100 or higher) overnight at 4 °C. (6)
Wash in cold PBS-Tween (3 × 5 min). (7) Incubate with anti-species A-IgG
(1:10 or 1:20 diluted with PBS containing 0.25% BSA) for 60 min at room
temperature. (8) Wash in cold PBS-Tween (3 × 5 min). (9) Incubate with
1:20 or 1:40 diluted PAP complex in 0.05 M Tris-HCl buffer, pH 7,6, containing 0.25% BSA. (10) Wash in cold PBS-Tween (3 × 5 min). (11) Further
as in steps (7b) to (10) described before.

Destruction of Endogenous Peroxidase

Staining of endogenous peroxidase in tissue sections may cause difficulties
in interpretation. This problem can be circumvented by blocking endogenous peroxidase activity by means of different methods. Prior to immunoperoxidase staining: (1) Incubation with a freshly prepared solution of
phenylhydrazine hydrochloride (1 g per 1000 ml) in 0.05 M phosphate buffer, pH 7.1, for 1 h at 37 °C (Strauss 1980) or (2) incubate with freshly pre-

pared absolute methanol containing 0.1% H_2O_2 for 15 to 30 min (Streefkerk 1972) or (3) incubate with 0.01 M periodic acid in distilled water for 10 min. Then wash off and treat the sections with 0.01% potassium borohydride in water for 10 min to block aldehyde groups (Isobe et al. 1977).

Controls. The unspecificity of staining caused by unspecific absorption of antibodies to sections or the occurrence of cross-reactions of the antibodies with unknown antigens is the main source of misinterpretation. Therefore, control experiments must be performed with care (see, Petrusz et al. 1976; Swaab et al. 1977; Buffa et al. 1979; Bergroth et al. 1980; Grube and Weber 1980). The following control experiments are possible. (1) Incubation of the antiserum with the antigen under test is the commonly used specificity control. After staining with the absorbed antiserum specific staining must not be observable. For this, antiserum is incubated overnight with antigen at 4 °C, or better with antigen coupled to agarose or carrier-antigen complex prepared with glutaraldehyde (see above) to avoid the formation of soluble immune complexes. Then the insoluble immune complex cannot stain tissue antigen. Cross-reactivity of the antiserum is ruled out (excluded) by absorption experiments with similar antigens, which could be present in the tissue. (2) Pre-immune serum or unrelated nonimmune serum of the same species can be used as a negative control. (3) Staining of the same antigen with different antisera as demonstrated for proctolin-like immunoreactivity in terminal ganglion of the cockroach *Periplaneta americana* (Figs. 1, 2). (4) Omission of one or more reagents. In this case the antisera are replaced by 1% BSA solution to prevent non-specific binding. In the near future the application of monoclonal antibodies with defined specificity will help to overcome some of these problems.

Multiple Staining Methods. Simultaneous visualization of two or more antigens in the same tissue section is often of significance for investigation of functional relationships. Different methods have been developed for this purpose (Lechago et al. 1979; Valnes and Brandtzaeg 1981; Gu et al. 1981; van Rooijen 1980, 1981). In our laboratory (Eckert 1977) we use Nakane's (1968) method for the demonstration of various corpora cardiaca antigens. After the initial staining sequence with DAB, the first layers of antibodies were removed from tissue sections by incubation in 0.2 M glycine-HCl buffer, pH 2.2, containing 0.5 M NaCl. The second staining procedure of the R-cell neurosecretions was performed with 4-chloro-1-naphthol as peroxidase substrate (Fig. 19). The sections were mounted in glycerol. More complete elution of antibodies from sections has been achieved by treatment with an acid potassium permanganate solution (Tramu et al. 1978) or by electrophoresis (Vandesande and Dierickx 1975).

CoCl$_2$-Iontophoresis and Indirect Immunofluorescence Method

There has been a significant development in various techniques allowing classification and matching of nerve cells to functional pathways (Strausfeld and Miller 1980). The biochemical characterization of these systems is of general importance. One morphological method is to combine such techniques with immunohistological methods. In our group this problem was adapted for CoCl$_2$-iontophoresis and immunofluorescence. Brains of the cockroach *Periplaneta americana* were iontophorized towards the brain via the nervus corporis allati I. The brain-corpora cardiaca-corpora allata complex was immersed in Grace medium (Bio. Cult. Lab., Glasgow). One corpus allatum was cut and CoCl$_2$ was applied for 16 h at 4 °C by means of a suction electrode (1.2 μA). After iontophoresis Co^{2+} was precipitated in insect Ringer solution by adding saturated $(NH_4)_2S$ (0.1–0.2 ml to 3 ml Ringer solution). Then the brains were fixed in Bouin's solution overnight at 4 °C, dehydrated, and embedded in paraffin. Tissue sections were silver-intensified according to the method of Timm (1958 a). Sections were incubated in darkness for 45 min at 55 °C with a solution consisting of 1 part of 1% aqueous AgNO$_3$ solution, 4.5 parts of an aqueous solution containing 0.86% citric acid, 0.34% hydroquinone, and 20% sucrose, and 4.5 parts of a 30% alcohol solution containing 60% gum arabic. Finally, careful rinsing in warm tap water and distilled water (55 °C) is an important step. Stock solutions are stable for some weeks in darkness at 4 °C, but the mixture cannot be used longer than 50 min. After this treatment, sections were stained with anti-neurosecretion serum using the indirect immunofluorescence method. The application of bright field excitation allows simultaneous viewing of the immunoreactive neurosecretory cells and cobalt-filled neurons (Fig. 20). Immunoreactive cells of the pars intercerebralis were not cobalt-filled, but many other cobalt-containing cells were observed in the contralateral pars intercerebralis. Results indicate that the neurosecretory cells of the pars intercerebralis terminate in the corpora cardiaca. In comparison, the immunoreactive lateral neurosecretory cells were stained simultaneously by cobalt precipitate. It is proposed that their axons go to targets outside the corpora cardiaca. In a similar manner, other techniques, e. g. autoradiography (van Rooijen 1981), retrograde transport of HRP (Bowker et al. 1981), retrograde transport of fluorescence stains (Sawchenko and Swanson 1981), and monoamine histofluorescence (McNeill et al. 1980; McNeill and Sladek, 1980) have also been used in connection with immunocytochemical techniques for tracing biochemically defined pathways.

Supplementary Methods

Immunostaining of Whole-Mount Preparations. Costa et al. (1980) invented a modified immunocytochemical staining technique for whole-mount preparations for the identification of polypeptides in peripheral autonomic nerves of mammals. This original technique can be used also in insects (Eckert et al. 1981). Best results were obtained after fixation with the formaldehyde mixture according to Stefani et al. (1967). There are differences in the penetration of FITC-labelled antibodies and larger PAP complexes. Nerve cells with proctolin-like immunoreactivity can be localized in the terminal ganglion of *Periplaneta americana* with immunofluorescence, but not with the PAP method. However, nerve terminals in the hindgut musculature are visualized with both techniques (Figs. 5, 6). In small preparations the method works excellently.

Protein A, Avidin, and Biotin. By reason of the extraordinary affinities of protein A to the Fc-fragment of IgG and of avidin for the small vitamin biotin, these molecules become suitable tools in immunocytochemistry (Notani et al. 1979; Guesdon et al. 1979; Hsu and Raine 1981; Hsu et al. 1981; Yasuda et al. 1981). Instead of the second antibody layer in indirect techniques protein A or protein A-conjugates can be used. The advantage of the biotin-avidin system is the high affinity (over one million times higher than most antigen-antibody interactions) of avidin, a 68,000 MW glycoprotein from eggs, for biotin. Biotin can be cross-linked with any protein, which can in this way react with avidin. On the other hand, avidin conjugated with fluorochromes or peroxidase can be used for the identification of biotinyl antibodies. These reagents are commercially available (Vector Lab. Inc., Burlingame, California). It appears certain that these new techniques will replace or complement existing methods.

Electron Microscopy

Electron microscopic immunocytochemistry is also a valuable tool for in situ identification of intracellular antigenic sites in insect nervous tissue. For this, different approaches have been worked out. In all cases, chemical fixation of the tissue must be performed. Glutaraldehyde and to a greater extent glutaraldehyde-formaldehyde mixtures are used, but one must consider that any fixation reduces immunocytochemical labelling (Hayat 1973). In the pre-embedding labelling technique, widely used in vertebrates, tissue fragments were incubated with the antibodies and other reagents prior to embedding in a epoxy resin. The main problem in these cases is the poor penetration of the large antibody molecules through cellular membranes. Penetration can be achieved either by freezing, by treatment with different detergents, or by sectioning the fixed material into $20-30$ μm thin slices by means of a vibrating microtome (Hökfelt and Ljungdahl 1972; Buijs and Swaab 1979; Pickel 1979; Pickel et al. 1979), which are then incubated. In several cases, small amounts of Triton X-100 are added to promote penetration of antibodies. Unfortunately, all these processes alter the cellular fine structure to some extent.

As an alternative method, postembedding labelling of ultrathin sections was developed, using different resins (Kawarai and Nakane 1970; Leduc et al. 1969), or polyethylene glycol or serum albumin (McLean and Singer 1970) as embedding media. Fortunately, in most cases the specific antigenicity survives dehydration and embedding in resin. Post-embedding labelling of antigenic sites in ultrathin sections was first performed by means of ferritin-conjugated antibodies (McLean and Singer 1970). However, this method was severely limited by a strong non-specific interaction between the ferritin molecules and the plastic material. The application of peroxidase as a marker, either coupled to the primary antibody (Nakane 1971; Nakane and Kawaoi 1974) or in the form of the PAP-complex (Sternberger et al. 1970) gives highly specific results and is therefore widely used both for vertebrate and invertebrate tissue.

As an alternative and complementary technique, Roth (1982) and Roth et al. (1978, 1980) developed a new postembedding staining method, taking advantage of the ability of protein A from *Staphylococcus aureus* to interact with IgG immunoglobulins from several mammalian species. In this method, colloidal gold is used as the electron-dense marker, though peroxidase can also be employed (Dubois-Dalq et al. 1977) permitting the detection of antigenic sites in electron microscopic ultrathin sections with a high specificity.

In the following, the preparation of insect nervous tissue for electron microscopic identification of antigenic sites (proctolin or proctolin-like peptides and their intracellular distribution) is described in detail for an alternative semithin-ultrathin section technique (Agricola pers. comm.) and for the protein A-gold method as well.

Preparation of Fixatives

Glutaraldehyde, commercially available as a 25% solution in water, gives good results mainly due to the fact that it reacts rapidly with proteins (Millonig and Marinozzi 1968; Schiff and Gennaro 1979). Being a dialdehyde, it stabilizes structures before extraction of any components can occur (Hayat 1970). Glutaraldehyde contributes little to the osmotic effect of the fixative, so that in fact the osmolarity of the buffer determines the effective osmolarity of the fixative mixture. One must take into consideration that certain constituents, e. g. lipids. are not fixed by the glutaraldehyde and may therefore be extracted during subsequent dehydration (Stein and Stein 1971).

Purification of some brands of glutaraldehyde is necessary, because of contaminants e. g. glutaric acid, polymerization and other products (Frigero and Shaw 1969), which show absorption at 235 nm (Trelstad 1969). Monomeric glutaraldehyde absorbs at 280 nm. Purification can be performed by charcoal filtration or by vacuum distillation (Glauert 1975). Purified glutaraldehyde can be stored for up to a year at $-20\,^{\circ}$C (Gillett and Goll 1972).

Formaldehyde reacts with proteins, but with less cross-linking than glutaraldehyde. However, it has some advantages, penetrating very rapidly and reacting with lipids in an unknown fashion (Millonig and Marinozzi 1968).

Since commercially available formaldehyde contains methanol, it must be freshly prepared by dissolving 10 g of paraformaldehyde in 25 ml of bidistilled water. The solution is heated to approximately 65 °C with continuous stirring. A few drops of 30% sodium hydroxide are added to clear the solution. The resulting solution contains 40% formaldehyde and is allowed to cool before mixing with other components of the fixative.

Preparing Glutaraldehyde-Formaldehyde Fixative

50 ml of 0.2 M phosphate buffer, pH 7.3 is mixed with 2.5 ml of 40% formaldehyde and 4 ml of 25% glutaraldehyde. Distilled water is added to make 100 ml.

The final solution contains 1% formaldehyde and 1% glutaraldehyde in 0.1 M buffer. For fixation of insect nervous tissue adjust the osmolarity to approximately 400 mOsm with glucose.

Fixation for Electron Microscopic Immunocytochemistry

It is advantageous to perform the initial fixation in vivo and then to complete fixation by immersing of small pieces in fixative.

The optimum fixation time depends on the size of the specimen. In most cases 2 h at room temperature are sufficient. Fixation at lower temperatures (4 °C in a refrigerator) requires a duration of up to 4 h. Longer periods result in extraction of materials by the buffer.

Rinsing Specimen. It is important to thoroughly rinse specimens after fixation to remove all unreacted fixative that would otherwise interact with the antisera during the incubation stages. At least three changes of washing solution are usual each, 1/2−1 h. Longer washing should be avoided because extraction may occur in spite of stabilization by the fixative.

The washing solution (the fixative buffer) is adjusted to a final osmolarity of 400 mOsm with glucose.

Dehydration. As a rule, tissues are dehydrated with a series of graded ethanol, but acetone can also be used.

Dehydration should be performed as rapidly as possible to avoid extraction. As an intermediate solvent before resin, propylene oxide is used. Dehydrate in 50%, 70% and 90% ethanol or acetone (10 min each), two changes of 100% ethanol or acetone (15 min each) and two changes of propylene oxide (10 min each).

Embedding. The most frequently used embedding media for electron microscopic immunocytochemistry, as well as for pre- and postembedding staining, are Epon or Araldite (Durcupan). In a few cases, Epon-Araldite mixtures are used. The embedding follows the general schedule:

Propylene oxide (PO)

PO: Durcupan mixture	3:1	1 h, room temp.
PO: Durcupan mixture	1:1	1 h, room temp.
PO: Durcupan mixture	1:3	1 h, room temp.
Durcupan mixture I	1 h, 50 °C	
Durcupan mixture II	1 h, 50 °C	
Embedding in mixture II		

All steps up to now, including fixation and dehydration, should be performed with continuous specimen agitation on a rotary shaker. During the embedding procedure this enhances penetration.

The Semithin-Ultrathin Technique

A very useful method for the electron microscopic localization of antigenic sites in insect nervous tissue is the semithin-ultrathin technique. It allows the identification of the same stained structure both at the light microscopic and at the electron microscopic level.

Procedure. The tissue is fixed in a mixture of 2% glutaraldehyde and 2% formaldehyde in 0.1 M phosphate buffer pH 7.3−7.4 for 2 h at room temperature. The osmolarity of the mixture was adjusted to 430 mOsm. Tissues were dehydrated in a series of graded acetone, transferred in propylene oxide and embedded in Durcupan ACM according to the general schedule.

Semithin sections (about 2 μm thick) are mounted on a carefully cleaned glass slide without any adhesive. Subsequent sections must be mounted with sufficient distance to preserve the possibility of reembedding the sections separately.

The embedding medium is dissolved by adding a drop of freshly prepared sodium ethoxide onto the section for 45 min (Na-ethoxide is prepared by dissolving 2 g NaOH in 100 ml ethanol. The solution must be filtered prior to use). After this, sections must be rinsed with ethanol and transferred to water.

Immunocytochemical staining is performed by covering the sections with the following solutions: (1) Normal goat serum 1:3, 15 min at room temp. (2) Primary antiserum 1:100, overnight at 4 °C. (3) Wash three times (5 min each) with PBS to which traces of Tween 80 have been added. (4) Goat-anti-rabbit IgG 1:10, 1 h at room temp. (5) Wash 3 times according to 3. (6) PAP 1:20 diluted with Tris 0.05 M, 1 h at room temp. (7) Wash 3 times according to 3. (8) DAB-reaction (5 mg DAB in 10 ml Tris buffer pH 7.6 plus 7 μl 30%

H_2O_2). (9) Wash with distilled water. (10) 1% OsO_4 in distilled water 1 h at room temp. (11) Wash 3 times with distilled water. Sections were then covered with glycerol and investigated under the light microscope (Fig. 27).

For further electron microscopic investigation, selected sections were transferred to water. Then part of a gelatine capsule is mounted over it with a little glue. The section is then dehydrated with ethanol, reembedded in Durcupan and sectioned for electron microscopy (Fig. 28).

Control experiments for the identification of non-specific antibody absorption are performed according the methods described for light microscopy (see p. 291).

Protein A-Gold Method

The protein A-gold method is another useful technique for the electron microscopic detection of antigenic material in ultrathin sections of aldehyde-fixed and resin-embedded insect nervous tissue. Protein A, molecular weight 42,000, is produced by *Staphylococcus aureus* as an integral component of the cell wall.

Its most conspicuous property is the ability to interact with mammalian immunoglobulins. This interaction involves the Fc-region of the IgG and does not affect the antigen-antibody reaction. On this basis Roth (1982) and Roth et al. (1978, 1980) developed a postembedding stain technique for the immunocytochemical detection of antigenic material using a complex of protein A and gold particles (see Fig. 29).

Preparation of Colloidal Gold. Twenty mg of tetrachloroauric acid ($HAuCl_4$) are dissolved in 200 ml of bidistilled water and brought to the boil. Since the reduction process is affected even by the smallest contaminants, special care must be taken to prepare the glassware used. It should be gently cleaned and siliconized. When boiling, 10 ml of 1% trisodium citrate, dissolved in bidistilled water, is rapidly added (Frens 1973). The solution becomes blue and after 5 min further heating it clears to orangish red. To 10 ml of this solution add 30 ml Tris buffer, 0.005 M, pH 7.6. Then add 40 μl Tween 80 and 10 mg sodium azide (NaN_3). The diameter of the gold particles in the final solution depends upon the amount of trisodium citrate in a constant volume of tetrachloroauric acid. In the present conditions it is about 15 nm.

Preparation of Protein A-Gold Complex. Fifty ml of gold suspension is rapidly mixed with 50 μg protein A which is available from Pharmacia Fine Chemicals and can be stored as a 0.1% solution at $-20\,°C$ for several months. The mixture is then centrifuged at 4 °C for 15 min (15,000 rpm). The supernatant, sucked off with a Pasteur pipette, was discarded and the pellet immediately suspended in 2 ml 0.005 M Tris buffer pH 7.6 to which

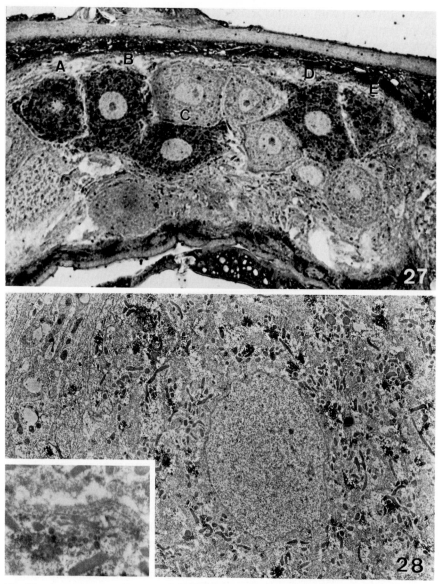

Fig. 27. Light microscopical morphology of the dorsocaudal part of the 6th abdominal ganglion of *Periplaneta americana.* From a 2-μm semithin section, resin was dissolved by means of sodium methoxide. Subsequently, the section was incubated for the immunocytochemical demonstration of proctolin, using the PAP-method. Proctolin-like immunoreactivity is limited to specific cells (*A–E*) and occurs in the form of dense complexes dispersed equally throughout the cytoplasm. × 800

Fig. 28. After reembedding in resin, the section has been processed for electron microscopy. The part shown is the central region of cell *C* in Fig. 27. Reaction product is limited to the numerous Golgi-stacks, distributed in the cytoplasm. It occurs as well in the

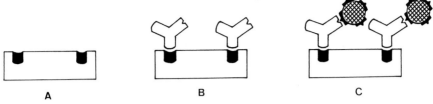

Fig. 29. Schematic representation of the protein A- gold technique. **A** Antigenic sites on the surface of an ultrathin section of resin-embedded tissue. **B** Binding of the specific antibody molecule. **C** After incubation with the pAg-complex, gold particles are bound by interaction of the protein A-molecules with the FC-region of the gamma globulin

1 μl Tween 80 per ml was added. To remove non-adsorbed protein, the complex must be washed at least twice, each time centrifuged 15 min at 4 °C.

Procedure of the Protein A-Gold Technique. Tissues were fixed with a mixture of 0.5 to 1% glutaraldehyde and 0.5 to 1% freshly prepared formaldehyde in 0.1 M phosphate buffer pH 7.4 for 1−2 h at 4 °C. Objects were then rinsed twice in phosphate buffer, rapidly dehydrated in a series of graded acetone (for 10 min each) and embedded in Durcupan ACM according to the general schedule. Silver to light-gold sections were mounted on uncoated nickel grids. Before use, grids were dipped into an approximately 0.5% solution of Zapon lac in order to enhance the adhesion of the sections. Grids were then placed face down onto drops of the following solutions (The drops of antibody-washing and protein A-gold solution should be placed on a plate of dental wax which is kept in a moist chamber throughout the procedure):

1% ovalbumin for 5 min at room temp. in order to block non-specific binding sites. Grids were then transferred without any rinsing to drops of anti-proctolin, 1:300 diluted with PBS, for 48 h at 4 °C. Then they are washed with 0.05 M Tris buffer pH 7.6 at least three times and placed on drops of protein A-gold solution for 12 h at 4 °C. For this, the stock solution (see above) is diluted up to tenfold with 0.005 M Tris buffer pH 7.6 to which Tween 80 (1 μl ml^{-1}) was added. Finally, grids are washed gently first with Tris buffer, then with bidistilled water and transferred to the staining solutions (5% aqueous uranyl acetate and lead citrate according to Reynolds (1963) (Figs. 30 and 31).

Washing the grids is performed as follows: Grids are placed onto drops of buffer on a wax plate which is fixed to a magnetic stirrer, working at low speed.

Golgi cisternae as in dense complexes, associated with smaller granules and resembling lysosomal structures (*inset*). Lysosomal elements are possibly involved in the maturing of peptide granules, as described by Smith and Farquhar (1966) in cells of the anterior pituitary gland. × 3600

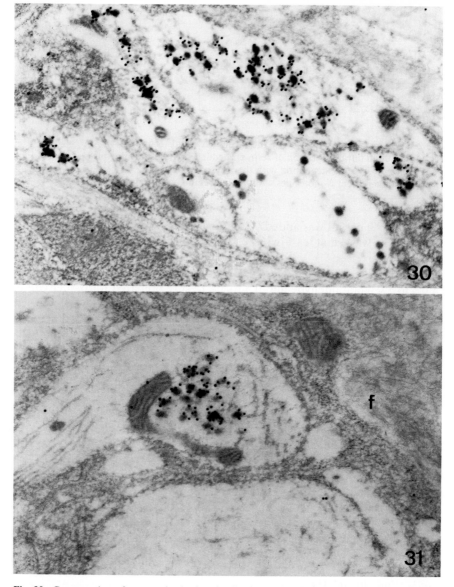

Fig. 30. Cross section of a nerve in the longitudinal muscle of the cockroach proctodaeum. The aldehyde-fixed tissue was embedded in Durcupan, and ultrathin sections were incubated with proctolin antiserum and protein A-gold solution. Most of the profiles exhibit a specific reaction with the pAg-complex, demonstrated by the gold particles, virtually restricted to the peptide granules. Note that the granules in the lower right profile are not labeled indicating the occurrence of a peptide other than proctolin-like in the proctodeal muscle. × 28 000

Fig. 31. Nerve terminal in close spatial relation to a muscle cell, separated from it by a fibrous acellular material (*f*). Note the localization of gold particles, restricted to the peptide granules. Only few non-specific gold particles occur. × 28 000

Conclusions

The avidin-biotin system and protein A method demonstrate that the development of more sensitive and specific immunocytochemical techniques, especially for immunoelectron microscopy, is not yet over. The introduction of monoclonal antibodies opens a new perspective in immunocytochemistry for the investigation of neuronal structures, mainly in two directions. First, monospecific antibodies with defined specificity against only one determinant group are suitable for more exact chemical identification of molecules of interest at a morphological level, e.g. neuropeptides. However, the problem of crossreactivity with identical determinant groups in other molecules persists. A disadvantage of the use of monoclonal antibodies to the application of normal antisera, containing a heterogeneous antibody population with different specificities for various determinant groups of the antigen or hapten, is the loss of immunoreactivity when the determinant group for the monoclonal antibody is changed by the fixation procedure.

Much improvement for neuroanatomical and neurodevelopmental research is expected by application of monoclonal antibodies to the identification of specific neuronal marker molecules, e.g. cell surface antigens with many different functions, and molecules related to neurotransmitter action. Crude extracts of neuronal tissue can be used for the immunization of mice or rats. Lymphocytes of the immunized animals must be fused with myeloma cells. The hybridoma secrete numbers of antibodies differing in their specificity (see Zipser and McKay, 1981a; Zipser 1982). Therefore, the hybridoma growth medium must be screened by means of immunocytochemical, radioimmunological, or radiochemical techniques (Barnstable 1980; Cohen and Selvendran 1981; Reichardt and Mathew 1982; Teugels and Ghysen 1983; Trisler 1982; Wulliamy et al 1981; Zipser and McKay 1981b) for a selection of hybridoma, producing interesting antibodies. In spite of its advantages, the production of monoclonal antibodies is not yet a routine method like the established techniques described here. It requires experience and is still restricted to specially qualified laboratories.

Detection of Serotonin-Containing Neurons in the Insect Nervous System by Antibodies to 5-HT

Nikolai Klemm

Fakultät für Biologie der Universität
Konstanz, F.R.G.

Introduction

The occurrence of serotonin (5-HT) in the CNS of vertebrates was first shown by Twarog and Page in 1953. Since then there has been abundant evidence that serotonin is a neuroactive substance (neurotransmitter and neuromodulator) in vertebrates and invertebrates (see Dismukes 1979; Evans 1980). Its presence in the insect CNS was first shown by Gersch et al. (1961) and its intraneuronal sites were demonstrated with aldehyde fluorescence (the Falck-Hillarp method) by Klemm and Axelsson (1973) and Klemm (1974, 1976, 1980).

Although the aldehyde-fluorescence method is very sensitive to catecholamines it is less so for 5-HT (see Ritzén 1967; Klemm 1980). This is because 5-HT is less reactive to formaldehyde (Ritzén 1967), and its fluorophore fluoresces less intensely than that of catecholamines, fading quickly when exposed to fluorescent light. The typical 5-HT fluorophore appears yellow in routine fluorescence microscopy and changes colour (orange, yellowish-brown) according to its concentration (Caspersson et al. 1966; Klemm and Axelsson 1973). In insects a yellowish-brown fluorophore can be detected, in addition to yellow fluorescence. Manipulation of the procedure of fluorophore formation revealed that neurons with a yellow fluorophore may change to yellowish-brown and vice versa (Klemm and Axelsson 1973).

Since the introduction of the immunocytochemical method by Coons et al. (1941), antibodies have been used to label larger molecules (e. g. enzymes) in vertebrates (see Nairn 1976; Coons 1971; Sternberger 1979) and invertebrates (Eckert et al. 1971 and later many others). Although many enzymes share common functions in different species they often have different immunogenic properties. Thus the application of antibodies is often limited to certain species. Small molecules, however, such as 5-HT are shared by almost all animals. Thus, antibodies to such small molecules lack species specificity and can be used regardless of the species studied. Steinbusch recently introduced a highly specific antibody against 5-HT (Steinbusch et al.

1978, 1983), which is now commercially available. Besides its universality, easy application and high specificity, it is very sensitive. Model experiments have shown that the immunofluorescence method for 5-HT is approximately 1000 times more sensitive than the formaldehyde fluorescence method (Schipper et al., 1981). Moreover, it can be applied for electron microscopy (cf. Pasik et al. 1982). Antibodies to 5-HT have also been successfully used on the insect nervous system (Klemm and Sundler 1983; Klemm et al., 1983; Nässel and Klemm 1983; Taghert et al., 1983; Bishop and O'Shea 1983).

General Considerations of Antibody Staining

Since its introduction by Coons et al. (1941) the original fluorescent anti-body staining has been much developed and modified (compare: Coons 1971; Sternberger 1974; Larsson 1981). The early "direct immunocytochemi-cal technique" (incubation of the tissue with fluorochrome-labelled anti-body) has been replaced by the more sensitive and more specific "indirect technique" (sandwich technique) (Nairn 1969; Coons 1971; Sternberger 1974; Vandesande 1978; Larsson 1981). Two procedures are most frequently used today: the indirect immunofluorescence technique (Coons 1971) and the "unlabelled antibody enzyme technique" (Sternberger 1979). The sensi-tivity of anti-5-HT in the insect nervous tissue is almost the same in both techniques (Klemm et al. 1983). The intensity of labelling is dependent on the tissue antigen concentration and the ease with which the antibody pen-etrates to the antigen. In addition, the tissue-antigen concentration is de-pendent on the physiological state of the tissue and is dependent on 5-HT retention in the tissue after fixation and processing the tissue (Vandesande 1978; Larsson 1981). In both techniques several steps are involved:

1. The tissue antigen (e.g. 5-HT) reacts with an antibody raised in species I (e.g. rabbit) directed against the antigen 5-HT.

2. Another antibody, raised in species II (e.g. sheep) and directed against the immunoglobulins (IgG) of the species I (sheep anti-rabbit) binds to the antibody of species I which thus acts as an antigen to the second antibody. In the immunofluorescence technique the second antibody (anti-immuno-globulin) is labelled with a fluorochrome and can be detected by fluores-cence microscopy. In the "unlabelled technique", this antibody is unlabelled.

3. In the "unlabelled antibody enzyme method" a third step is involved. In the most widely used technique, a soluble peroxidase-antiperoxidase complex (PAP) is added. The antiperoxidase compound of the complex should be from the same species as the primary antibody (e.g. rabbit). When given in excess, one binding site of the bivalent secondary antibody binds with the primary antibody (e.g. anti-5-HT) while the other remains free to combine in the third step with the antiperoxidase component (raised in species I) of the PAP complex. The peroxidase component of the PAP compound is enzymatically active. In the presence of hydrogen peroxidase

and diaminobenzidine tetrahydrochloride (DAB) it oxidizes DAB into an insoluble brown polymer which can be visualized in the light microscope. Because the polymer is osmiophilic it also can be studied under the electron microscope (Pasik et al. 1982).

The Immunofluorescence Technique (Figs. 1—5)

Several fluorochromes are available as markers in immunocytochemistry (e. g. fluorescein isothiocyanate — FITC, tetramethyl-rhodamine isothiocyanate — TRITC) (Nairn 1969; Coons 1971). The procedure used in this laboratory is adapted and slightly modified from Hökfelt et al. 1978; Larsson 1981; Steinbusch et al. 1982 (see also Klemm et al. 1983).

Fixation: A proper fixation is the most crucial step in antigen preservation.

1. Anaesthetize the insect slightly with ether and quickly dissect out the nervous system.

2. Transfer the specimen into freshly prepared ice-cold fixation fluid (insect Ringer solution) or other physiological solutions, add 4% paraformaldehyde, pH 6.9—7.4 (see a below). Continue fixation for 3—12 h at 4 °C.

3. Rinse the tissue for 48 h at 4 °C in phosphate-buffered saline (PBS) or other physiological solutions to which 15%—25% sucrose is added (b). The tissue can be stored at 4 °C in the rinsing solution for many weeks provided the solution is replaced by a fresh one every 3 days to prevent bacterial growth.

a) To prepare the fixation fluid: add paraformaldehyde to the buffer, warm to about 55 °C, and stir until the paraformaldehyde is completely dissolved. In case it does not dissolve completely, add a few drops of dilute NaOH to the fixative (Larsson 1981), if necessary titrate to pH 6.9—7.4.

b) To prepare PBS: mix the following two solutions. Solution $A-Na_2HPO_4$ 1.4 g/100 ml distilled water or $Na_2HPO_4 \cdot 2 H_2O$ 1.78 g/100 ml distilled water. Solution $B-NaH_2PO_4 \cdot H_2O$ 1.4 g/100 ml distilled water. To achieve a solution of pH 6.9—7.2 mix 84.1 ml of solution A and 15.9 ml of solution B and add 8.5 g NaCl and 900 ml distilled water.

Mounting and Sectioning

Remove excess liquid by touching with blotting paper. Mount the specimen on a cryostat chuck or directly to the cryostat stage using either tissue paste (e. g. homogenized vertebrate brain tissue) or Tissue-Tek II (Lab-Tek products, Naperville, Ill, USA) and make sections of 8—14 μm with a cryostat microtome (knife temperature −25 °C). Adhere the sections to a warm (room temperature) microscope slide pretreated with chrome alum (see c below). Store at −20 °C. Before proceeding further let the specimens warm to room temperature.

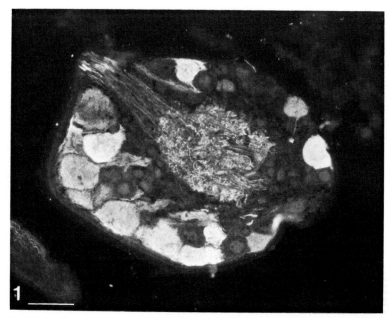

Fig. 1. Section of the frontal ganglion of the locust, *Schistocerca gregaria* Forsk. Anti-5-HT reacts in cell bodies and neuropil. Bar = 50 μm

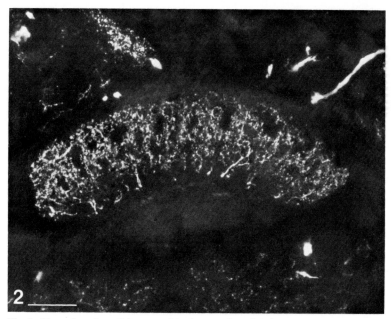

Fig. 2. Frontal section of the central body complex of *Schistocerca gregaria*. Copy from colour slide. 5-HT$_i$ (immunoreactive) fibres exclusively in the central body (fan-shaped body). Bar = 50 μm

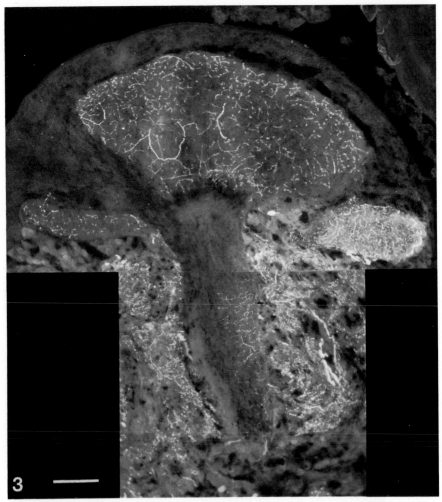

Fig. 3. *Periplaneta americana.* 5-HT$_i$ innervation of the peduncle and the calyx.
Bar = 50 μm

c) To prepare and treat slides with chrome alum-gelatine: 1 g gelatine and
0.1 g chrome alum [CrK(SO$_4$)$_2$ · 12 H$_2$O are dissolved in 200 ml distilled water and slightly warmed up. Before use the solution should be stored at room
temperature for 24 h. The cleaned microscope slides are dipped into the solution and subsequently dried in an oven at 50 °C for 12 h.

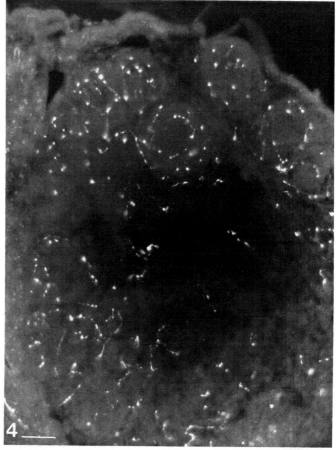

Fig. 4. *Apis mellifica.* 5-HT$_i$ fibres in the olfactory glomeruli. Bar = 50 μm

Antibody Staining

Rinse the sections in PBS at room temperature for 20 min, blot with filter paper.

1. Incubate warm (room temperature) specimens in diluted primary antiserum (anti-5-HT) (see *d* below) by placing drops on the sections. Store the preparations in a moist chamber either for 3–4 h at room temperature or 12–24 h at 4 °C (*e*).
2. Rinse the sections at room temperature in PBS. Shake gently. 2 × 15 min
3. Blot the sections by carefully touching them with filter paper.
4. Incubate with diluted FITC-labelled secondary antiserum in a moist chamber at room temperature (*f*). 1 h

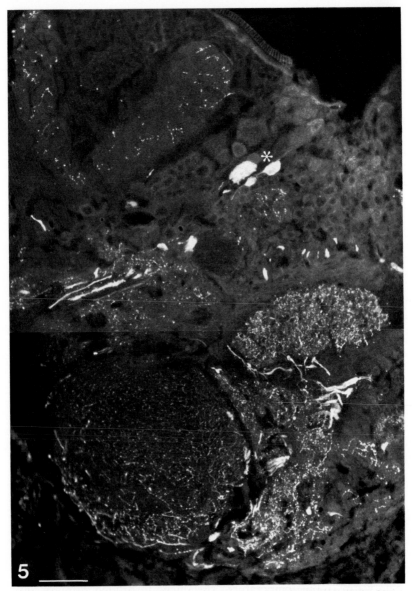

Fig. 5. *Periplaneta americana.* Innervation of the central (fan-shaped) body. Lower left to the central body is the *β*-lobe, upper left is the calyx. *asterisk:* cell bodies. Bar = 50 μm

5. As step 2. 2 × 15 min
6. Mount in a mixture of PBS and glycerin (1 : 1 − 1 : 3 v/v) and cover with a cover slip.
7. For permanent mounts see p. 312.

d) To prepare the primary antiserum solution: The following antisera were used (concentrations in brackets): 6−7 Ser (Steinbusch) (1 : 160), 7−6-Ser (Steinbusch) (1 : 600−1 : 1000). N-Ser 2 (Immuno Nuclear Corp. Stillwater, MN USA) (1 : 800). Lyophilized antisera can be stored at −20 °C. All antisera were brought to room temperature before dilutions were made. PBS containing 0.25% of the detergent Triton X-100 and 0.25% of human (or bovine) serum albumine was the dilution medium. One part of the antiserum was dissolved with 5−10 parts of the buffer. The diluted antiserum was stored at −20 °C in quantities sufficient for one series of experiments. Before use, the antisera were diluted with the same buffer to the required final concentration. The optimum concentration was determined by labelling the specimens with different concentrations of the antibody. Each batch of antiserum and all new tissue should be tested in a broad range of dilution of the antiserum. Since frequent freezing and thawing of the antibody may damage the proteins, the final concentration of the antiserum should be kept at 4 °C (up to 2 weeks).

e) Temperature: The longer cold incubation period has the advantage that the concentration of the antiserum can be reduced by 50% to 20% of the concentration used in short-term incubations. Higher dilution reduces the unspecific (background) staining and is more economical.

f) The secondary FITC-labelled antiserum: use sheep antirabbit antiserum (ShARb)/immunoglobulin (Ig)-FITC (Statens Bakteriologiska Laboratorium, Stockholm, Sweden). The initial antiserum solution was made with PBS (1 : 15) and stored at −20 °C. Before use the antiserum was brought to room temperature and diluted 1 : 1 (v/v) with PBS containing 0.5% Triton X-100 to achieve the final concentration of the detergent of 0.25%. The final solution (1 : 30) can be stored at 4 °C for approximately two weeks.

Comments: The FITC-fluorophore does not diffuse and is relatively photo-stable. The method is easy to perform. The endogenous peroxidase activity can be neglected.

Shortcomings: The FITC-fluorophore fades under fluorescent light. However, Giloh and Sedat (1982) report that the presence of the antioxidant n-propyl gallate (2%−5% w/v in glycerol) in the mounting medium reduces the fading rate by a factor of 10. As the anti-oxidant is detrimental to tissue, the authors recommend rinsing the specimens in PBS and remounting them for storage. Johnson et al. (1981) reduce fading by adding phenylene diamine to the mounting medium. The background staining in FITC preparations is higher than in the PAP method, but can be reduced by appropriate filter combinations. The preparations can be stored at −20 °C for several months. FITC preparations can be post-treated according to the PAP meth-

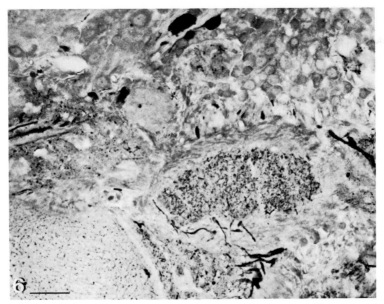

Fig. 6. Same preparation as Fig. 5 post-treated with PAP. Bar = 50 μm

od to achieve permanent mounts (see p. 312). Their quality, however, is usually less good than in straight-forward PAP preparations (Figs. 6 and 7).

Fluorescence Microscopy and Photography

Several types of fluorescent microscopes are available which employ either a xenon or a mercury-type light source. I prefer a Zeiss Standard microscope with epi-illumination using a mercury lamp (HBO 50). A suitable filter set is mandatory for FITC fluorescence, since FITC is excited between 450−500 nm (ex. max: excitation maximum 490 nm) and emits light of 500−550 nm (em. max: emission max 520). As ex. max and em. max are close, the secondary filter set must include a filter sharply cutting off above 500 nm (Zeiss FT 510) to prevent excitation light from being mixed with emission light. A selective filter set with narrow band filters (Zeiss filters: BP 585, FT 510, LP 520) provides good contrast and sufficient intensity. For less intensive reactions, a filter set consisting of BP 450−490, FT 510, LP 520 or BP 515−565 can be used (BP 515−656 cuts off the possible occurrence of reddish fluorescence). Black and white photos can be made with Kodak Tri-X or Ilford HP5 (both usable between 400 and 1600 ASA) or even with "slow" films as Kodak Panatomic X (64 ASA). For colour photography Ektachrome High Speed (400 ASA) was preferred. The exposure time for these films is between 10 s and 2.0 min.

The Unlabelled Antibody Enzyme Method for Sections
(PAP Method According to Sternberger 1974) (Fig. 7)

For this method, the specimens can be fixed and further handled in two different ways:

For cryostat sectioning: fixation, rinsing, mounting and sectioning is the same as for FITC technique (see p. 307).

For paraffin sectioning: deep freeze and subsequently freeze dry the sample.

A) Deep freezing: freeze a mixture of propane-propylene [or Freon-22 (Larsson 1981)] in a metal beaker in liquid nitrogen. Kill the animal, remove the sample and dip for 5–10 s into the mixture. Transfer the sample into a metal holder immersed in liquid nitrogen (immediate dipping into the liquid nitrogen would result in a layer of vaporized nitrogen on the surface of the sample and thus prevent fast cooling. Fast cooling is essential to prevent the formation of ice crystals which would damage the tissue). Tissue samples covered with liquid nitrogen can be stored for several days.

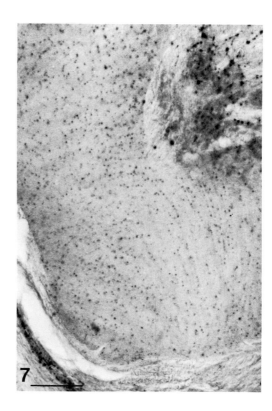

Fig. 7. *Periplaneta americana.* Freeze-dried paraffin-PAP preparation. Perpendicular 5-HT$_i$ fibres in the α-lobe and β-lobe. Bar = 25 μm

Freeze drying: many suitable freeze dryers are commercially available. The procedure should be performed at 10^{-2} Torr and at low temperature ($-80\,°C$). I prefer the "cold finger" type freeze dryer (see Klemm 1980).

B) Paraformaldehyde treatment: Store deep freeze specimens in a desiccator and allow them to come to room temperature. Place them in a vial containing 5 g paraformaldehyde/1000 ccm. Cover the vials tightly and heat in an oven for 1 h at 80 °C.

C) Allow the specimens to cool to room temperature. Embed in paraffin in vacuo (see Klemm 1980).

D) Section at $6-14\,\mu m$. Place the sections on microscope slides covered with glycerol-gelatine (see g, p. 313). Blot excess glycerol-gelatine with filter paper.

E) Treat the sections with paraformaldehyde vapors as in (b). Allow them to come to room temperature.

F) PAP-staining for paraffin sections:

1. Deparaffinize the sections in xylene − 3 × 5 min
2. Place in absolute alcohol − 2 × 5 min
3. Treat in absolute methanol containing 0.03% H_2O_2 to inactivate endogenous peroxidase activity (Larsson 1981) − 30 min
4. Rehydrate in 96% alcohol and 70% alcohol − each 5 min
5. Place in 0.9% NaCl − 5 min
6. Rinse in PBS (see p. 304) − overnight
7. Blot specimens with blotting paper
8. Incubate in primary antiserum (see instruction 4 p. 307) in a moist chamber at 4 °C − 12−24 h
9. Rinse in PBS with agitation at room temperature − 2 × 15 min
10. Blot specimens with filter paper
11. Incubate with unlabelled goat anti-rabbit antiserum (see h below) in a moist chamber at room temperature − 30 min
12. Rinse in PBS (to which 0.25% Triton X-100 can be added) with agitation at room temperature − 2 × 15 min
13. Blot the sections
14. Incubate with rabbit peroxidase-antiperoxidase (PAP) complex at room temperature (see i below) − 1 h
15. Rinse in PBS (to which 0.25% Triton X-100 can be added) at room temperature and with agitation − 2 × 15 min
16. Rinse in 0.05 M Tris, pH 7.6 (see j below) − 5 min
17. React with DAB (see k below) − 1 h
18. Rinse with distilled water − 1 min
19. Postfix in 4% glutaraldehyde in 0.05 M Tris pH 7.6 or PBS − 10 min
20. Dehydrate in 70% alcohol, 96% alcohol, absolute alcohol and xylene − each 5 min
21. Counterstain
22. Mount in Permount (Fisher USA) or DePeX (Gurr USA)

G) PAP treatment for cryostat sections:

1. Rinse the section in PBS − 10 min
2. Dehydrate in 70% alcohol, 96% alcohol, twice in absolute alcohol and place in xylene − each 5 min
3. Place in absolute alcohol − 5 min
4. Treat in absolute methanol containing 0.03% H_2O_2 − 30 min
5. Rehydrate in 96% alcohol and 70% alcohol − each 5 min
6. Place in 0.9% NaCl − 5 min
7. Rinse in PBS − 15 min
8. Proceed according to PAP-staining for paraffin sections (F, p. 312).

H) Post-treatment with PAP for FITC sections for permanent mounts (Fig. 6): rinse FITC preparations in PBS and remove the cover slip − 5−10 min. Proceed according to (F) stage 10−22.

g) To prepare glycerol-gelatine jelly: Dissolve 1 g gelatine in 100 ml distilled water at 40 °C. Cool the solution to room temperature and filter. Add 15 ml glycerol. A small amount of camphor or thymol prevents bacterial growth. This mixture can be stored at 4 °C for a long time. Before using, warm to approximately 40 °C. Remove surplus liquid on the microscope slide with filter paper.

h) To prepare unlabelled goat anti-rabbit antiserum (Statens Bakteriologiska Laboratorium, Stockholm, Sweden or Nordic, Tilburg, The Netherlands): Dilute 1:30 with PBS including 0.25% Triton X-100, 0.25% BSA.

i) To prepare rabbit peroxidase-antiperoxidase (Dakopatts, Copenhagen, Denmark): Dilute with PBS including 0.25% Triton X-100 and 0.25% BSA to 1:90−1:160. Steinbusch et al. (1982) recommend dilution only with PBS.

j) To prepare Tris: 6.05 g Tris ($C_4H_{11}NO_3$)/1 l distilled water. Adjust pH to 7.6 with conc. HCl.

k) To prepare DAB: dissolve 60 mg 3,3′-diaminobenzidinetetrahydrochloride (DAB) (Sigma) in 100 ml 0.05 M Tris, pH 7.6. Filter and add 0.01% H_2O_2 (ca. three drops 30% H_2O_2/100 ml). Take safety precautions when weighing out DAB (BIOHAZARD!)! Use fume hood!

Comments: The method is easy and has certain advantages although it is more time-consuming than the FITC method. There is low background staining. Paraffin sections have better tissue preservation than cryostat sections. Mounts are permanent. The PAP method can also be applied for electron microscopy (Sternberger 1979; Larsson 1981; Pasik et al. 1982 − and see Chap. 14).

A Whole-Mount Method for Antibody Staining (After Taghert et al. 1983)

The method was developed by Taghert et al. for staining cell bodies, neurites and processes of 5-HT-containing neurons in locust embryos. The method is as follows (see Fig. 8)

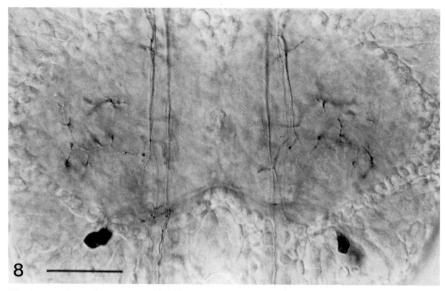

Fig. 8. Anti-5-HT staining (PAP whole-mount method) of cells S1 and S2 in the meso-thoracic ganglion of a 70% developed locust embryo (Nomarski interference phase contrast: photomicrograph by courtesy of C. Goodman). Bar = 50 μm

1. Fix in 2% paraformaldehyde in slightly hypotonic saline, 1 h.

    To prepare saline:
    150 mM NaCl
      3 mM KCl
      2 mM CaCl$_2$
      1 mM MgSO$_4$
      5 mM TES buffer (Sigma) adjusted to pH 7.0

2. Rinse the embryo (up to 70% development from fertilization) in saline, 1 h.

3. Incubate in primary rabbit anti-5-HT (N Ser 2; Immunonuclear Corp) 1:500 dilution with 0.25% Triton-X or 0.2% Saponin and 0.25% BSA for 12 h at 4 °C.

4. Rinse in saline 1 h at 4 °C.

5. Incubate in unlabelled goat anti-rabbit IgG (diluted 1:20) 12 h at 4 °C.

6. Rinse in saline 1 h at 4 °C

7. Incubate in HRP-labelled rabbit-antiperoxidase (diluted 1:40) 12 h at 4 °C.

8. Rinse twice in saline for 1 h starting at 4 °C raising temp. to 20 °C by completion.

9. Incubate 1 h in 50 mg 3,3'-diaminobenzidine in saline.

10. Develop 0.5−1 h in the same solution containing 0.006% H$_2$O$_2$.

11. Rinse in saline, clear in glycerol and view under Nomarski interference optics (Fig. 8).

After rinsing in saline tissue can also be counterstained and dehydrated and embedded in a suitable medium under cover slip.

Specificity of Anti-5-HT Labelling

Specificity of antibody labelling is a crucial problem in immunocyto-chemistry (see Vandesande 1979). Non-specific positive reactions can be caused by reagents used in the procedure. They can be reduced by applying highly purified antibodies in low concentration. Positive results must depend on the dilution of the primary antiserum, and no staining should be obtained in the absence of primary antibody or its replacement with normal serum of the same species (Petrusz et al. 1976). For the PAP technique, inactivation of the endogenous peroxidase activity is advised (Straus 1971; Larsson 1981). Serotonin immunogen was made by coupling 5-HT to a carrier protein (BSA) with formaldehyde. The antiserum therefore contains primary anti-bodies directed against both 5-HT and BSA. The antisera used are purified from BSA antibodies by affinity chromatography (Steinbusch et al. 1978, 1982). To test whether the coupling of anti-5-HT is specific the following two procedures were used to absorb antibodies with antigen:

A. The antiserum of the working solution was absorbed with $200-1000 \mu g$ 5-HT/1 ml antiserum and left in room temperature for $12-24$ h. The mixture can be stored at -20 °C.

B. The mixture of antigen and antiserum was shaken at 37 °C for 1 h and kept at 4 °C for 12 h. The precipitate was removed by centrifugation at $20,000 g_{max}$ for 15 min (Steinbusch et al. 1978). The mixtures can be stored at -20 °C and should be shaken before use.

The primary antiserum and the absorbed primary antiserum were applied on successive sections at step 1 for FITC technique and at step 8 in the PAP technique. Short-term reactions at room temperature ($2-3$ h) gave better results than long-term reactions. Too much of the antigen in the absorbed antiserum, however, may fail to inhibit antibody coupling (see Moriarty 1976; Sternberger 1974; Larsson 1981). This test assumes that the antibody does not differentiate between the antigen in tissue and in solution. Immobilization of small water soluble substances in tissue is critical. For example 5-HT requires fixation in formaldehyde to prevent the antigen from dislocation or elution. Fixation in such highly reactive substances may chemically modify the antigen. Schipper and Tilders (1982) recommend gelatine models, fixed and incubated in antiserum and treated identically to the tissue in order to study specificity. According to these authors, the 5-HT antibodies, which were used in this study have higher affinity to 6-hydroxy-tetrahydro-β-carboline than to 5-HT. 6-Hydroxy-tetrahydro-β-carboline is condensated when 5-HT reacts with formaldehyde (Whaley and Govin-dachari 1981) which could be formed during the synthesis of the im-munogen (Steinbusch et al. 1983) and during the fixation of the tissue. Nevertheless, the immunocytochemical procedure is specific for serotonin due to specificity of the antiserum and to specificity of fixation (see Stein-busch et al. 1978, 1983; Schipper and Tilders 1982).

Acknowledgement. I thank Renate Wieskirchen for photographic assistance.

Chapter 16

Monoaminergic Innervation in a Hemipteran Nervous System: A Whole-Mount Histofluorescence Survey

Thomas R.J. Flanagan

Department of Biology, Wesleyan University
Middletown, Connecticut U.S.A.

Introduction

A variety of monoamines known to function as neurotransmitters in vertebrate nervous systems have been identified in homogenates of insect ganglia (for reviews see Evans 1980; Robertson and Juorio 1976; Smyth 1977; Walker and Kerkut 1978). The cellular distribution of several of these has been described from formaldehyde vapor-treated sections of insect brains (Klemm 1976; Klemm and Sundler 1983). Considerable differences in neuroanatomical design and monoaminergic patterning, in particular, have been noted among brains of different insect species and no conserved pattern of monoaminergic innervation of the insect brain has yet been detected.

The ventral nerve cord has rarely been the focus of neuroanatomical study (Power 1948; Björklund et al. 1970). However, as it is less highly evolved than the brain (Lane 1974), the ventral nerve cord may be organized in a simple pattern that is common to insects.

The ventral nervous system of the blood-feeding hemipteran, *Rhodnius prolixus,* is intermediate between the fully fused system of highly evolved Diptera and the more primitive segmental system of Dictoptera. *Rhodnius* nervous system has a segmental (prothoracic) ganglion, fused meso- and metathoracic ganglia, and fused cephalized sub- and supraoesophageal ganglia (the brain). Here, technical modifications are described that allow rapid determination of the distribution of monoaminergic somata and some monoaminergic projection patterns in insect ganglia whole mounts. A summary of such distribution is described for the prothoracic ganglion and parts of the brain of *Rhodnius.*

Materials and Methods

1. Dissections

The central nervous systems of fifth instar *Rhodnius prolixus* Stal (Cambridge strain), from a laboratory colony were dissected into either a physiological saline described by Maddrell and Gee (1974) for physiological incubations or a Ca^{2+}-free, Mg^{2+}-rich (25 mM) saline, the latter restricting release of endogenous monoamine prior to histofluorescent and immunocytological analysis.

2. Histofluorescence

Glyoxylic acid (GA) methods have been developed for visualization of monoamines in whole-mount preparations (de la Torre and Surgeon 1976; Lindvall and Björklund 1974; Lent 1982). Ganglia from *Rhodnius* were incubated in 5% GA (Mallinkrodt Co.), 0.2 M sucrose in Ca^{2+}-free, Mg^{2+}-rich saline for 60 min at room temperature or for 8 to 12 h at 12 °C. The relatively high concentration of GA was necessary to penetrate the neural sheath barrier. After GA incubation, tissues were dried and then heated to clear the tissue and catalyze the second step of the GA reaction.

Whole mounts were gently blotted dry and stored over calcium carbonate for 3−5 days in a dark chamber at room temperature. Dried preparations were then covered with a drop of light mineral oil and flash-heated on copper blocks at 85 °C for 2 min. During this step, air was driven from the tracheae that heavily invest these ganglia (for a description of staining tracheae in an insect ganglion see Burrows 1980). If air remains trapped within the tracheae and hinders fluorescence illumination, a second flash-heating is done. Prolonged heating increases background fluorescence and damages tissue.

At neutral pH GA reacts well with catecholamines and has a 10^{-17} M threshold in reactions with dopamine (Moore and Loy 1978). This low threshold makes it probable that the most reactive catecholaminergic cell types are dopaminergic cells. Serotonin (5-HT) is most reactive with GA at pH 5. However, the insect ganglia used in this study are not really suited for prolonged incubations at this pH. Preparations that were immediately processed after a 5−10 min incubation at pH 5 displayed serotonin-like fluorescence within the neurohemal organs, but cells were poorly stained in the ganglia. For this reason, catecholaminergic cells were studied at pH 7 under conditions which permitted regular, but not routine, staining of the serotonergic cells. The identity of serotonergic cells was then corroborated by an immunocytological method.

After GA treatment, fluorescence was visualized with a Leitz dark-field fluorescence microscope equipped with KGl, BG38, and BG3 excitation filters and a K_{490} barrier filter. Black and white photographs were taken using Ilford Pan-F ASA 50 film. A BG12 excitation filter was substituted for the BG3 filter, and image resolution was improved at the expense of yellow-green color distinctions. Catecholaminergic reaction products proved stable for over a week at 12 °C.

For further information on the specificity, sensitivity, limitations, and alternative applications of the histochemical methods used in this study, the reader should consult the following: Lindvall and Björklund (1974); Loren et al. (1980); Lent (1982); Moore (1980); Moore and Loy (1978); Bolstad et al. (1979); Axelsson et al. (1973); Furness and Costa (1975); de la Torre (1980); Fuxe and Johnsson (1973); de la Torre and Surgeon (1976).

3. Monoamine Loading

Preloading monoaminergic cells with precursors (Kerkut et al. 1967), fluoro-genic monoamine (Gershorn 1977; Myhrberg et al. 1979; Berger and Glowinski 1978), or fluorogenic nonmetabolizable monoamine analogues (Klemm and Schneider 1975) facilitates the histofluorescent visualization of monoaminergic cells. Monoaminergic cells within ganglia from *Rhodnius* display selective cell loading when tissues are incubated in 0.1−0.5 mM dopamine for 20−60 min. High monoamine doses seem to be required due to the neural sheath of intact, isolated preparations (Pichon 1974; Treherne and Pichon 1972; Bernard et al. 1980). No accumulation of this metaboliz-able monoamine was seen in tracheae.

4. Immunocytology on Whole-Mount Preparations

Immunocytological methods developed to detect serotonin in crustacean ganglia whole mounts (Beltz and Kravitz 1983) were adapted for insect ganglia. Tissues were fixed in situ with 1.4% lysine monohydrochloride, 0.2% sodium metaperiodate, and 4% paraformaldehyde in 0.1 M phosphate buffer at pH 7.2 (McLean and Nakane 1974) for 3 to 6 days at 12 °C in a dark chamber. Fixed tissues were washed and then permeabilized in a solution of 0.3% Triton X-100 in 0.1 M phosphate with 0.1% sodium azide for 24 h at 12 °C in the dark. Rabbit anti-5-HT antiserum (Immuno Nuclear Co.) was diluted 1 : 100 with permeabilizing solution and incubated with tissues for 24 h at 12 °C in the dark. Tissues were washed for 24 h in Triton X/phos-phate to remove unbound rabbit antiserum. Fluorescein (FITC)-conjugated goat antirabbit IgG (Boehringer Mannheim Biochem.) was diluted into Triton X-100 phosphate 1 : 20 and incubated with tissue for 9 h at room tem-perature. Excess goat IgG was removed in a 5-h wash at room temperature. Tissue was immersed in 80% glycerol with 4 mM $NaHCO_3$ to enhance the signal from the fluorescein. Fluorescence was observed using a Leitz dark-field fluorescence microscope equipped for catecholamine histofluorescence.

5. Specificity of Immunocytological Reaction

Rabbit anti-5-HT antiserum was preabsorbed with BSA-5-HT conjugate (Immuno Nuclear Co.) to demonstrate the specificity of the immunocytological reaction. Antiserum diluted 1:100 and absorbed against 100 μg 5-HT-conjugate per 1 ml diluted antiserum for 12 h at 12 °C failed to stain any recognizable cellular elements within the ganglia.

6. Pharmacological Treatments

Monoamine stores were depleted with reserpine (for review see Carlson 1966). Isolated ganglia were incubated for 60 min in 1 mM reserpine (Serpasil, CIBA) which had been sonicated into physiological saline. Ganglia were washed briefly in saline after incubation and processed for histofluorescence analysis.

Vinblastine was used to block axoplasmic transport. Ganglia were removed from insects 3 and 7 days after they had been given an effective dose of 5×10^{-5} M vinblastine in 2% ethanol. This treatment has been previously shown to block axoplasmic transport of neurosecretory products in *Rhodnius* (Berlind 1981).

7. Neutral Red Vital Staining

The vital stain and pH indicator neutral red (Lillie 1969) was prepared as a 0.1% stock solution in distilled water, and was filtered and diluted to 0.0004% in physiological saline prior to use. Tissues were stained for 3−12 h at 24 °C or for 20 h at 12 °C. Ca^{2+}-free saline was used to reduce precipitation of dye (Baker 1958) onto the ganglion sheath without altering the staining pattern.

Results

Overview

Rhodnius possesses a typical hemipteran central nervous system composed of a brain, suboesophageal ganglion (SOG), prothoracic ganglion (PG), and a fused meso- and metathoracic ganglion (MG). Glyoxylic acid (GA) treatment of fresh whole mounts demonstrates fluorogenic neurons in all ganglia. The brain contains about 120 fluorogenic cells (excluding a cell set within the optic lobes); the SOG, about 9; the PG, exactly 11; and the MG, exactly 19. These cells fall into two distinct classes; green (catecholaminergic) cells and rapidly fading yellow (serotonergic) cells. Green cells can be subdivided into intensely fluorescent and weakly fluorescent catecholaminergic cell types, possibly reflecting differences between intensely reactive dopaminergic and less reactive noradrenergic cells. All fluorogenic responses of these cells are eliminated by incubating the tissues in 1 mM reserpine in

Table 1. Monoaminergic cell types within the prothoracic and suboesophageal ganglia

Segments	Cell types					
	IFC	YC	SC	LSC	F	DM
Prothoracic	3	4	4	6+4[a]	2	about 8
Suboesophageal	3	6+8[b]	4	uncertain[b]	2	about 8

[a] Variable population of cells within a certain cell set.
[b] Population of cells in a certain cell set that indicates a tendency to form a discrete subset of perikarya.
IFC = Intensely fluorescent catecholaminergic cells.
YC = Yellow cells.
SC = Less fluorescent catecholaminergic cells (possibly noradrenergic).
LSC = Lesser satellite nf-MA cells.
F = Flanking nf-MA cells.
DM = Dorsomedial nonfluorogenic monoaminergic cells staining with neutral red. Presumed octopaminergic.

physiological saline for 60 min prior to histofluorescence analysis. Additional cell types appear after the ganglia have been incubated in and loaded with dopamine because monoaminergic cells opportunistically sequester exogenous monoamine. Newly fluorescent cells are referred to as nonfluorogenic (nf-MA) cells and constitute a third general class of cells described within this study. The distribution of cells by cell type within each ganglion is summarized in Table 1. Monoaminergic patterning of these cell types is described in detail for the prothoracic ganglion and the brain.

Prothoracic Ganglion

The prothoracic ganglion is the only segmental ganglion in *Rhodnius* that is not fused, and distributions of monoaminergic cells within it are taken to represent primitive organizational patterns within this insect nervous system. The PG contains 11 fluorogenic cells and 18 nonfluorogenic (dopamine-sequestering) monoaminergic cells. Six distinct monoaminergic cell sets are identifiable in this ganglion.

The most conspicuous cell type in this and all other ganglia is the intensely fluorescent cell (IFC) type (Fig. 1). The IFCs are interganglionic interneurons that display intensely green, catecholaminergic fluorescence (Flanagan, in prep.). There are three dorsomedially distributed IFCs in the PG; two lateral and one medial. Only the medial IFC shows an extensive and highly diffuse innervation of the PG. In vinblastine-treated animals, these processes are conspicuously beaded (Fig. 2). Interganglionic projections from IFCs will be described in detail elsewhere (Flanagan, in prep.).

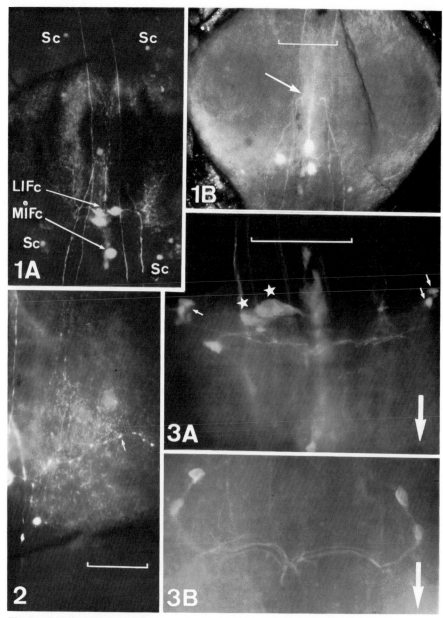

Fig. 1 A, B. Photographs of the three intensely fluorescent cells (IFC) and their primary neurites in the prothoracic ganglion showing lateral (*LIFc*) and medial (*MIFc*) subtypes and also the less fluorogenic satellite cells (*Sc*). **B** emphasizes the local pattern of the IFC projections. *Arrow* indicates the dorsomedial convergence of the primary neurites from the three cell types. Bar = 50 μm

Fig. 2. *Arrow* indicates the transverse branch of the medial IFC subtype after vinblastine treatment. Note varicosities along the fiber. Bar = 50 μm

The PG also contains an anterior and posterior pair of small (5 μm) cells (satellite cells: Sc) situated at the periphery of its dorsal surface. Their green fluorescence is typical of a catecholaminergic reaction product, but they are distinctly less fluorescent than the IFC. Even in heavily dopamine-loaded preparations, no distant projections arise from these cells, and they appear to be strictly local.

A third group of cells (yellow cells: YC) display a serotonin-like, rapidly fading yellow histofluorescence in fresh whole mounts and are also resolved by immunocytology for serotonergic cells. The PG has four YCs, which always occur as doublets at the posterolateral margin of the ganglion. Their processes, which can be seen after dopamine loading (Fig. 3A), project medially to converge with proximal processes from the IFC and then continue contralaterally to innervate anterior PG neuropil. Immunocytological staining shown in Fig. 3B corroborates this pattern. Terminal YC fields overlapping with terminal fields of medial IFCs show that much of the neuropil receives both serotonergic and catecholaminergic innervation.

Eight rapidly fading yellow cells are also found as doublets in the anterior and posterior fused meso/metathoracic ganglion. They project medially in close association with fibers from the IFCs (Fig. 4), and recapitulate the pattern of YCs found in the PG. Yellow cells are more persistent pattern elements than IFCs and flanking nf-MA cells in the MG. There is no evidence of any additional YCs (of abdominal origin) having been recruited into the fused MG.

Results of Exogenous Dopamine Loading. When the PG are loaded with exogenous dopamine, three additional cell sets become fluorogenic (Fig. 5). The most conspicuous of these normally nonfluorogenic monoaminergic (nf-MA) cell sets is situated dorsomedially and contains six to eight large ($>10 \mu$m) cells. Fibers from this dorsomedial nf-MA cell set project anteromedially, converge, and branch bilaterally into a transverse "common" tract before leaving the PG through the major ganglionic root. These are the only monoaminergic cells in the PG that clearly project to the periphery. Figure 6 also shows some results of dopamine loading on the fused meso/metathoracic ganglion.

A pair of cells flanking the IFC also become fluorogenic in dopamine-loaded preparations. They lie in a focal plane just below the dorsomedial nf-MA cells, isoplanar with the IFC. They project anteriorly, cross the midline, and continue in contralateral tracts into the anterior ganglion. Flanking

Fig. 3. Dopamine-loading allows photographic recording of the yellow cells (indicated with *arrows* in **A**) and their primary neurites. Immunocytological staining with antisera directed against serotonin stains these same cells (**B**). Note the medial convergence of the YC projections. In **A**, two dopamine-loaded nonfluorogenic dorsomedial cells are lightly stained and indicated with *asterisks*. Heavy arrow points anteriorly. Unless so indicated anterior is up in other figures. Bar = 50 μm

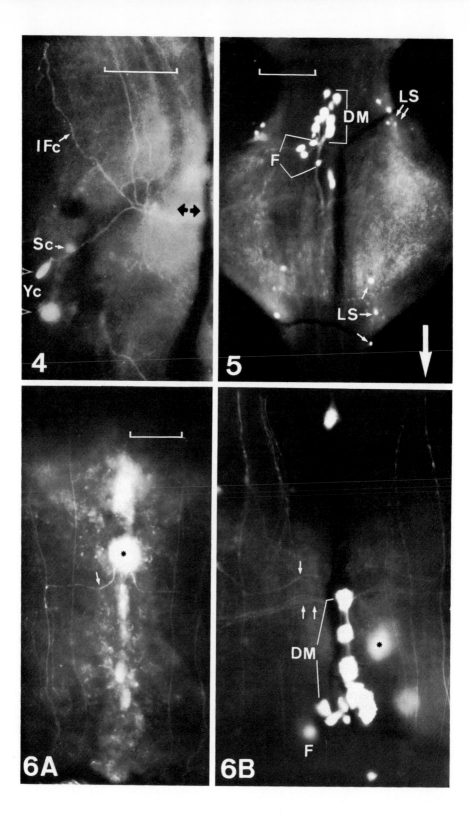

cells, like yellow cells, have only contralateral projections. They are inter-ganglionic monoaminergic interneurons. Their native transmitter is un-known.

The third set of nonfluorogenic monoaminergic cells visible in dop-amine-loaded PG resembles the satellite cell set. A group of six small (5 μm) cells distributed within the dorsoanterior aspect of the PG are regularly de-tectable. An additional pair of these cells is often seen adjacent to each yel-low cell set. They all appear to lack remote projections. Their native trans-mitter is unknown and they are referred to as lesser satellite, nonfluorogenic monoaminergic cells.

All identifiable monoaminergic cells are dorsally distributed within the PG (Table 1). Cellular positions and soma size correlate with the extent of cellular projection. Three dorsomedial sets of large cells, and only these three sets, project to remote target sites; whereas only lateral cell sets project exclusively to target fields within the PG. The distribution of monoamin-ergic somata can be compared with distributions of histologically recogni-zable neurosecretory cells and of cells that stain with the vital dye neutral red (see Fig. 7) which selectively stains monoaminergic cells in some in-vertebrate systems (Stuart et al. 1974; Lent 1982; Evans 1980; O'Shea and Adams 1981; Chap. 13). In *Rhodnius*, the dorsomedial nf-MA cells appear stained, but staining is not detectable in any other monoaminergic cell type. Neutral red appears to stain a combination of neurosecretory and mono-aminergic subsets in this insect.

Summary of Suboesophageal Ganglion Cell Distribution

Ganglia anterior to the thorax become specialized for cephalic functions. The suboesophageal ganglion (SOG) retains most of the pattern themes seen in the PG but also displays some unique designs.

Fig. 4. Medially projecting YC fibers and peripherally projecting IFC fibers travel through a shared path (*double-headed arrow*). *Yc* yellow cells; *Sc* satellite cells; *IFc* intensely fluores-cent cells. Bar = 50 μm

Fig. 5. Dopamine-loaded prothoracic ganglion displays two novel cell types near the IFC: flanking nonfluorogenic monoaminergic cells (*F*), and dorsomedial nonfluorogenic monoaminergic cells (*DM*). A third class of nonfluorogenic monoaminergic cells revealed by this method are called lesser satellite (*LS*) cells. Bar = 50 μm

Fig. 6A, B. The single intensely fluorescent cell of the mesothoracic ganglion is indicated by an *asterisk* in both **A** and **B**. The transverse arch of the IFC is indicated by a *single ar-row*. Fibers descending from prothoracic and suboesophageal IFC are also apparent. In dopamine-loaded mesothoracic ganglia (**B**), flanking cells (*F*) and dorsomedial (*DM*) cells are also apparent. The transverse arch of the DM cells is indicated with *double arrows*. Bar = 50 μm

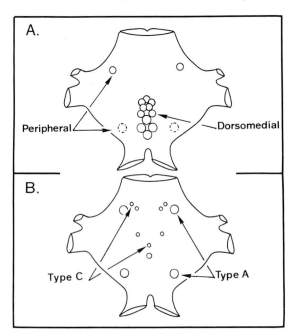

Fig. 7. Neutral red staining of prothoracic cell types (**A**). Broken lines indicate ventral cells Neurosecretory cell staining of prothoracic cell types (**B**); redrawn after Baudry (1968)

The SOG contains three IFCs analogous to those seen in the PG, two lateral and one medial. The projections from these cells resemble those seen in the PG (Flanagan, in prep.). Serotonergic yellow cells are detected immunocytologically and in dopamine-loaded SOG (Fig. 8). In this ganglion, YCs lying lateral to the IFCs appear as dispersed triplets, rather than the doublets seen in thoracic ganglia. However, their projection pattern is typical of yellow cell projections in the PG and MG. As in other ganglia, a single satellite cell is associated with this YC. An anterior satellite cell is also seen in the SOG, making the fluorogenic monoaminergic pattern in the SOG highly reminiscent of the PG pattern.

In the monoaminergic innervation of the SOG the basic PG pattern is retained and augmented with additional monoaminergic cell types, including cells contributing to catecholaminergic commissures (Fig. 9). The serotonergic doublet distribution pattern described in the PG appears in the SOG also, but serotonergic triplets suggest a bias towards an increased cell number in some yellow cell sets in this ganglion.

The Brain

The insect brain is formed from a fusion of embryonic ganglia and in adults has three major divisions: the proto-, deuto- and tritocerebrum. A paucity of

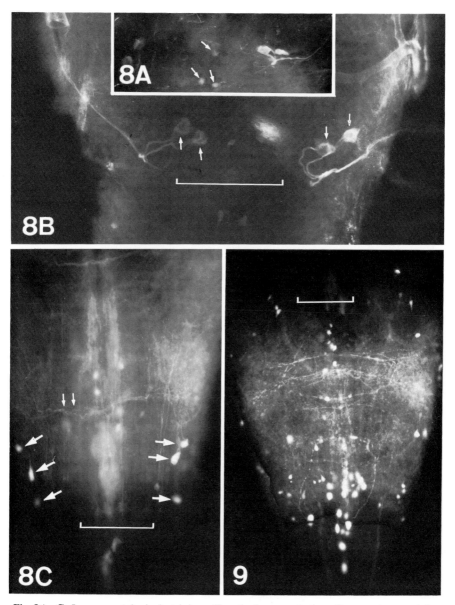

Fig. 8A–C. Immunocytological staining offers the best resolution of serotonin-containing yellow cells. Four giant YC and four YC (**A,** *arrows*) lie in the anterior of the suboesophageal ganglion. The giant cells (**B,** *arrows*) clearly project out of the central nervous system. More posteriorly, YC triplet sets are detected (**C,** *arrows*). The transverse arch within which these fibers project is marked with *double arrows*. Bar **B** = 100 μm; **C** = 50 μm.

Fig. 9. Dopamine-loaded suboesophageal ganglion displays numerous transverse commissures. Bar = 50 μm

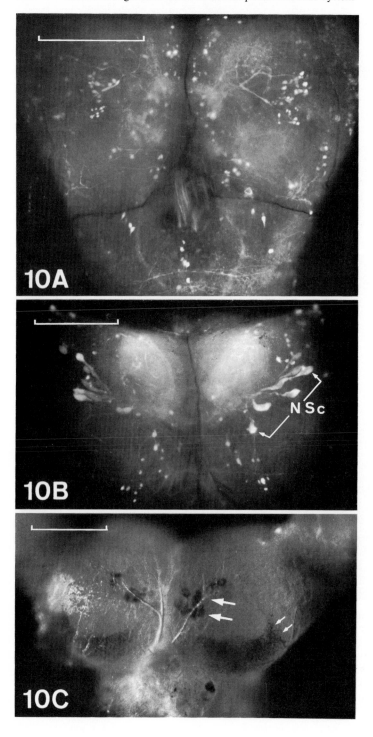

internal landmarks within whole-mount preparations and known differences in anatomical organization of insect brains across species makes direct monoaminergic comparison between *Rhodnius* and other species difficult. Likewise, pronounced differences of organization do not allow simple comparisons between brain and ventral ganglia. Dopamine-loaded brain whole mounts offer greatly reduced resolution of nonfluorogenic monoaminergic cell sets. Given these constraints, pattern elements detectable within both brain and ventral ganglia include (1) doublet sets, (2) local and remote innervations, (3) ipsilateral, bilateral and contralateral projection types, (4) similarities of soma size, number, and, in some cases, position, and (5) some limited resolution in differences of fluorescent intensities. Although the brain is architecturally more complex than the ventral ganglia, its fluorogenic cell sets conveniently project as groups, thereby facilitating the mapping of monoaminergic and other projections (Fig. 10). Furthermore, starvation or vinblastine treatment can enhance the fluorescence in certain peptidergic cells such as the median neurosecretory cells in the pars intercerebralis (Fig. 10 B) while depressing fluorescence of other normally fluorogenic cells. These cephalic neurosecretory cells are also unusual in that they stain with neutral red (Fig. 10 C).

Discussion

In the prothoracic ganglion in *Rhodnius* six cell-type sets are distinguishable (Table 1): (1) serotonergic yellow cells (YC); (2) the intensely fluorescent catecholaminergic cells (IFC), which are probably dopaminergic, and (3) the less fluorescent catecholaminergic cells (SC) which are probably noradrenergic. The dorsomedial (DM) nonfluorogenic monoaminergic cells (4) correspond in size, position, approximate number, projection pattern, and affinity for neutral red to previously described octopaminergic cells in cockroach and locust ventral ganglia (for review see Evans 1980). The dorsomedial nf-MA cells are therefore inferred to be octopaminergic. The transmitters of the flanking (F) nf-MA cells (5) and the lesser satellite (LS) nf-MA (6) cells are unknown.

The brain differs markedly from the presumably more primitive monoaminergic innervation pattern of the PG. Small (5 μm) weakly fluorescent catecholaminergic (noradrenergic?) cells are totally confined to a single, densely packed set within each optic lobe and the number of cells per cat-

Fig. 10 A−C. Fluorogenic cells and fibers in a brain whole mount (**A**). Starved or vinblastine-treated animals display a decreased fluorescence of most of these cells and an enhanced fluorescence of medial cephalic neurosecretory cells (*N Sc*; **B**). These conditionally fluorescent NSCs and neuropils also stain with the vital dye neutral red (**C**) (*double arrows*). Bar **A** = 100 μm; **B, C** = 50 μm

echolaminergic cell cluster is generally much greater in the brain than in the ventral ganglia. In addition, these cells project to target sites more circumscribed than do analogous cells within the ventral ganglia. The brain of *Rhodnius* also contains ventral catecholaminergic cells (as has been previously reported in histofluorescent studies of the cockroach brain; Frontali 1968) and cells that display unusual catecholaminergic responses to pharmacological treatments. The medial cephalic neurosecretory cells are the only neurosecretory cells in *Rhodnius* that displayed monoaminergic properties. The types of monoamine-handling cells are greater within the cephalic ganglia than within the thoracic ganglia.

The present method allows a rapid and simple screening of monoaminergic cell bodies in whole mounts of an insect central nervous system. Applied to the central nervous system of *Rhodnius* it showed that the catecholaminergic innervation of the brain was more abundant, i.e. the number of cells per cell set was greater than in the ventral ganglia. Furthermore catecholaminergic projections of the brain converge towards multiply innervated, circumscribed target sites. Such a feature may be interpreted as indicating a divergence in the evolution of the prothoracic- and brain-patterning strategies. The strategy of the ventral ganglia may reduce, whereas that of the brain may increase, the potential for further changes in patterning.

Acknowledgements I would like to thank Dr. B. Beltz for suggestions in determining fixation parameters appropriate for whole-mount immunocytology of serotonin in invertebrate ganglia. Funding for this study was provided by NSF grant PCM 76-80236 to A. Berlind and by Grants in Support of Scholarship from Wesleyan University. I also thank Renate Weisskirchen, European Molecular Biology Laboratory, for preparing the final plates.

Chapter 17

Identification of Neurons Containing Vertebrate-Type Brain-Gut Peptides by Antibody and Cobalt Labelling

Hanne Duve and Alan Thorpe

School of Biological Sciences, Queen Mary College, University of London
London, England

Nicholas J. Strausfeld

European Molecular Biology Laboratory
Heidelberg, F.R.G.

Introduction

Different chemical compounds such as cobalt chloride, heme proteins and fluorescent dyes (e.g. Procion yellow and Lucifer yellow) have proved invaluable for tracing neurons. They do not, however, reveal the identity of the chemical contents, whereas immunocytochemistry does. Immunocytochemical methods are inferior, however, for revealing the complete dendritic domain and axonal pathway of a particular cell due to a variety of reasons. First, with the thin sections $(5-10\,\mu m)$ obligatory for this technique, a curved and branched axonal pathway cannot be followed very far since the plane of section does not usually coincide with the axonal trajectory. Second, there are considerable variations in the metabolic activities of peptidergic neurons that are reflected in a changing distribution of material in both the perikarya and the axons. Factors such as age, diet, sex and, possibly, behavioural history are very important in determining how much of a particular pathway is stainable at any particular time (see also Chap. 13).

With these considerations in mind we thought it would be useful to combine immunocytochemical techniques with a method for labelling entire neurons. This would reveal the nature and distribution of the particular peptidergic type of material in a totally stained neuron within the central and peripheral nervous system.

A combined method must satisfy the following criteria: (1) The nervous system should be reasonably well-known in terms of its general organisation and the origins and projections of peripheral nerves. (2) Backfilling into axons should be aimed at neurons known to arise from a brain area that contains neurosecretory cell bodies. (3) Secondary labelling of antibodies must be compatible with the backfilling histochemistry. For example, FITC cannot be used in conjunction with methods that require dehydration, and PAP cannot simply be combined with cobalt-silver intensification (see Chap. 6).

In this chapter we demonstrate simultaneous labelling with a gastrin/cholecystokinin (CCK) antibody (L:112) and cobalt-silver of neurons be-

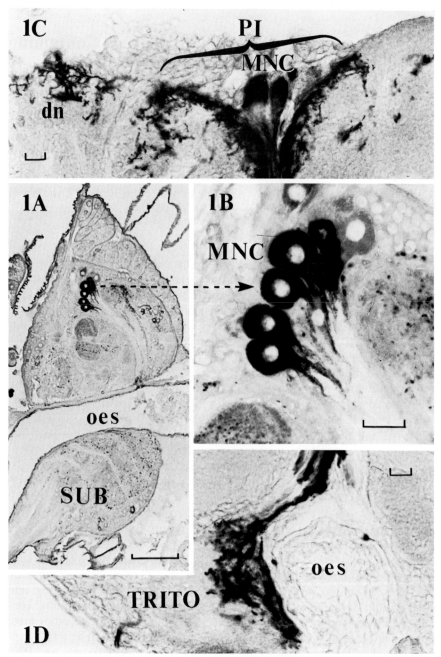

Fig. 1. Median neurosecretory cells (*MNC*) can be chemically defined by antibody stain-ing (**A, B**) but cannot be visualised in their entirety. Here only the cell bodies and axons are shown up by labelling with an antiserum to gastrin-CCK. On the other hand, cobalt back-filling into the cardiac-recurrent nerve, just proximal to the corpus cardiacum, reveals that

longing to the median neurosecretory cell group (MNC) in the pars intercerebralis of the blowfly, *Calliphora erythrocephala.* (A similar method, independently developed by Eckert and Ude in studies on another species, is described in Chap. 14.)

A variety of antisera raised against mammalian peptide hormones have shown cross-reacting material within specific neurons in insects (see Chaps. 12, 13, 14). The peptides within insect neurons may be variously related to the original mammalian antigenic material used to raise the antiserum (see Duve et al. 1982; Dockray et al. 1981). Several large neurons and some smaller ones in the MNC group are known to contain material that is immunoreactive to antibodies of the gastrin/CCK family (Duve and Thorpe 1981). It is also known from earlier "classical" neurosecretory studies on the MNC that at least some neurons arising in the pars intercerebralis project via the median bundle to the anterior part of the corpus cardiacum in the thorax (Thomsen 1965). The neuroanatomy and neurochemistry of these neurons suggested to us that they would make an ideal experimental model for developing a double-labelling technique. We wanted to examine back-filled neurons and the distribution of immunoreaction products in the axons of the median bundle and in the neurons of the MNC. Comparison should not be *indirect*, as in Fig. 1A−D, but *direct* using the same preparation. Thus, it was decided to introduce the cobalt chloride solution at the anterior tip of the corpus cardiacum in order to backfill axons originating in the brain and then to carry out the immunocytochemical treatment, hoping that cobalt filling, sulphidation, and the final silver intensification treatment would not prevent visualisation of the immunoreaction products.

Method

Flies must be old enough to allow preparation and isolation of minute peripheral nerves without the presence of excess haemolymph. The age, sex and diet of the flies must be controlled since these factors can influence the amount of immunoreactive material in the neurons. We used male or female blowflies, at least 3 weeks old (posteclosion) which had been fed on a diet of sugar, water and protein-rich Ovaltine.

Each fly was immobilised dorsal-side up with its head tilted forwards and down on a wax bed. The corpus cardiacum-corpus allatum complex was exposed by removing cuticle from the neck and prothorax. This complex is

---

many neurons arise in the pars intercerebralis (*PI*). **C** and **D** illustrate such a preparation photographed immediately after silver intensification (water immersion Zeiss×25 Neofluar). Although cobalt-silver precipitate shows up entire cells in the PI including their dendrites (*dn*) and their projections to the tritocerebrum (*TRITO*), it provides no information about the neuron's chemical identity. *SUB* suboesophageal ganglion in sagittal section; *oes* oesophagus. Bar in **A** = 100 μm; **B**, 25 μm; **C, D**, 10 μm

immediately recognisable in older flies from its pale metallic blue sheen (of massed neurosecretory materials) in incident light. After the gut was pinned to one side, the corpus cardiacum was supported on a thin (100 μm) sliver of black plastic (the thinnest type of plastic trash bag does nicely) which was slipped between it and the aorta. The area was flooded with O'Shea-Adams TES Ringer (see Chap. 7) and the cardiac-recurrent nerve was gently teased free of connective tissue. In other experiments where we filled the abdominal and cardiac nerves, a similar routine was used prior to isolation of the nerve in a vaseline trough. The Ringer solution was then removed and a vaseline trough was made around the corpus cardiacum from a small-bore syringe and tamped onto the black plastic. The trough was flooded with distilled water to check for leaks and the water was replaced by 5% aqueous cobalt chloride or by 5% cobalt chloride complexed to lysine (Görcs et al. 1979), prepared by adding 5 ml 10% $CoCl_2$ to 5 ml saturated lysine (or proline) dissolved in distilled water. To test for free cobalt a drop of this solution was added to a weak solution of ammonium hydroxide. If a pink flocculent precipitate appeared (indicating free cobalt ions), another 0.5 ml lysine was added to the solution.

After the trough was flooded with cobalt solution, the corpus cardiacum was severed from the cardiac-recurrent nerve and the trough sealed with a square of Parafilm. Each preparation was left for 4−6 h at 4 °C in a damp chamber.

After cobalt diffusion the head capsule was snipped open between the eyes and dropped into 0.01% $(NH_4)_2S$ in cold 70% ethanol to precipitate cobalt sulphide. After 3 min the head was washed twice for 5 min in cold 70% ethanol and fixed in aqueous Bouin's fluid for 3−6 h. Afterwards the brain was removed, dehydrated and rapidly (over 2 h) embedded in paraffin and serially sectioned at 5−10 μm. All further treatment was carried out on the slide.

For immunocytochemistry we used an anti-gastrin/CCK, COOH-terminus-specific antiserum L:112 (kindly provided by Prof. G. J. Dockray, Dept. of Physiology, University of Liverpool), which has previously been used in immunochemical and immunocytochemical studies on *Calliphora* (Dockray et al. 1981; Duve and Thorpe 1981). The antigenic determinant is exclusively within the common COOH-terminal tetrapeptide of gastrin and CCK. The basic methodology for the immunocytochemical study of insect neuropeptides has already been described (see Chap. 13). In this particular study, the primary antibody was used overnight at a concentration of 1:1000 at 4 °C and was followed by a secondary sheep anti-rabbit, rhodamine-conjugated fluorescent antibody applied at a concentration of 1:200 and incubated for 1−2 h at 4 °C. A rather large amount of rhodamine-conjugated immunoreaction product was necessary for it to fluoresce through heavily cobalt-silver-labelled neurons. Excess fluorescent antibody was removed in several washes of Tris-HCl buffer, pH 7.2, and the slices were mounted in buffer for preliminary examination by means of UV illumination. Fluores-

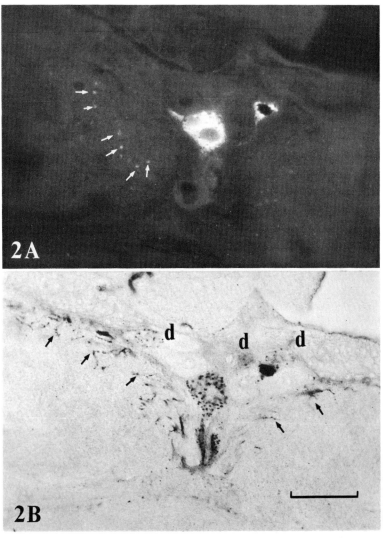

Fig. 2 A, B. Double labelling with L:112 antiserum and cobalt. After silver intensification, tissue can be briefly checked for successful fills with a low-power water immersion lens as in Fig. 1 C, D. However, it is convenient to use immunocytochemistry straight away. Here two MNCs are shown by direct illumination to contain silver precipitates within the cell bodies, axons and dendrites. Interestingly, the fluorescent blebs in the protocerebrum (*arrowed* in **A**) do not correspond to cobalt-silver filled dendrites (*arrowed* in **B**). They are derived from another gastrin-CCK cell system that resides exclusively in the brain. The nonfluorescent cobalt-filled cell bodies (*d*) belong to MNCs without gastrin-CCK that go to the corpus cardiacum. Antibodies to other peptides, such as α-endorphin, are known to reside in certain MNCs (Duve and Thorpe 1983). (Reproduced by kind permission of the Journal of Neurocytology) Bar = 25 μm

cent cells matched the distributions previously described by PAP labelling (Duve and Thorpe 1981). Figure 2A shows one large and one small cell within the MNC group containing immunoreactivity towards L:112 antibody.

To visualise cobalt distribution in neurons, the cobalt sulphide deposits within the cells were silver-intensified according to the method of Tyrer and Bell (1974). The silver-intensified slides were dehydrated, mounted in Fluormount and examined by normal tungsten illumination or by UV illumination or a combination of both.

Interpretation of the Results

Figure 2B shows two fluorescent cell bodies, partly labelled with cobalt-silver. Certain cells within the group do not show any immunoreactivity or cobalt uptake whereas some others contain cobalt but are not immunoreactive.

Visual comparison of the same cells after two separate procedures − immunocytochemistry and silver intensification of cobalt sulphide − can be done only when the number of cells is limited and their distribution is not too complicated. These restrictions can be circumvented by examining the tissue using the two methods simultaneously. This was achieved by carrying out immunostaining and silver intensification of cobalt before microscopic examination of the tissue (Fig. 3A−D). When such a preparation is examined under UV illumination (Fig. 3A), certain large and small cell bodies in the MNC group are revealed to be immunoreactive to gastrin/CCK COOH-terminal-specific antibody. Some of the cells showed bright fluorescence (Fig. 3A, a and b), others showed faint fluorescence (Fig. 3A, c) and some showed none at all (Fig. 3A, d). On the other hand, some of the immunoreactive cells did not show any cobalt-silver content (Fig. 3B, a) indicating that they contribute to peptide-containing systems confined to the brain (Duve et al. 1983). A notable result of backfilling the MNCs is that the cobalt-silver deposits can vary considerably. Thus, within the group, some cells are very densely filled (Fig. 3B, c) and some are sparsely filled (Fig. 3B, b).

Direct comparison of the results of cobalt filling with those of immunostaining can be quite difficult for such a complex group of cells, even when one switches quickly from one form of illumination to another. We found it useful to combine epi-ultraviolet illumination and normal tungsten illumination from below. This visually superimposes immunoreactivity (fluorescence) and cobalt-silver deposits and provides a quick and easy check of whether a particular cell is both immunoreactive and backfilled. This is demonstrated in Fig. 3C in cell group b and to some extent c, where the very heavy deposits of cobalt-silver almost obscure the fluorescence.

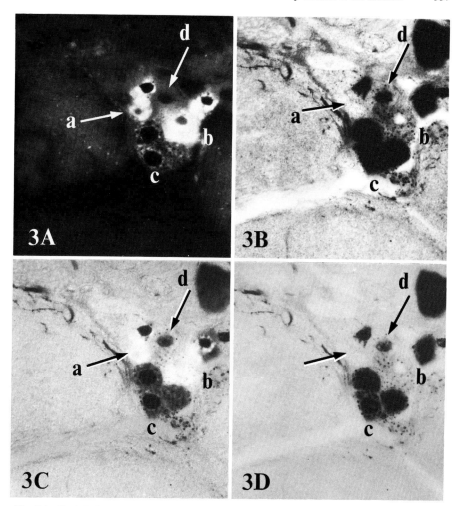

Fig. 3 A–D. MNCs are difficult to match with immunocytochemically stained neurons in some parts of the pars intercerebralis simply because there are so many cells present. Double marking resolves this problem. In this series of photographs of the same preparation there are three kinds of gastrin-CCK immunoreactive neurons. Those that contain no cobalt (*a*) and are exclusive to the brain; those that contain low amounts of cobalt (*b*) and those that contain large amounts of cobalt (*c*) that almost mask the fluorescence. Other cells (*d*) contain cobalt but are not reactive to gastrin-CCK antiserum and still others contain neither cobalt nor antigens related to gastrin-CCK. Presumably these belong to other peptide-containing systems that do not leave the brain. **A** Epi-UV illumination. **B** Tungsten illumination. **C** Combining tungsten illumination and fluorescence epi-UV illumination allows a direct check of cobalt- and rhodamine-labelled neurons. **D** Red filter with tungsten illumination eliminates brown background "noise" incurred by silver-reduction reaction products which can cause a rather grainy appearance (**B**). (Reproduced by kind permission of Journal of Neurocytology) Bar = 25 μm

Conclusions

The structure and function of the insect central nervous system has been much better understood since the development of neuronal filling techniques. These methods, however, contribute little to our knowledge of neurochemistry. Conversely, immunocytochemical techniques reveal something of the nature of substances in some parts of a neuron without saying very much about its projections, the dendritic domain or its inputs. The combined application of both techniques may help to bring together these two important facets of neurobiology, particularly if they can eventually be used in combination with other double markers (see Chap. 6). In so doing, they may help to elucidate intriguing problems about the specific functions of the newly discovered neuropeptides within systems of identified neurons.

Acknowledgements. We are pleased to acknowledge the support of an EMBO short-term fellowship to H. D. and A. T. for work at the European Molecular Biology Laboratory's Neurobiology group. We gratefully acknowledge support of the Science and Engineering Research Council of Great Britain in the form of a grant to A.T. (GR/B88136). Renate Weisskirchen (EMBL) printed the photomicrographs.

Chapter 18

Interpretation of Freeze-Fracture Replicas of Insect Nervous Tissue

STANLEY D. CARLSON, RICHARD L. SAINT MARIE
and CHE CHI

Department of Entomology, University of Wisconsin
Madison, Wisconsin U.S.A.

Introduction

Freeze-fracture (FF) replication is an important method for visualizing, at high resolution and in three dimensions, ultrastructural topography of membranes and organelles in hydrated cells without the intervention of embedding media and with little or no chemical fixation. The tissue replicas provide extensive vistas of cleaved membranes. This remarkable technique, which splits unit membranes, also permits resolution of macromolecular architecture of biological membranes particularly with reference to the location of membrane proteins. When correlated with thin section EM, FF confirms and greatly extends our knowledge of specialized intercellular contacts. Observations of the latter provide essential data for cell biology: intercellular adhesion, low resistance metabolic and electronic coupling, chemical transmission and paracellular occluding barriers. The FF technique has also proved invaluable for understanding the nature and location of electrical and chemical junctions.

The following section briefly summarizes the history of freeze fracture. Vignettes follow on four critical phases of this process: fixation, cryoprotection, freezing and fracturing. Following this background, a stepwise "recipe" is presented for freeze-fracture replication of insect nervous tissue as performed in our laboratory. The results of these procedures are illustrated from the visual neuropils and some are correlated with thin-sectioned material. Particular diagnostic features are demonstrated, with a commentary on the interpretation of freeze-fractured organelles and membrane specializations as seen in replicas. The chapter ends with a brief discussion of current trends and future prospects in FF technology.

Historical Perspective

The present state of freeze-fracture (FF) technique owes much of its success to the development of the freeze-fracture/-etching instrument (a freezing microtome within a vacuum evaporator) designed by Moor et al. (1961). This design was later licensed for commercial sale by Balzers. Some other less complex (and far less costly) instruments (e.g. Bullivant and Ames 1966) are enclosed in a bell jar that can be fitted onto most conventional vacuum evaporators. A current model freeze-fracture/-etch instrument is shown in Bullivant (1973), and a double replica maker used by our laboratory is supplied by Balzers. For more precise accounts of the development of FF technique see Mazur 1970; Steere 1973; Bullivant 1973; Stolinski and Breathnach 1975; Sleytr and Robards 1977; Rash and Hudson 1979.

Fixation

Both chemical fixation and cryoprotection can be eliminated if freezing is sufficiently rapid. Using a unique rapid-freeze machine, Heuser et al. (1975) have produced an artefact-free, freeze-fracture picture of synaptic vesicle exocytosis without the use of any tissue pretreatment. Unfortunately, good cell preservation is limited to an area no deeper than about 20 μm from the point of contact with the freezing source. Additional details about this method are discussed in later sections.

For most FF applications a minimal period of aldehyde prefixation is necessary to minimize cell necrosis during tissue preparation (see: Procedure-Tissue preparation p. 342). Problems arising from chemical fixation prior to FF are discussed by Stolinski and Breathnach (1975) and, Satir and Satir (1979). The use of cryoprotectants without prefixation are discussed by McIntyre et al. (1974).

Cryoprotectant

While the use and value of fixatives can be argued, the importance (and necessity) of cryoprotectant for most tissue is obvious. Two major cryoprotectants are glycerol and dimethyl sulfoxide (DMSO) — in that order of popularity. The latter compound (in buffer) has a limited shelf-life and we have not used it on insect nervous tissue. Stolinski and Breathnach (1975) cite DMSO for reducing possible freezing injury without interfering with etching at concentrations less than 40%.

Glycerol has long been a cryoprotectant of choice in FF work. This is not surprising given the number of insects that are able to supercool in cold environments because of the naturally occurring glycerol and glycerates in their bodies (Gilmour 1965). Glycerol is compatible in the phosphate-buffered solutions we use during the permeation prior to freezing. Lane and coworkers (e.g. Skaer and Lane 1974) have obtained excellent FF results of ventral nerve cords from locust and cockroach by incubating tissue in cacodylate buffer with 0.2 M sucrose and various glycerol concentrations (up to 20%) at room temperature for 30 min.

Freezing

The fundamental rule for freeze fixation is that it be accomplished rapidly. Nervous tissue, as is the case in other animal tissues, contains over 80% water and the formation of large ice crystals is ruinous to cell fine structure. At least two important conditions for successful tissue preservation are obtained if tissue is rapidly frozen: (1) the heat of crystallization is minimized, thus reducing the transient rise in temperature which permits the formation and migration of large ice crystals within the cells, and (2) the so-called "vitreous" ice is formed which has a density similar to liquid water and thus causes minimal or no tissue damage.

We freeze specimens by plunging them into liquid Freon-22 that is itself cooled to $-150\,°$C by liquid N_2. This indirect use of cooling capacity eliminates gas formation that occurs when a warm specimen contacts liquid N_2 directly. Vaporization at the specimen surface creates a shroud of insulation which decreases the tissue cooling rate to perhaps a tenth of the optimum rate at which vitrification should proceed. The use of an intermediate cooling agent, such as liquid Freon-22 (melting point $-150\,°$C), achieves an adequate cooling rate of $600\,°$C to $1000\,°$C/sec. In general practice, liquid-N_2-cooled isopentane and liquid propane may substitute for Freon although they are more dangerous to work with.

A case for even more rapid cooling can be made and a recent example (Maul 1979) shows differences in intramembrane particle distribution when cooling rates are suboptimal. Possibly the answer to the fast cooling rate problem is in the ultrafast freezing machine developed by Heuser et al. (1975). In this process, tissue is rapidly propelled onto a cold, (pure) copper block which is cooled to the temperature of liquid helium ($-269\,°$C). The tissue is then immediately (without chemical fixation) ready for the balmier climate of the FF apparatus.

Fracturing

Cryotome knives were first used to fracture specimens. However, in more recent instruments, thin, inexpensive "throw-away" razor blades are employed. The chilled blade chips into the frozen tissue and compresses it for a short distance before a cleavage plane is set up which then carries completely across the specimen. The fracture plane runs in an unpredictable fashion through the cytosol, but when it approaches any membrane system, the fissure then parallels or splits the membrane. Some disadvantages of knife fracture are: (1) the fracture plane cannot be controlled with any precision; (2) knife marks obliterate fine structure; (3) compressional forces and the release of frictional heat also alter fine structure. More detailed information about knife speed and position during various operations and its use as a cold trap during etching can be found in Hudson et al. (1979).

More recently, hinged double-replica holders have been employed to fracture tissue by applying tension to the specimen. This method eliminates knife marks along with compressional and thermal effects. Another clear advantage is the production of complementary surfaces. The complementary surface is lost when a knife is used. Drawbacks of the double-replica maker are that less relief is developed on the replica and the cost of this small, brass, hinged device is considerable. Since both the compression and extension techniques provide some H_2O and CO_2 contamination of the fractured surface, a choice cannot be based on this criterion alone.

Procedure

The following is a general protocol for successful freeze fracturing with modifications applicable to insect nervous tissue using a Balzers Apparatus BA 301 or BA 510. Although much of this methodology is borrowed, the steps presented below have yielded our best results. This is not to say, however, that modifications will not be called for in the future as ability and techniques evolve. The procedure below is presented in sufficient detail as to provide a useful starting point for the novice.

1. Tissue Preparation

1.1 Dissect out a larger than needed portion of the insect nervous system or, better yet, cut the insect open and pin it out to expose the desired area.

1.2 Submerge the tissue in a buffered (pH 7.2−7.4) physiological solution containing 2.5% glutaraldehyde.   1−2 h

1.3 Rinse twice in buffered physiological solution without glutaraldehyde.
                                                                    30 min each

1.4 Soak in buffered physiological solution containing glycerol:

10% glycerol	15 min
20% glycerol	15 min
30% glycerol	30 min

Use a rotating stage to keep the viscous glycerol agitated.

1.5 Mount tissue on the specimen holder.

The duration of the fixation period, and the temperature and composition of the fixative solution may each be varied depending upon the thickness, density and osmolarity of the particular tissue being investigated. In general, even for specimens of high cell density, such as the compound eye of *Musca*, one hour of fixation is often sufficient for adequate preservation throughout the tissue. Fixation periods of up to 3 h, however, have been used with no apparent adverse effects. Because glutaraldehyde diffuses slowly into tissue, we routinely fix eyes at 0 °C to retard tissue necrosis during fixation. One should be aware, however, that cooling tissue during preparation may, itself, introduce artefacts, i.e. aggregation of intramembranous particles at phase transitions within the membrane. Such temperature-dependent particle segregation has been observed within nuclear and mitochondrial membranes of certain mammalian cells (Wunderlich et al. 1974; Maul 1979). If fixation at room temperature is desired, it may be accomplished using buffered paraformaldehyde (2%), alone or in combination with glutaraldehyde (2.5%). Our own experience indicates that buffered glutaraldehyde, by itself, is preferable for studying intercellular junctions. When paraformaldehyde is incorporated into the fixative solution, the intramembranous particles associated with certain membrane specializations tend to fracture, more or less randomly, onto either of the two membrane faces. This indifferent partitioning makes subsequent detection and identification of intercellular junctions much more difficult (see also Staehelin 1973).

To minimize tissue damage resulting from dissection, final trimming of the tissue should be performed during the buffer washes following fixation, when the tissue is less pliable. At this stage, it is advisable to shape and orient the tissue block to fit the dimensions of the particular specimen holder being used. During trimming, it is also important to remove as much chitinous substance (tracheae, exoskeleton) as possible without disrupting the structures of interest. This additional care will become important during subsequent cleaning of the tissue replica.

When mounting the tissue on the specimen holder, care should be taken to avoid the formation of air pockets which could result in poor heat exchange during freezing as well as inadequate anchorage while fracturing. Adhesion of the specimen to its holder can be enhanced further by scoring the surface of the holder each time it is used. After mounting, excess fluid should be removed from the holder with a wedge of filter paper. The tissue itself, however, should remain moist and not be allowed to dehydrate or shrink before freezing. It is best to freeze immediately after mounting the tissue.

2. Freezing

2.1 Using clamped forceps, rapidly plunge the specimen holder and attached tissue into freezing (but not frozen) Freon-22. Swirl vigorously for 10 s.

2.2 Remove and quickly draw off excess Freon from specimen holder and forceps with filter paper. Store frozen specimen in liquid nitrogen (N_2).

Freon gas will condense in a metal cup if surrounded by liquid N_2. Special, double-chambered, insulated flasks are available commercially for this purpose or such vessels can be constructed using a wide-mouth Dewar vessel (see Chap. 5, Fig. 3, Vol. 1). Freon cooled in this manner will eventually solidify. Immediately before freezing a specimen, a pool of liquid Freon is made by applying a (clean) blunt metal object to the surface of the frozen Freon. The specimen should be plunged as rapidly as possible into the pool before Freon refreezes to prevent cold air, immediately above the Freon, from causing premature freezing of the tissue.

When removing excess Freon from the specimen holder, care should be taken not to let the tissue warm above $-100\,°C$. The safest procedure is to remove the specimen holder from the liquid Freon and place it immediately into the liquid N_2. The Freon will form an icy frost around the specimen holder. Remove the specimen holder from the N_2. As the Freon around the specimen holder begins to melt, draw it off rapidly with filter paper. Resubmerge the specimen holder in liquid N_2 and repeat as necessary until frost no longer forms.

Since specimens can be stored in liquid N_2 indefinitely (Stolinski and Breathnach 1975), they can be prepared in advance for several freeze-fracture runs.

3. Preparing the Freeze-Fracture Apparatus

3.1 Prepare and install carbon and carbon-platinum electrodes (Fig. 1 C, D)
3.2 *For microtome fracture only.* Clean a fresh knife blade with acetone to remove any oils. Install with razor edge parallel to the surface of the specimen stage.
3.3 Cover the empty specimen stage with aluminum foil.
3.4 Evacuate the specimen chamber.
3.5 Begin cooling the specimen stage with liquid N_2 when vacuum reaches 2×10^{-5} Torr.

Electrode holders should be cleaned of excess or flaking carbon deposits. A fine-wire brush or emery cloth may be used for this purpose. Fine particles or carbon dust may be removed with a jet of compressed air. Check and set spring tension of electrode holders. Do not handle electrodes, electrode holders, microtome, or the internal apparatus of the vacuum chamber with bare hands, as skin oils will contaminate the vacuum. All metal surfaces are cleaned with acetone.

Prepare carbon and carbon-platinum electrodes as illustrated in Fig. 1 A−D. Alternate electrode configurations may be used (see Moor 1959; Stolinski and Breathnach 1975). Use minimum length of carbon rod required for holder as this will result in a more thorough outgassing when the rods are heated prior to specimen fracture and replication.

The amount of platinum used must be determined empirically for each instrument. We generally use 6.5−8.0 cm of (0.1 mm diam.) platinum wire; evaporated at a distance of about 14 cm from the center of the specimen

Fig. 1 A–D. Carbon-platinum electrode. **A** Sharpen two carbon rods to a point (about 0.5 mm in diameter). **B** Pass platinum wire through an alcohol flame to clean and stiffen. Loop platinum wire at the junction of the two carbon rods. **C** Wrap half the wire tightly around the tip of one rod (cover an area of 2–3 mm from the tip). Use forceps to wrap wire; do not touch rod or wire directly with fingers. **D** Wrap second half of wire around the tip of the second carbon rod. Be sure that individual coils contact each other and the carbon rod

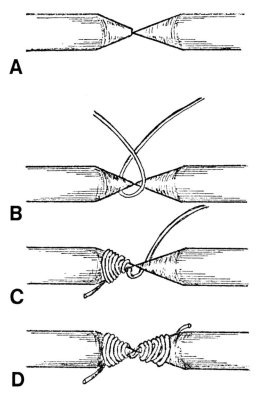

stage. Since this distance may vary from instrument to instrument, the amount of platinum used should be calculated in proportion to the square of the distance. An alternate method for platinum evaporation is the electron beam evaporator (available from Balzers Corporation). Such devices allow evaporation at a lower temperature than that of conventional evaporating electrodes, thus resulting in less radiant heat damage to the fractured surface (Moor 1971, 1973). Other advantages are that less platinum is required and the quality of replication can be consistently reproduced. Such devices, however, are costly and, in our opinion, the quality of the replica (in terms of grain size) is less satisfactory than that of the best replicas obtained using conventional evaporating electrodes.

To obtain a strong replica it is important that the carbon electrode be installed directly above the specimen stage so that carbon is deposited normal to the specimen surface. Conversely, the carbon-platinum beam should be presented to the specimen surface at a 45° angle to accentuate the surface relief. Some workers (Stolinski and Breathnach 1975) advocate an evaporation angle as steep as 60° to increase the amount of tissue surface receiving the platinum deposit. In our opinion, however, by casting shorter shadows, the three-dimensional topographic features produced at 60° may be less prominent, and important fine-structural details (e.g. individual intramembranous particles) may be less easily resolved.

4. Loading the Specimen Stage

4.1 Vent vacuum and open specimen chamber when specimen stage reaches −150 °C. Vent with dry nitrogen gas or air (dried with silica gel) to reduce water condensation on the cold, stage surface.

4.2 Remove aluminum foil and paint the specimen stage with liquid Freon. This will insure good thermal contact with the specimen holder and prevent accumulation of water condensate.

4.3 Quickly mount and orient specimen holder(s) onto the stage using forceps with tips precooled in liquid N_2. Position a piece of white cardboard near the stage to monitor film thickness (see subsection 6 − Fracture and Replication).

4.4 Evacuate specimen chamber immediately to minimize vapor condensation onto specimen stage.

5. Preparing to Fracture

5.1 When vacuum is at 2×10^{-5} Torr, slowly raise the temperature of the stage to −100 °C. The vacuum will deteriorate at this point as the Freon on the stage evaporates and any water condensate sublimes. Leave specimen stage at −100 °C until ready to fracture. This step is necessary to reduce specimen contamination during the fracture-replication phase. *Do not allow specimen stage to rise above −100 °C* at any time or ice within the tissue may recrystallize causing damage to the specimen.

5.2 *For microtome fracture only.* Begin cooling microtome with liquid N_2 when vacuum again reaches 2×10^{-5} Torr.

5.3 Outgas carbon electrode when vacuum reaches 2×10^{-6}. This is done by heating the electrode until it glows, but not enough to cause carbon evaporation. Heating will release gases trapped in the porous carbon rods and thus reduce contamination of the vacuum when the electrode is subsequently heated for replication. Outgas electrode until vacuum begins to improve again (usually 1−2 min). Repeat this step for the carbon-platinum electrode.

5.4 Lower specimen stage temperature to −115 °C and maintain. Begin fracture and replication procedure at $1-2 \times 10^{-6}$ Torr.

6. Fracture and Replication

6.1 *For microtome fracture only.* When microtome is at liquid-N_2 temperature (−190 °C), begin to shave away the surface of the specimen(s) using slow, gentle strokes. Lower microtome approximately 5 μm for each stroke until the desired specimen depth has been reached.

6.2 *For microtome fracture only.* For the final cut, lower microtome 5 μm and fracture rapidly. After fracturing the specimen(s) completely, halt the microtome arm immediately above the specimen stage. The microtome arm, which is colder ($-190\,^\circ$C) than the specimen stage ($-115\,^\circ$C) will function as a cold sink, trapping residual gases around the specimen surface and thus improving the local vacuum. *Do not permit the microtome blade to touch the specimen surface after the final fracture has been accomplished.*

6.3 Heat carbon-platinum electrode to initiate evaporation. (A) *For microtome fracture:* move microtome arm away from specimen surface. (B) *For double-replica device:* open hinged specimen holder 180°. Continue evaporation until desired amount of platinum has been deposited (usually 5–10 s).

6.4 Heat carbon electrode and evaporate an appropriately thick layer onto specimen surface (5–10 s).

6.5 Allow electrodes to cool for a few minutes then vent vacuum with dry air or dry nitrogen gas. Maintain the specimen stage at $-115\,^\circ$C.

The most critical phase for specimen contamination occurs in the interval between fracture of the specimen and replication of the fractured surface. Contamination can be reduced, to some degree, by the cold trapping capacity of the microtome arm. Nevertheless, shadowing and replication should begin as soon as possible following specimen fracture. When initially heated to the point of evaporation, the carbon-platinum electrode will release potential contaminants into the vacuum chamber. It is thus advisable to continue to shield the specimen surface with the microtome arm for the first 1–2 s of carbon-platinum evaporation. If a hinged, double-replica device is used, the device should be opened 1–2 s after evaporation has begun. This source of specimen contamination can be reduced (but not eliminated) by outgassing the electrode prior to fracturing (refer to step 5.3). Evaporation of the carbon electrode (for replication) should be initiated immediately following the deposition of the platinum shadow to insure good adhesion between the two films.

The amount of platinum deposited onto a fractured specimen is important. If too little platinum is deposited, the replica will appear pale in the electron microscope and many details will be lost. If too much is applied, fine resolution will suffer. Moor (1973) recommends depositing a layer of carbon-platinum, 10–15 Å thick, for adequate contrast. Film thickness can be estimated during evaporation by using a quartz crystal monitor. Such devices are highly accurate, but expensive. We use a much cheaper (and less accurate) method which involves attaching a small (2 × 3 cm) piece of white cardboard to the specimen stage, near the level of the specimen. The card is positioned (during step 4.3) so that one part of the card is shaded from the platinum beam by the specimen stage. In this manner, film thickness can be visually estimated (empirically) during evaporation by comparing the darkening of the unshaded portion of the card to that of the shaded area. Either method (card or monitor) can also be used to determine the thickness of the subsequent, pure-carbon film. A layer of carbon 150−200 Å thick is recommended for structural stability of the replica (Moor 1973).

7. Freeze Etching

Although some authors use the terms *freeze fracture* and *freeze etch* interchangeably, the two procedures are not identical. Freeze etching is a process whereby water is allowed to sublime (under vacuum) from the surface of frozen tissue. The procedure may be employed following freeze fracture, before shadowing and replication are carried out, to expose the surface of underlying cell structures such as organelles or unfractured plasma membrane, not revealed by the fracture plane. Etching of fractured tissue is not required nor is it recommended for most freeze fracture work. The increase in water vapor, released by the prolonged sublimation process, can be a serious source of specimen contamination if allowed to recondense on the cold, fractured surface before it is replicated. For proper etching to take place, glycerol should not be used as the cryoprotectant (Staehelin and Bertaud (1971). For details concerning the use of freeze-etch replication and its associated problems consult Moor (1971), Staehelin and Bertaud (1971), Benedetti (1973), Rash and Hudson (1979).

If etching is desired, use the following procedural modifications:

7.1 Eliminate step 5.4 and maintain the specimen stage at −100 °C for tissue fracture at $1-2 \times 10^{-6}$ Torr.

7.2 After the final fracture of the tissue (step 6.2), maintain the tissue under vacuum (up to 1 min) before applying the platinum shadow. Continue to shield the specimen(s) with the cold microtome arm throughout this period. If a double-replica device is used, the specimen(s) must be protected by a liquid-N_2-cooled shroud during the etching phase that follows tissue fracture.

7.3 Shadow and replicate the specimen(s) (steps 6.3 and 6.4). Subsequent steps are the same as for conventional freeze-fracture.

8. Cleaning the Replica

8.1 Remove specimen holder(s) from the cold specimen stage and place in a small plastic Petri dish containing frozen methanol or ethanol (100%).

8.2 Leave Petri dish (covered) at room temperature for 30 min to melt and dehydrate the tissue.

8.3 Using a single-hair brush, trim away excess replica from around the fractured specimen surface and gently tease the specimen from its holder if it has not already floated free. These manipulations are done in the now-liquid alcohol.

8.4 Transfer the specimen (replica side up) to a drop containing a 50:50 solution of methanol (ethanol) and 5.25% sodium hypochlorite (commercial bleach) for 1–5 min. Do not handle the specimen with forceps or you may damage the thin replica. Instead, transfer on a small piece of filter paper. *Keep the exposed surface of the replica clean and dry after removing from 100% alcohol.* If the replica surface becomes wet, it will submerge in the cleaning solutions and be difficult to retrieve once the tissue has dissolved.

8.5 Transfer specimen to a drop of full strength, commercial bleach (5.25% sodium hypochlorite), again using a piece of filter paper. Leave in bleach for 24 h to completely dissolve the chitinous structures. Keep in a covered, humid chamber to prevent evaporation of the bleach.

8.6 Transfer cleaned replica using a wire loop or a fire-polished, Pasteur-pipette tip through three changes of distilled water.

8.7 Replica may be retrieved from the surface of the final drop of distilled water onto a Formvar-coated, slot grid, bringing the grid down from above, or from below using a mesh grid. Formvar-coated grids may be reinforced with a thin carbon film before the replica is mounted. However, the carbon film must be applied to the back of the Formvar film because the replica will not adhere to a carbon surface.

Techniques for handling and cleaning replicas vary widely among investigators and are dependent, to a large extent, on the tenacity of the tissue being digested. Single-cell suspensions, for example, can be thawed and digested in a matter of minutes. Replicas of most animal tissues can be cleaned in less than 2 h using full-strength, commercial bleach. Insect tissues, on the other hand, often contain chitinous material (cuticle, tracheae) which is resistant to conventional treatment. Our experience is that prolonged cleaning periods (up to 24 h) or increased temperatures (up to 60 °C for 2–4 h) are required to remove this troublesome chitin. If such procedures prove to be inadequate, a longer cleaning duration or the use of strong acids and bases may be employed (Steere 1973; Stolinski and Breathnach 1975).

Large tissue blocks frequently expand when thawed under vacuum, in the air, or in aqueous solutions, resulting in fragmentation of the surface replica. Such breakage may be minimized by using the frozen alcohol procedure (described above) which allows the specimen to thaw slowly while being dehydrated. The use of alcohols has the added advantage of removing lipid and oils from the tissue, and dehydrating the surface of the replica. If the replica surface is kept dry following the alcohol treatment, it will float on the surface of subsequent aqueous solutions where it is more easily handled. A transfer from the 100% alcohol into 50% alcohol: 50% bleach prevents the specimen from expanding during rehydration. Presumably, the bleach breaks down cell integrity before osmotic pressures can develop within the tissue.

Interpretation of Replicas

General Principles. The last step in FF technique is the observation and photography of the replica with the electron microscope. One then sees the faithful, (down to 20 Å) three-dimensional metal casting of the fractured tissue surface. Though many problems associated with interpretation of replicas have been solved, the practitioner still needs a detailed knowledge of the preparation conditions to anticipate particular kinds of artefacts. Several rules follow to aid in evaluating the present figures but the student of this technique will want to read more detailed sections on this topic by Benedetti (1973), Bullivant (1973), Orci and Perrelet (1975), Stolinski and Breathnach (1975), and da Silva (1979).

1. Branton (1966) determined that the two types of membrane faces were derived from cleavage at the hydrophobic midregion of membranes. Thus, the P (or protoplasmic)-face refers to the hydrophobic face of that leaflet closest to the cytoplasm. The E-face (or exoplasmic) refers to the inner face of the outer membrane leaflet. The polar surfaces of membranes (those which actually contact the cytoplasm or extracellular space) are rarely revealed in replicas without etching of the fractured surface.

2. In insects (like vertebrates) the P-face usually possesses most of the intramembranous particles (IMPs). This feature, however, does vary with regard to certain membrane specializations (i.e. gap junctions) and is dependent on the degree of fixation (discussed above in Procedure). As in other animals, the lipid surfaces themselves are generally smooth.

3. As the fissure progresses through the cell during cleavage, cytoplasmic areas are naturally broken in a random, cross-fractured way. Such fracture surfaces reveal little of the cytoplasmic ground substance that is exposed, although the cleaved remnants of the organelles suspended in the cytosol are frequently seen.

Freeze-Fractured Neural and Glial Tissue

We are studying the fly retina and its optic lobes. Its nerve and glial tissue demonstrate the usefulness of FF technique. Micrographs were selected to provide both P- and E-face views. In most cases, correlative thin-section micrographs are provided to orient the reader with regard to key ultrastructural features. To insure proper three-dimensional perspective, all FF figures should be viewed with platinum deposit coming from below. Encircled arrows are included in each FF micrograph to indicate the direction of the shadowing material.

The Cleaved Cell: A Survey

Figure 2 reveals two apposed retinula (photoreceptor) cells in the housefly retina and illustrates features of a freeze-cleaved and replicated cell. The E- and P-faces of the adjacent plasma membranes are readily apparent as is the intervening extracellular space. The replica also demonstrates the random nature of cleavage in the cytosol resulting in its coarse appearance. Mitochondria in this figure are cross-fractured. The cleavage plane also reveals portions of the nuclear membrane and smooth endoplasmic reticulum (ER).

Organelles

Most, if not all, organelles in insect neuronal and glial cells can be seen in the freeze-cleaved state. Examples of microtubules, smooth ER (Fig. 3A) and a cell nucleus (Fig. 4A) are shown. In the first case, a broad expanse of longitudinally-oriented microtubules and smooth ER fill a glial cell found between the retina and first optic neuropil of the fly eye. The outlines of these parallel-running organelles correspond to that seen in the companion thin section (Fig. 3B) of a similar cell with a common orientation. In Fig. 4A two features of the neuronal cell nucleus are especially noteworthy. (1) The P-face of the inner nuclear membrane is seen underlying the E-face of the outer nuclear membrane of this double-membraned organelle. As is the case for all double-membraned organelles (mitochondria and nuclei), the membrane leaflets apposed to the intermembranal space are referred to as the E-faces. Those apposed to the cytosol (protoplasm) and the lumen of the organelle (in this case the nucleoplasm) are referred to as the P-faces. The sequence of faces is thus P-E, E-P; the same sequence as that of two apposed plasma membranes. For single-membraned organelles (i.e., ER, Golgi, vesicles, etc.) the leaflet abutting the lumen is denoted as the E-face and that apposed to the cytosol as the P-face (Branton et al. 1975). (2) Nuclear pores are randomly and rather uniformly distributed and a concept of the frequency of these pores is easily gained from such a replica. Nucleoplasm is not exposed

---

Fig. 2. Replica of two retinula cells demonstrates key cytological features. The fracture plane has cleaved a nuclear membrane (*N*) in the left cell, and traversed the cytoplasm (*CY*), cross-fracturing several mitochondria (*M*) and some smooth endoplasmic reticulum (*SR*). Of main importance is the appearance of plasma membrane leaflets of the neighboring cells. *P* P-face; *E* E-face of adjacent cell. Between the seemingly juxtaposed leaflets is the intervening extracellular space (*ECS*). *Encircled arrow* direction of platinum deposit in this and all other FF micrographs. × 53,000

Fig. 3. A Replica of glial cell in the distal portion of the lamina. Cleavage is along long axes of parallel microtubules (*m*). Scattered cisternae of smooth endoplasmic reticulum (*sr*) run parallel to the microtubular array. × 60,000. **B** Thin section correlate of **A**. × 60,000

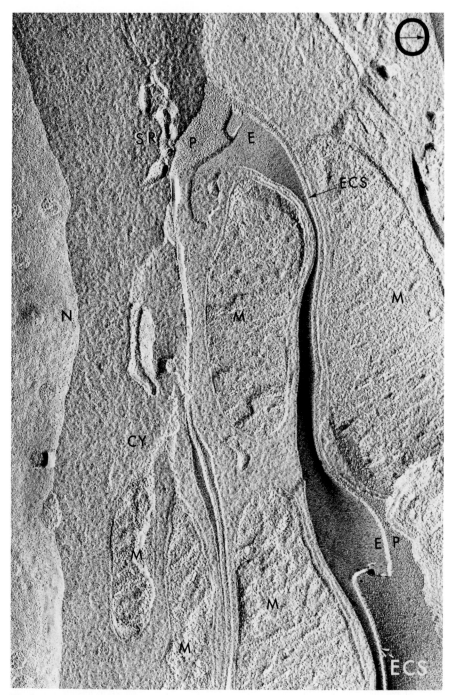

Fig. 2. Legend on page 351.

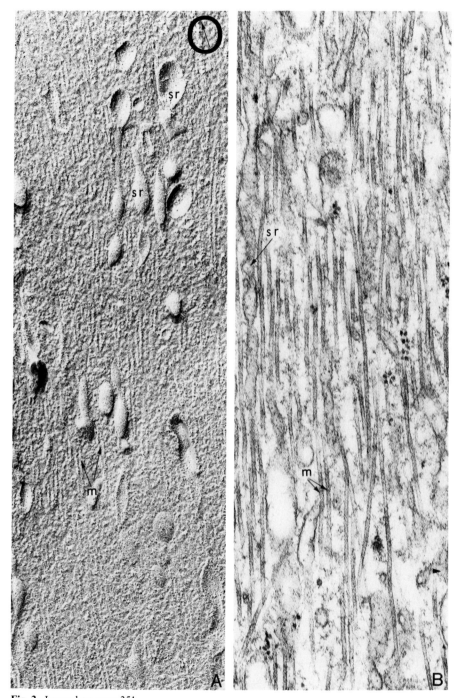

Fig. 3. Legend on page 351.

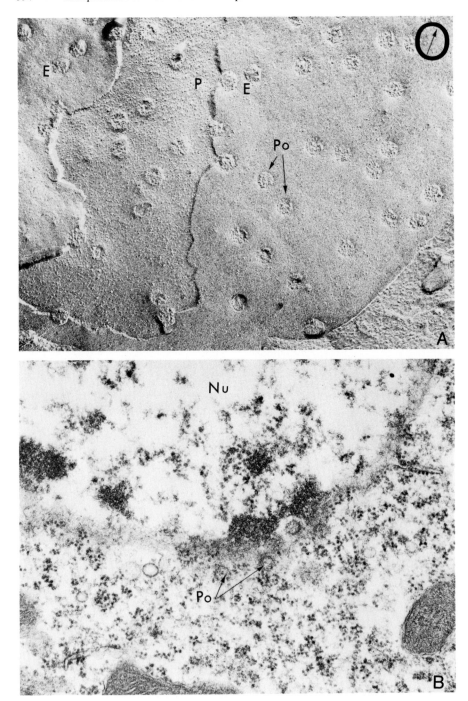

in this particular replica but it is usually featureless when revealed by fracture. In the counterpart thin section (Fig. 4B) nucleoplasm is visualized but even the fortunate tangential sectioning of the nuclear membrane has revealed only a relatively few pores. In nearly all of the replicas illustrated here a variety of organelles are seen to greater or lesser extent; e.g., pigment granules (Fig. 8A); mitochondria (Figs. 2, 9B); rhabdomeric microvilli (Fig. 5A).

Lest the reader be confused as to where many of these organelles are located or their relative sizes, the major cellular components of the distal ommatidium are presented in a survey replica (Fig. 5A). The cleavage plane carries across most of the optical components (lens, corneal and large pigment cells, Semper cells, pseudocone and photoreceptor cells). The periodic array of the rhabdomeric microvilli (photoreceptor organelle) is evident at this magnification. This and other such surface elaborations are discussed in detail at the end of this section. The structural features correspond to those seen in thin section (Fig. 5B).

Membrane Specializations

Perhaps the main advantage of freeze fracture is that it reveals great vistas of cleaved membranes. If special areas of cell contacts are present in the fracture plane, these will be readily apparent. They can be unequivocally identified by the particular array, alignment or formation of intramembranous particles on a given membrane leaflet. Several of the more prominent junctions found in the peripheral retina and first optic neuropil are shown in Figs. 6–9.

Gap Junctions. The presence of gap junctions is generally considered to be strong morphological evidence for ionic and metabolic coupling between cells. Such junctions are common in the retina and lamina between certain neurons and between glial cells. Gap junctions occur between the same and different kinds of glial cells indicating syncytial connections between them.

Details about gap junctions between identified neurons and glia in the fly lamina can be found in Chi and Carlson 1980a, b; Saint Marie 1981; Shaw and Stowe 1982; Saint Marie and Carlson 1983a, b. A gap junction (Fig. 6A, inset) is identified by the edentation of the membrane surface and

---

Fig. 4. A Replica of Semper cell nucleus in the distal retina. Nuclear pores (*Po*) are randomly situated. The double-membraned nature of the nuclear envelope is revealed. The P-face (*P*) lies above the nucleoplasm. The fracture plane also passes through the outer nuclear membrane, exposing its E-face (*E*). × 57,000. **B** Thin section of a nucleus (*Nu*) in a monopolar neuron. The nuclear envelope is sectioned obliquely so that no crisp membrane outlines are seen. Nuclear pores (*Po*) are thus sectioned en face. × 51,000

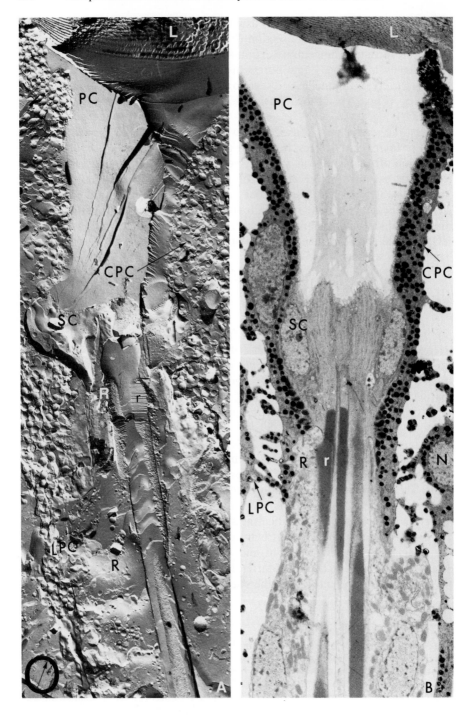

by the presence of small plaques of unordered, intramembranous particles on the E-face of the plasmalemma. Such preferential partitioning of gap-junctional particles onto the (usually particle-free) E-face in arthropods greatly facilitates their detection in FF. In Fig. 6A, the fracture plane has passed through the prominent perikaryal rind of the first optic neuropil, revealing a series of gap junctions between two monopolar neuron cell bodies (see also the small (5-particle) gap junction connecting adjacent retinula axons in Fig. 11A). In thin section (Fig. 6B), the presence of gap junctions between Type I and Type II monopolar interneurons is confirmed. Note that the thin section offers a very limited sampling of surface membrane and provides little information about the prevalence or distribution of gap junctions.

Tight Junctions. Tight junctions have recently been reported in a variety of insect tissues (reviewed in Lane and Skaer 1980). Of particular interest are findings which implicate tight junctions within the insect perineurium as the structural basis of the insect blood-brain barrier (Lane et al. 1977; Lane and Swales 1978a, b, 1979). Our own research indicates that tight junctions are widespread in the optic lobe of *Musca* where they occur between neurons, between glial cells, and between glia and neurons (Chi and Carlson 1980a; Saint Marie 1981; Saint Marie and Carlson 1983b; see also Lane 1981). One such example can be found between the marginal glial cells (Fig. 7). Cleaving the surface membrane of this type of glial cell exposes a system of aligned P-face ridges and E-face grooves which represent the FF view of focal tight junctions (see inset for thin-section view of these punctate, membrane fusions). These elongate particle ridges (grooves) usually run parallel to one another, and some form loops.

Tight junctions occur in the marginal cell layer as part of a more pervasive barrier system which completely surrounds the lamina and which isolates individual optic cartridges within the neuropil (Saint Marie 1981; Saint Marie and Carlson 1983a, b). Such a barrier system has been implicated in permitting inhibitory field interactions within the optic lobe (see also Laughlin 1974, 1981; Shaw 1975, 1979, 1981).

Septate Junctions. This membrane specialization is believed to function in intercellular adhesion although some reports suggest they form a partial occluding barrier (reviewed in Lane and Skaer 1980; see also Saint Marie 1981; Saint Marie and Carlson 1983a, b). Satir and Gilula (1973) and Lane and Skaer (1980) present three-dimensional illustrations of the septate junction

---

Fig. 5. A Survey of retinal epithelium of the housefly. The corneal lens (*L*) Pseudocone (*PC*) rests proximally on four Semper cells (*SC*) and is flanked by corneal pigment cells (*CPC*). Rhabdomeric microvilli (*r*) of peripheral retinula cells (*R*) are directly below Semper cells. Large nucleated (*N*) pigment cells (*LPC*) laterally shroud each ommatidium. × 3450. **B** Longitudinal thin section correlate of **A**. × 2800

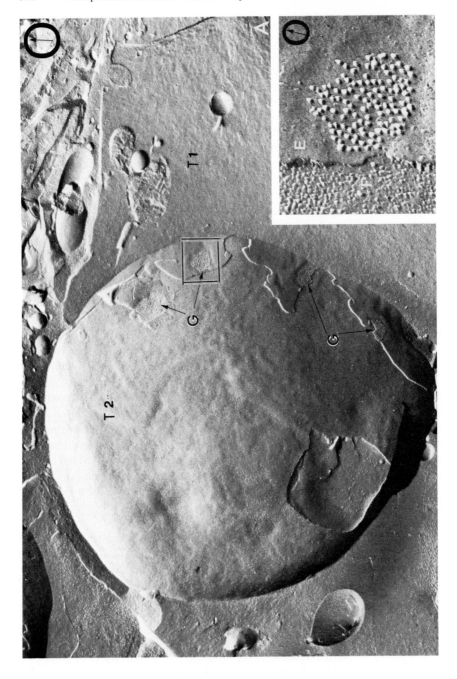

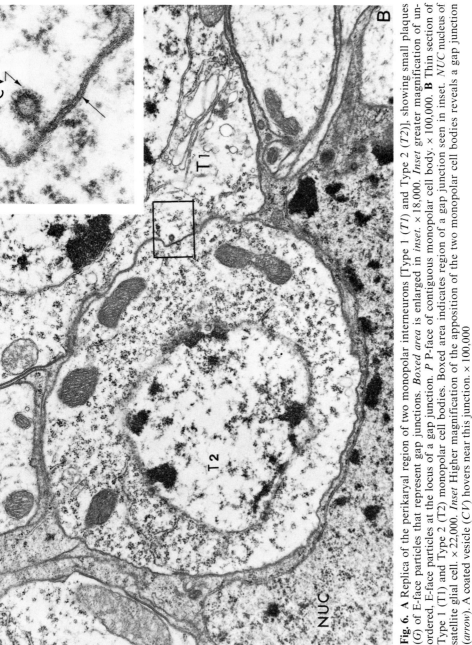

Fig. 6. **A** Replica of the perikaryal region of two monopolar interneurons [Type 1 (*T1*) and Type 2 (*T2*)], showing small plaques (*G*) of E-face particles that represent gap junctions. *Boxed area* is enlarged in *inset*. × 18,000. *Inset* greater magnification of unordered, E-face particles at the locus of a gap junction. *P* P-face of contiguous monopolar cell body. × 100,000. **B** Thin section of Type 1 (T1) and Type 2 (T2) monopolar cell bodies. Boxed area indicates region of a gap junction seen in inset. *NUC* nucleus of satellite glial cell. × 22,000. *Inset* Higher magnification of the apposition of the two monopolar cell bodies reveals a gap junction (*arrow*). A coated vesicle (*CV*) hovers near this junction. × 100,000

in which the pleated sheet (thin section) appearance of this junction is related to the parallel rows of P-face particles (freeze fracture). They propose that the intercellular septa "... are attached to the membranes at points coincidental with the rows of particles seen in freeze fracture" (Lane and Skaer 1980). In our example (Fig. 8) of the septate junction we turn to the base of the pseudocone where, four Semper cells are linked to each other and to corneal pigment cells by extensive septate junctions (Fig. 8 A). Also there are relatively large, laterally projecting digitiform processes which extend from each Semper cell and form a "tongue-in-groove" construction with neighboring Semper cells. The extracellular tracer, lanthanum, penetrates these septate junctions and in unstained thin sections (Fig. 8 B) outlines its "pleated sheet" construction.

Retinula Cell Junction. Retinula cells in the retina of *Musca* are joined to each other by a membrane specialization dubbed unimaginatively (Chi et al. 1979) the retinula cell junction (Figs. 9 A−C). This particular junction is probably a variant of the reticular septate and scalariform junctions, which have been extensively reviewed by Lane and Skaer, (1980). Fractured leaflets containing this junction reveal ridges of fused, P-face particles and (Fig. 9 B) E-face grooves. These ridges (grooves) are elongate, more or less parallel, and occasionally form circular enclosures (Chi et al. 1979). Though they resemble freeze-fractured tight junctions, the ridges and grooves of ap-

---

◀ **Fig. 7.** Replica of P-face (*P*) of marginal glial cells. Ordered membrane particles form ridges which frequently loop back on themselves (*single arrows*). Grooves (*G*) on the E-face (*E*) correspond to P-face ridges (*double arrows*). × 25,500. *Inset* Thin section of focal tight junctions (*arrows*) between marginal glial cells that correspond in location to the elongate parallel ridges seen on the P-face in the replica. × 150,000

Fig. 8. A Replica of cleavage plane through the base of the pseudocone (*PC*) revealing the ▶ P-face (*P*) of two Semper cells (*SC*) and E-face (*E*) of a neighboring corneal pigment cell (*CPC*). Wavy, parallel rows of membrane particles on Semper cell P-face are extensive septate junctions between *SC* and overlying corneal pigment cell. Note interdigitating processes (*I*) of the Semper cells and pigment granules (***) in adjacent corneal pigment cell. × 22,500. **B** Thin section, unstained. Lanthanum has penetrated through obliquely sectioned septate junction (*SJ* pleated sheet material) and is found between a pigment (*CPC*) and Semper cell (*SC*). Note adjacent gap junction (*GJ*) between the two cells. (***), osmophilic pigment granule. × 84,000

Fig. 9. A Thin section of retinula cell junction. Apposed plasma membranes of two retinula cells. Faint septa span the intercellular space. Mitochondria are parallel to, and near the plasma membrane. Tannic acid preparation. × 40,500. **B** Replica of retinula cell junction (counterpart to A) Numerous intramembranal particles on the P-face (*P*) some of which form elongate ridges. E-face membrane shows non-corresponding grooves. Abundant cross-fractured mitochondria (*M*) are aligned along the cell border. × 42,000. **C** Thin section of retinula cell junction. A central retinula cell (*Cc*) between two peripheral retinula cells. Note the scalloped borders and discontinuous cell contact characteristic of a junction slightly shrunken in mildly hypertonic fixative. × 16,000

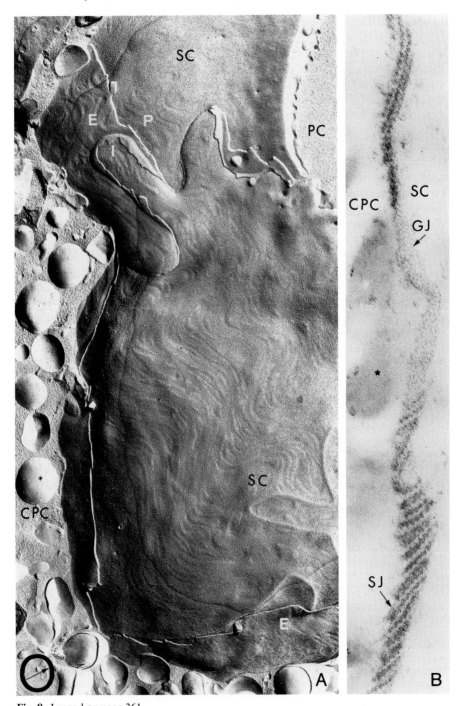

Fig. 8. Legend on page 361.

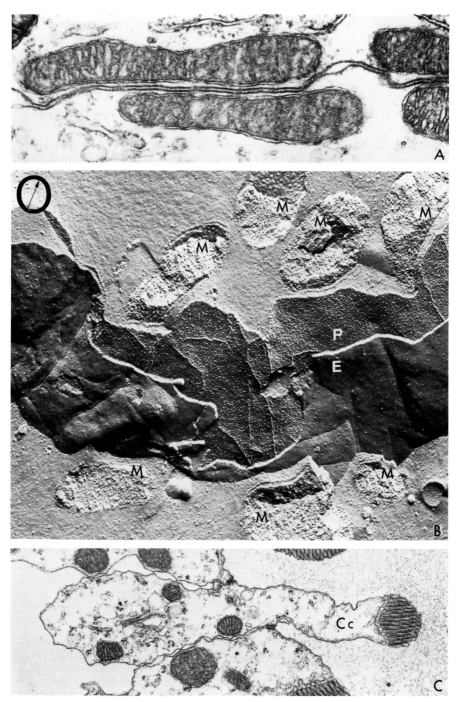

Fig. 9. Legend on page 361.

posed retinula cells rarely correspond (Fig. 9B). Similar junctions also lacking complementarity have been reported in the insect rectum where they are referred to as reticular septate junctions (Lane 1979a, b; Lane and Skaer 1980). It is problematic whether this junction has any adhesive qualities. When slightly hypertonic solutions are used during fixation, retinula cells shrink slightly and pull away from their neighbors (Fig. 9C). They take on a scalloped appearance and the distances between these membrane creases correlate with the spacings exhibited by adjacent P-face ridges. Chi and Carlson (1981) showed that lanthanum can penetrate between retinula cells.

Retinula cells exhibit a second type of intercellular junction which runs with the particle ridges of reticular septate junctions. Evidence of this specialized membrane can be found within the uniform (10−12 nm) cleft between thin-sectioned retinula cells (Fig. 9A). This junction contains very fine, barely distinct septae as well as a close association with mitochondria. In freeze fracture (Fig. 9B) the P-face exhibits a high density of IMPs. These features are similar to the mitochondrial-scalariform junctional complex found in insect tissues involved in ion transport (reviewed in Lane and Skaer 1980). It is possible that the retinula cell junction is involved in ion-pumping activities necessary for maintaining the extracellular cation pool.

Cell Surface Modifications

Several conspicuous modifications of the cell surface which are present in the fly retina and optic lobe are illustrated in Figs. 10−12. These stable features often considerably increase cell surface area and their visualization in freeze-cleaved cells adds structural (and sometimes functional) insights unobtainable by other EM techniques.

Gnarls. This term is applied to spherical glial processes that appear to be "pinched off" in series (Campos-Ortega and Strausfeld 1973). These invade only the β-processes of the optic cartridge (Burkhardt and Braitenberg 1976). In FF views (Fig. 10), IMPs appear to be associated with both the P- and E-face of this bulbous process. The correlating inset is of a tripartite gnarl entering a β-fiber. Dense particulate matter fills the space between glial process and nerve fiber. The gnarl, whose function is unknown, serves as a useful anatomical marker in pinpointing β-fibers in a replica. Another

→

Fig. 10. Replica of an epithelial glial cell intruding into the neuronal process of a β-fiber. This particular glial extension is called a "gnarl" (*GN*) several of these appear to bud from each other serially. The fracture plane reveals both faces [exterior (*E*) and interior (*P*)] of the gnarl. × 73,000. *Inset* thin section. Gnarls (*GN*) of a β-fiber: dense particulate material fills the interspace around it. Note presynaptic structure in α-fiber (*arrow*) close to gnarl. × 37,000

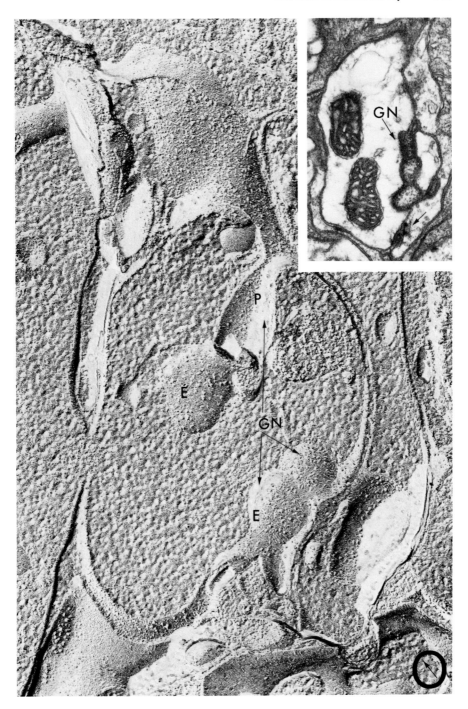

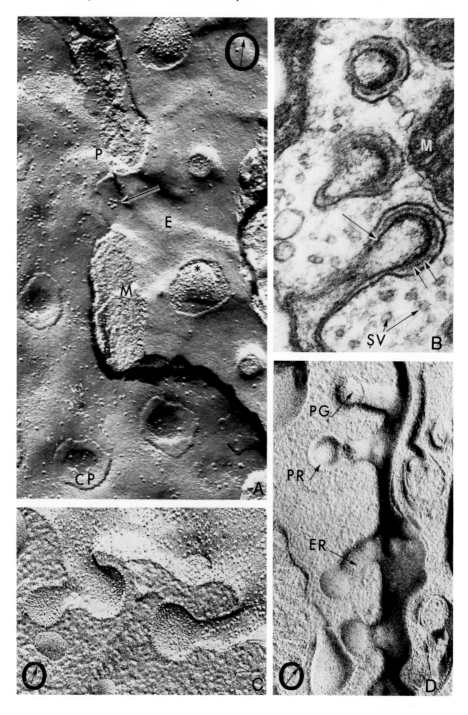

peculiarity is the rarely observed presynaptic T-bar (chemical synapse) opposite the gnarl in the inset. This so-called gliapse was first depicted by Boschek (1971).

Capitate Projections. These distinctive mushroom-shaped glial processes insert into retinula axons. Capitate projections were first described from thin sections by Trujillo-Cenoz (1965). Figure 11 A, C, D represents the FF views of these peculiar glial-neuronal intimacies. In the usual fracture, which follows the membrane plane of the axon terminal (Fig. 11 A), capitate projections are sheared off at their bases leaving "pockmarks" along the retinula axon cell surface, a convenient diagnostic FF marker for identifying the latter cell. These "pockmarks" may contain remnants of either P- or E-face membrane derived from the epithelial glial cell which gave rise to the capitate projection. A close look at the "pocks" shows the particle-rich nature of these tiny glial intrusions. IMPs are found on the E-face (Fig. 11 A) as well as the P-face (Fig. 11 A, D) of the glial cell at these loci. Rarely, cross-fractures show longitudinally exposed capitate projections (Fig. 11 C, D) revealing their profile and interior. Small P-face particles on the tip of the capitate projection (Fig. 11 C–D) may correspond to the electron densities observed in thin-sectioned material (Fig. 11). In the latter representation, an occasional capitate projection is longitudinally sectioned and seen in its entirety, whereas adjacent kindred structures are sectioned obliquely.

The Chemical Synapse

Perhaps the most important intercellular junction found in nervous tissue is the chemical synapse. Using a rapid freezing technique first proposed by Van Harreveld and Crowell (1964), Heuser and Reese (1979, 1981) and Heuser et al. (1979) identified active sites of synapses and isolated the various stages of vesicle exocytosis and recovery at stimulated frog neuromuscu-

---

Fig. 11. A Replica of two photoreceptor axons. Fragments of the E-face of an epithelial glial cell usually attach to the P-face (*P*) of an R cell membrane when capitate projections (*CP*) are cross-fractured. The P-face of the epithelial glial cell is exposed where capitate projections insert (*) into the E-face (*E*) of an axon. Membrane particles are numerous at these points. *Arrow* points to site of close contact between photoreceptor cells. A small cluster of five gap junction particles is seen on the retinula cell E-face. × 81,000. **B** Thin section. Three capitate projections invaginate a retinula terminal. Note increase in cytoplasmic density at hemispherical cap (*arrow*). A band of filamentous material (*double arrows*) is in the space surrounding the knob. Mitochondria (*M*) and synaptic vesicles (*SV*) are nearby. × 100,000. **C, D** Several longitudinally cleaved capitate projections showing stalk and cap invaginating into the retinula axon. **C** P-face of retinula axon. Small particles at the projection tip probably are insertion points of interspace filamentous material seen in the thin section. **D** Both E-(*ER*) and the P-face (*PR*) of the retinula axon and the P-face (*PG*) of the intruding glial cell. **C** = × 76,000; **D** = × 51,000

lar junctions. More recently, FF replication has been used to study the effect of stimulation at the photoreceptor synapse in the housefly (Saint Marie 1981; Saint Marie and Carlson 1982). Figures 12−13 show fracture faces of light-stimulated and unstimulated axon terminals from housefly photoreceptor cells at the sites where they make inhibitory (chemical) synapses onto monopolar interneurons in the first optic neuropil.

In the (unstimulated) presynaptic membrane (Fig. 12) the P-face is studded with the glial remains of sheared-off capitate projections. Of functional importance, however, are the "bowtie"-shaped active zones of the chemical synapses made by the retinula axon onto the second-order monopolar cells. The P-face of the retinula axon shows that these uniquely formed particle aggregations correspond well in location, overall size, distribution and en face geometry to the base of the presynaptic density as seen in thin section (Fig. 12 inset). This view of the retinula cell-monopolar cell synapse illustrates the power of correlative (FF and thin section) studies in identifying particular structures. In addition, data on the random orientation (note bars orthogonal to the long axis of the active zones) and distribution of these active sites would be difficult to obtain from thin sections. As a result of these studies, we estimated that each retinula axon makes a minimum of 175 such synapses (Saint Marie 1981; Saint Marie and Carlson 1982; see also Nicol and Meinertzhagen 1982; Shaw and Stowe 1982).

Active zones from light-stimulated retinula axons have small, P-face depressions on the axon just lateral to the active zone cluster of particles (Fig. 13). These punctate depressions, which rarely occur in unstimulated axons, resemble the vesicle attachment sites observed at the frog neuromuscular junction (Heuser et al. 1974, 1979; Heuser and Reese 1977, 1979). Saint Marie (1981) has also shown that the E-face of presynaptic membranes at stimulated photoreceptor synapses exhibit "dome-shaped protuberances" which have a central, crater-like depression indicative of cross-fractured vesicle attachment sites. Thus pit and elevation can be correlated as ultrastructural evidence of synaptic transmission between two neurons. Figure 13 also shows that intramembranous particles have accumulated within the extrasynaptic regions of the membrane in this stimulated axon, particularly in the region immediately surrounding the elevated plateau of the active zone (compare with Fig. 12). Similar results have been observed at the frog neuromuscular junction and the addition of vesicle membrane components

→

Fig. 12. Replica of unstimulated retinula axon terminal (P-face) showing numerous randomly oriented active zones. Each *bar* is at right angles to the long axis of a bowtie-shaped active zone. Note also nearby cross-fractured capitate projections *(CP)*. × 65,000. *Inset* Retinula cell, presynaptic structure is sectioned en face relative to the presynaptic density. In this presentation the presynaptic bar has the bowtie shape but only the "knot" and one side ("bow") are in the plane of section. Several synaptic vesicles (arrows) are present at the edge of the presynaptic density. Three postsynaptic processes (*) are also seen. × 130,000

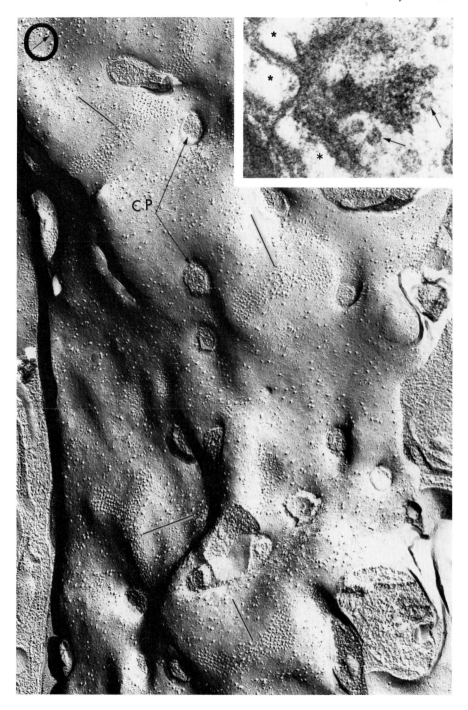

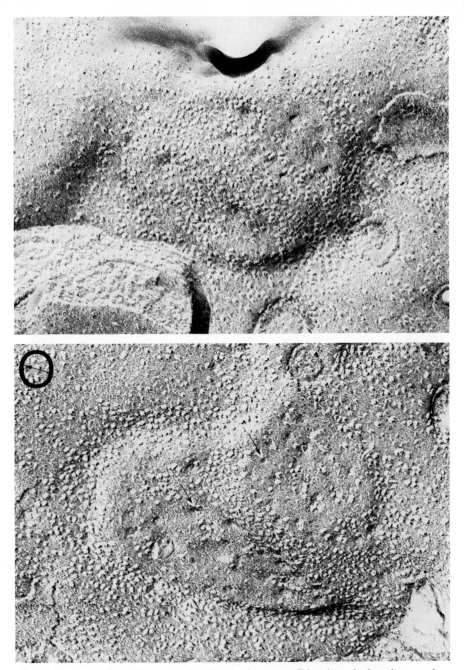

Fig. 13. Replicas of two light-stimulated retinula axons. Disc-shaped elevation on the P-face denotes a synaptic locus. *Arrows* point to pits (on both sides of the cluster of active zone particles) where synaptic vesicles are assumed to have fused with presynaptic membrane and opened, releasing neurotransmitter. Collected membrane particles along the perimeter of this plateau are absent in unstimulated axons (see Fig. 12). × 115,000

(used vesicles) to the presynaptic membrane has been cited as the most likely source of these additional membrane particles (Heuser et al. 1979; Heuser and Reese 1979, 1981).

Figure 14A reveals the P-face of a monopolar interneuron underlying the E-face of a retinula cell terminal (again note the diagnostic sheared-off capitate projections). The convergence of three synapses is indicated by three segments of fused, P-face particles on the dendritic membrane. Although these ridges resemble tight junctions, no corresponding ridges or grooves are evident on the presynaptic membrane, nor have any tight junctions been found at such synapses in thin-sectioned tissue (Saint Marie 1981). Except for the fused, P-face particles around the perimeter of the postsynaptic site, the postsynaptic membrane appears similar to that of the extrasynaptic membrane.

In another replica (Fig. 14B), we see a narrow dendritic process between two retinula axons. Again, fused, P-face particle ridges lie adjacent to the sites of synaptic contact. In the correlating thin section (Fig. 14C), two dendrites lie adjacent to a retinula cell synapse. The subsurface cisternae (arrows) that underlie the postsynaptic membrane at this dyad synapse and the presynaptic T-bar are features that could perhaps be brought out in a replica if etching were performed.

The Optic Cartridge

Recent FF views of the lamina ganglionaris (Chi and Carlson 1980a, b; Lane 1981; Saint Marie 1981; and Saint Marie and Carlson 1982, 1983b; Shaw and Stowe 1982) have both confirmed and extended our knowledge of

---→

Fig. 14. A Replica of contact area between a retinula axon terminal and dendrite trunk of a large monopolar neuron. Fused particle ridges (*arrows*) mark the convergence of three chemical synapses on the P-face (*P*) of the postsynaptic neuron. No complementary grooves are seen on the E-face (*E*) of the retinula axon. × 42,000. **B** Replica of postsynaptic monopolar cell spine (*L*) between two retinula cell (*R*) terminals. This dendritic spine bears ridges of fused, P-face particles (*arrows*) situated next to the synapses. × 76,000. **C** Thin section of retinula cell (*R*) synapsing onto two monopolar cell processes (*L*). Synaptic vesicles are present in the presynaptic retinula cell near the T-bar (*T*). Each postsynaptic process contains a flattened "bag" (*arrow*) opposite the T bar. × 100,500

Fig. 15. A Replica of a pair of (L1, L2) monopolar cells flanked by photoreceptor axons (*R*). Only the E-face of one monopolar cell remains overlying the P-face (*P*) of its companion cell. The E-face exhibits two clusters of particles (single arrows) which are sites of gap junctions. E-face grooves may indicate tight junctions or E-face figures of postsynaptic sites (see Fig. 14A). Capitate projections (*CP*) seen as large "pockmarks" on the P-face of the retinula axons. Circular patches (*) delineate presynaptic sites based on bowtie aggregates of particles (see also Fig. 12). × 22,000. **B** Thick (HVEM) section of complementary longitudinal section of the lamina. Type I monopolar interneurons (*L*) project in parallel between banks of photoreceptor axons (*R*). Epithelial glial cells send small capped processes [capitate projections (*CP*)] into the retinula cells. ×22,000

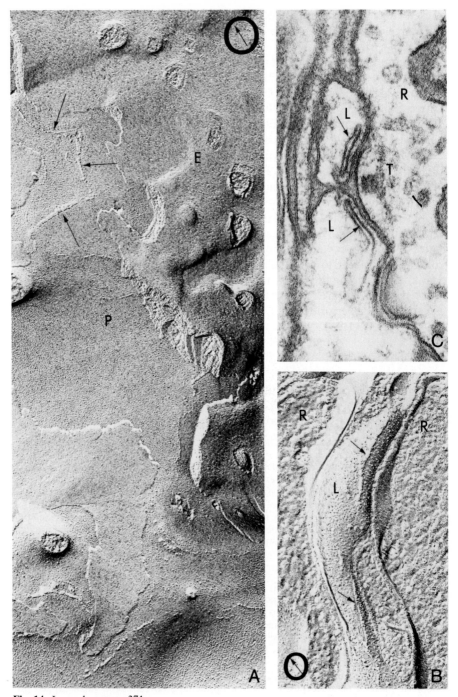

Fig. 14. Legend on page 371.

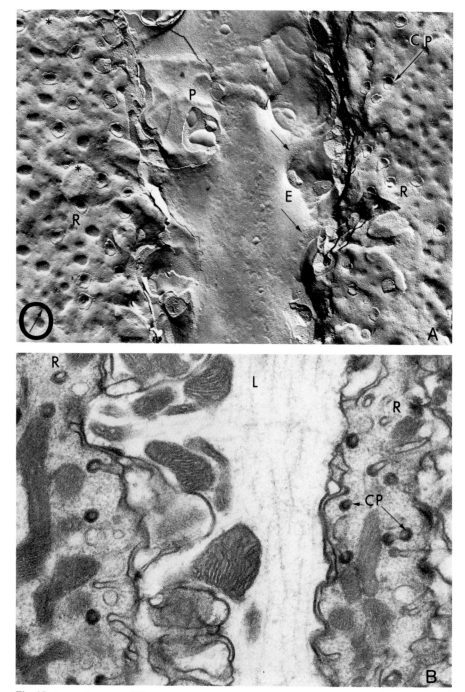

Fig. 15. Legend on page 371.

the optic cartridge (For summary of earlier thin section studies see Straus-feld and Campos-Ortega 1977; Shaw 1981; Strausfeld 1983). In Fig. 15 A the relatively smooth-surfaced membrane face is from a Type I (either L1 or L2) monopolar interneuron at the core of an optic cartridge. The "pockmarked" (from sheared-off glial capitate projections) membranes to either side are those of retinula axons. Presynaptic specializations were found on the P-face of a similar retinula axon (Fig. 12). Gap junctions are seen far better in this replica than in the corresponding longitudinal thick (HVEM) section.

The accompanying thick (0.4 μm) section shows the thin glial covering from which arise capitate projections that insert into retinula axon terminals. In these thick sections all membranes and organelles are present and in focus within the depth of the section. The outlines are less crisp. However, more of the organelles are seen.

Recent Advances and Future Prospects

The future of the FF technique is unquestionably bright for a variety of rea-sons. There have been quantum leaps in the development of solid-state cir-cuitry leading to precisely regulated and programmable FF instrumentation. Greater vacuums, faster and deeper freezing, shorter durations between frac-turing and replicating, computer processing (for reconstruction and enhance-ment of replica images), and improved stereoscopy of replicas reduce artefacts and help to evaluate the results.

Rapid Freezing

Mention has been made of the Heuser et al. (1975) device that plunges tissue against a pure copper block cooled to liquid helium temperature (4 K). This eliminates fixation and cryoprotection. Tissue is fractured and processed in the conventional manner.

While the rapid-freeze method works well on cells that are superficially situated, areas deeper than 20 μm are not well preserved. High pressure freezing may solve this and Moor (1977) briefly describes a procedure (see also Wolf et al. 1981) which would permit rapid vitrification of tissue 200 μm from the surface. High pressures (up to 2000 bar) are developed in a specially constructed chamber prior to freezing.

Improvements of fracturing which reduce specimen contamination and other artefacts can be accomplished by using the method of Gross et al. (1978), i.e. fracturing at $-196\,°C$ (77 K) and using an ultrahigh vacuum (UHV) which is about 3 log units greater (i.e., 2×10^{-9} Torr) than "con-ventional vacuums." Gross (1979) states "... it is possible to carry out freeze fracturing under UHV, practically contamination-free replication ... in-dependent of the specimen temperature." UHV keeps the specimen surfaces clean by eliminating hoar frosting and other gas-related artefacts which arise before and during coating. It also lowers heat damage during replication and lessens deformation in the course of fracture. The success of UHV is judged by excellent E- to P-face correlation of IMPs in yeast and in "purple" bacterial membranes, using digital image processing to visualize periodic structures (Kübler and Gross 1978).

Steere et al. (1979, 1980) reduced contamination and artefacts by using a double shroud cooled by liquid nitrogen to cover the specimen followed by fracturing at −196 °C. The simplicity of this method compares well with UHV technology which is intricate and costly.

The time spent between fracturing and replicating has always been an agonizing moment when one knows that ice chips on the knife, the contamination of backstreaming pump gasses, direct hoar frosting on the specimen, and water sublimation of the fracture face, etc., all increase the possibility for artefacts. A partial solution to those problems was reported by Ellisman and Staehelin (1979) who describe a gun shutter that has only a 1-sec delay between fracturing and platinum shadowing and which reduces heat absorbed by the specimen surface.

Since 1973, Steere and co-workers (e.g. Steere 1973, Steere and Rash 1979) have labored on "high resolution three-dimensional complementarity." The importance of stereo pairs of replicas for precise evaluation of ultrastructural relief seems to be little appreciated if only because stereo pairs rather seldom appear in published reports. Goniometric orientation of the replica faces with proper tilt angles greatly enhances the information content of FF micrographs (Steere and Rash 1979) and demonstrates the value of this fairly simple procedure.

Acknowledgements. We gratefully acknowledge support from the N.I.H., National Eye Institute, EYO 1686, University of Wisconsin Graduate School (Project 101451) and from the College of Agricultural and Life Sciences, University of Wisconsin-Madison, Hatch Project 2100. We thank Dr. Philippa Claude of the UW Primate Research Center, Dr. John E. Heuser, Washington University, St. Louis and Dr. Thomas S. Reese. N.I.N.C.D.S.-N.I.H. for training in freeze fracture technique. The work at the Primate Research Center was supported in part by Grant RR 00167 from the N.I.H.

Chapter 19

High-Voltage Electron Microscopy
for Insect Neuroanatomy

CHE CHI

Department of Entomology, University of Wisconsin
Madison, Wisconsin U.S.A.

Introduction

In 1962 Dupouy and Perrier described a transmission electron microscope with an accelerating voltage of one million volts (one megavolt = 1 MV). That achievement possibly set a standard for HVEM and today "high voltage" is regarded as that in the megavolt range. Voltages in the 100 to 1000 kV range have been used for neurobiological material (e.g. Shelton et al. 1971) with some success and a 500 kV scope was built in the U.K. as early as 1952 (Hirsch 1974). The world leader in megavoltage appears to be the 3 MV instrument in Toulouse, France although Meek (1976) mentions that the range extends to 10 MV. Nine other countries (six of them European) have megavolt microscopes. Present commercial models have been made possible by advances in electron optics and generators and accelerators largely provided by laboratories in France, the U.K. and Japan. Currently 55 such microscopes are in service, 14 of which are located in the U.S.A. (King et al. 1980).

The evolution of HVEM for the better viewing of biological specimens began with the expectation that the greater penetrating power of the highly accelerated electrons through thick sections could overcome the ultrathin sectioning problems experienced in the late 1940's and early 1950's when voltages were in the 40−100 kV range. When excellent ultramicrotomes became available, the need for HVEM by biologists was not so acute and interest waned. Obviously the high cost of MV microscopes also lessens enthusiasm and this insurmountable factor restricts their marketability. This is not overly troublesome in the U.S. as HVEM's in Boulder, Colorado and Madison, Wisconsin are designated National Research Resources by the National Institutes of Health. Qualified users are given free (or low cost) access to these instruments.

The experimental results (e.g. Fig. 1) in this paper were obtained on the HVEM at the University of Wisconsin, Madison. It is instructive to provide some basic data about the size and cost of this typical microscope. The model is an A.E.I., EM7 operating at a peak accelerating voltage of 1.2 MV with a lowest voltage of 100 kV. It has a magnification range of 63× to 1,250,000×. Resolution is specified as 5 Å at 1000 kV or 1250 kV. The electron generator-accelerator was manufactured by the Haefly Co. Basel, Switzerland. With generator-accelerator, the elongate column and allied instrumentation require a room 10 m high and 17 × 17 m in area. Scope weight is 25,500 kg and it rests on a 60,000-kg cement block that floats on bags filled with compressed air (see Pawley 1980). Today's costs for the microscope with generator-accelerator (only) would probably exceed one dollar per volt. Other ancillary instrumentation e.g. TV monitors and taping capability for the fluorescent screen, scanning densitometers, etc. appreciably add to the costs.

Rationale for HVEM for Biological Research

There are numerous reasons for biologists to use HVEM and in no way can the instrument be considered faddish. The writer cannot agree with Meek (1976) who wrote "the amount of significant biological information which has been published has been disappointingly small and could in the main have been obtained at 100 kV, though with the expenditure of greater effort." HVEM has unique advantages which transcend labour-saving considerations.

1. Electron penetration through the specimen is increasingly enhanced as accelerating voltage is augmented. In this way either thick specimens per se or thick sections of specimens can be visualized with good resolution. This aspect of HVEM is invaluable if serial sections are desired and tissue and reconstruction objectives are to be met.
2. At higher voltages, wavelength of the electrons decreases and resolution increases. (The theoretical resolving power increases 245% from 100 kV to 1 MV). Enhanced resolution also adds to magnification capabilities.
3. There is supposedly less interaction of the electron beam with the specimen, diminishing heating and ionization effects that would otherwise cause specimen damage. However, to quote King et al. (1980), "the same factors that lead to lessened radiation damage *per incident electron* also require one to use a *greater number* of *electrons* to obtain an *image* of the same information content".
4. At higher electron energies chromatic aberration decreases (but spherical aberration increases).
5. With all these advantages, (especially low beam damage to specimens) even living specimens, suitably encapsulated, can be examined with the HVEM without mortal harm.

There are at least two major difficulties with HVEM with regard to ultrastructural studies of animal cells. First, because focus is usually good throughout the depth of the cell examined there may be "too many" (overlapping) membrane profiles for each to be easily distinguished. This obscuring of cellular detail in neural tissue was discussed by Scott and Guillery (1974). Computer-assisted reconstruction techniques can help in this respect as well as the judicious selection of section thickness.

A second difficulty relates to the low specimen-beam interaction. Because electrons penetrate tissue so rapidly there is relatively little deflection by the heavy metal atoms. As a consequence, image contrast is greatly reduced. In response, accelerating voltage can usually be reduced with a corresponding greater image contrast albeit with a reciprocal loss in resolution. An optimum accelerating voltage must be determined by trial and error 'Cossletts Rule' ("resolution obtainable from a specimen of density = 1 is about one-tenth of the specimen thickness"; Meek 1976) can be a rough guide in striking a balance between given section thickness and desired resolution. Increasing specimen thickness or the use of dark field technique in which the scattered electrons create the ultimate image are marginal tactics to improve image contrast.

Three-dimensional visualization of selected fields can be easily accomplished on the HVEM as stereo pairs of thick, well-stained material are easily made and are a dramatic improvement over stereoscopy of thin sections. Tilt angles for the present stereo pair (Fig. 1) were determined from Hudson and Makin (1970).

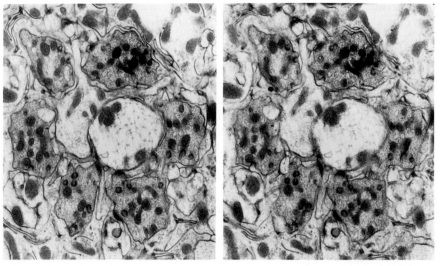

Fig. 1. Stereo pair of a cross-sectioned optic cartridge. Among other neurons are two electron-lucent, Type I, monopolar interneurons surrounded by six (electron-dense) photoreceptor axons. Epithelial glial cells enclose these neurons. Total angle of tilt is 39°. ×12,000

HVEM has been advantageous in elucidating fine structure in insects as demonstrated by the following: fly first optic neuropil (Chi and Carlson 1976); cockroach campaniform sensillum (Moran and Rowley 1975); beetle sperm (Ris and Allen 1975); fly satellite glia (Griffiths 1979); retinal pigment cells (Chi and Carlson 1976b) and chemoreceptor dendrites (Norris 1979). And although not of arthropod origin no better cytological advertisement for HVEM is know by this author than the magnificent high-voltage depictions of the microtrabecular lattice of cultured rat cells (Porter and Tucker 1981; micrograph by R. L. Anderson). The depth of focus and tridimensional appearance of this ground substance is truely a revelation. [See Heuser and Kirschner (1980) for an alternate viewpoint on cytoskeletons prepared by freeze-drying and use of conventional acceleratory voltages].

Method

In some ways preparation of insect neural or other tissue differs little from preparation for conventional electron microscopy. There are two exceptions; support film and staining techniques.

The illustrations (Figs. 1−6) presented here were based on the following procedure. The compound eye with intact lobes was dissected in 0.1 M phosphate buffered (pH 7.2) glutaraldehyde (2.5%), maintained in this solution for 3 h, and then rinsed overnight in buffer alone. Postfixation was with 2% osmium tetroxide dissolved in 0.1 N phosphate buffer for 3 h. A 30-min rinse in buffer was followed by a fast ethanol dehydration series up to 70% alcohol. The tissue remained in uranyl magnesium acetate (UMA) in 70% ethanol overnight before final dehydration. Tissue was embedded in Spurr[®] after a Spurr-alcohol series. The blocks were left overnight under vacuum at 20 °C and then at 65 °C for 24 h. Serial thick (0.25−0.30 μm) sections were picked up on copper slot grids covered with a durable carbonized Formvar film. This type of support is essential because the passage of highly accelerated electrons greatly stresses the film. These films must be well carbonized for stability and durability under the beam. Breakage of the support is far more commonplace than under beams of conventional (60−100 kV) voltage.

For serial thick sections, copper slot grids are preferred but this is not a requirement if an unobstructed field can be smaller and mesh grids are appropriate.

The problem of low specimen interaction with the highly accelerated electrons and the resultant low contrast can be somewhat overcome by an altered staining procedure, i.e. successive 6.5-h exposure to UMA (7.5%) at 50 °C followed by 0.5% lead citrate at room temperature (21 °C). Grids should then be rinsed in running double-distilled water and dried overnight. Moran and Rowley (1975), however, report that thick sections of the cuticular portion of an insect mechanoreceptor required 10 min of Reynolds' lead citrate (full strength) and an additional 10-min exposure to alcoholic 1% uranyl acetate. It is difficult to judge whether such abbreviated times would be sufficient to stain all of the cell's organelles. Moran and Rowley's sections are mostly through the distal (cuticular) portions of the campaniform sensillum, at a level above that of the neuronal perikaryon and axon.

Griffiths (1979) gives few staining details for his HVEM micrographs (taken in Madison) and the reasonable electron densities he obtained are at least partly the result of the greater tissue thickness ($1-2 \mu m$) employed. In addition, Griffiths achieved a very well-defined "staining" of the glial endoplasmic reticulum by cytochemically marking acid phosphatase with the substrate p-nitrophenol phosphate and thus avoided the use of conventional heavy metal stains. In other cases the tissue was "stained lightly" with uranium acetate and lead citrate. The enquiring reader is referred to pp. $41-43$ (general information about staining thick and thin sections for HVEM) in the comprehensive review by King et al. (1980).

HVEM of Insect Neurons

I have used HVEM for: (1) obtaining sufficient resolution in serial thick sections for identifying sites of chemical synapses (Fig. 2). This has enabled us to further elucidate patterns and distributions of synaptic connections. (2) With fewer serial thick sections (Fig. 3) taken in several planes we can begin to reconstruct individual neurons of the optic cartridges in the lamina and thus gain insight into the functional organization of this visual neuropil.

To a large extent, HVEM permits us to realize the detailed structure of neuropil (see Chi and Carlson 1976a). Even HVE-micrographs of the peripheral retina (Figs. $4-6$) provide reasonably clear information about cuticular features. The ommatidia are attenuated in the thicker sections but even in $0.3 \mu m$ sections of individual rhabdomeres, microvilli can just be discerned, as can certain membrane specializations between large pigment cells (Fig. 4). Again, the unexpectedly good resolution combined with added section thickness shows the full extent of the cells (Fig. 4) whereas parts of them would be out of the plane of an ultrathin section.   ·

Thick sections from the first optic neuropil are most informative if photographed as stereopairs. A three-dimensional view is shown of a transected optic cartridge in Fig. 1. Such views complement three-dimensional images derived from computer graphics based on serial thick sections. At 16,000× (Fig. 2) chemical synapses are also clearly visible so that relatively low-power electron micrographs can be used to achieve a "wiring diagram" of a region of neuropil.

Neural tissue associated with the insect cuticle has always proven difficult to section because of the cuticle's sclerotization, not to mention problems associated with embedding. This is particularly true for ultrathin sections with their tendency to wrinkle. Thicker (e.g. $0.5 \mu m$) sections have

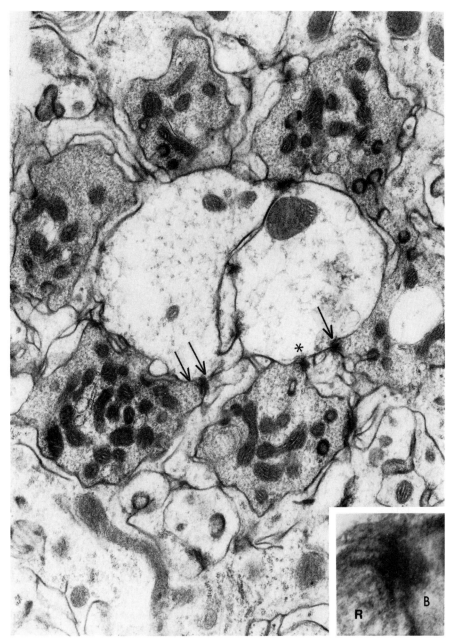

Fig. 2. Cross section of a synaptic cartridge. Synaptic locus (*single arrow*) between Type I interneuron and R axon. Contact (*double arrows*) between α-β-processes and R axons. *Inset* Enlargement of area with *asterisk* showing T-shaped synaptic ribbon in R axon (*R*) presynaptic to β-cell (*B*). ×16,000 and 70,400 (*inset*)

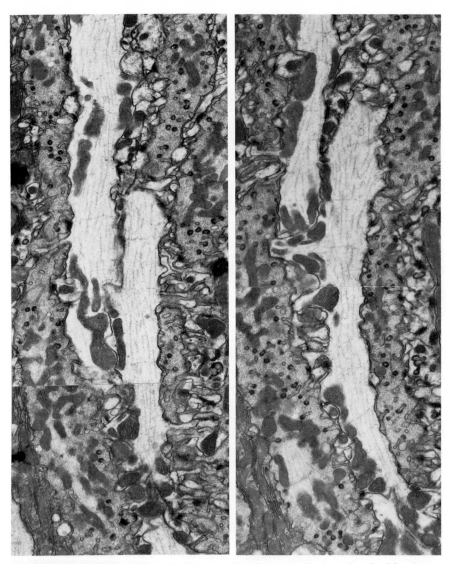

Fig. 3. Serial (thick) sections of the optic cartridge in longitudinal section. In this case an intervening section was omitted so that these two sections are about 0.6 µm apart. The electron-lucent pair of neurons are the Type I monopolars. Photoreceptor axon terminals surround and are on each side of these two interneurons. Scores of capitate projections (dark capped inclusions) enter the photoreceptor axons from the ensheathing epithelial glial cells. ×8,500

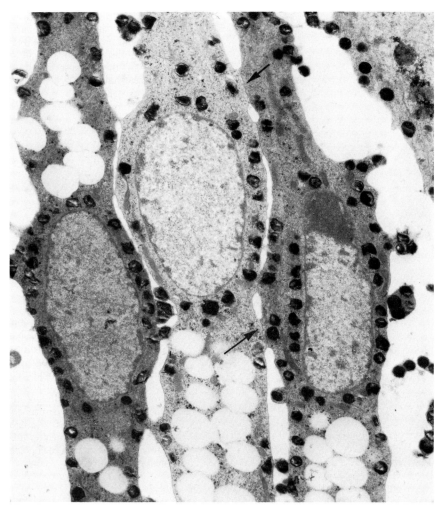

Fig. 4. Longitudinal section showing large pigment cells and the considerable contact between these cells. Two types of cell apposition are seen: direct contact with neighboring
cells and via digitiform processes. The membrane specializations (*arrows*) are probably
gap junctions since this type has been revealed in the same eye region through lanthanum
tracer studies and freeze fracture replicas. ×8,250

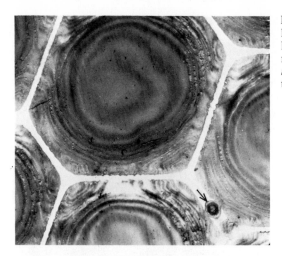

Fig. 5. Cross-sectional corneal lens facets showing the apparent helicoidal arrangement of protein microfibrils. At the lower right (*arrow*) is the cross-sectioned interfacetal hair. ×3,100

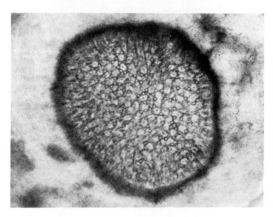

Fig. 6. At higher magnification the base of the interfacetal hair shows the tubular body (dendrite terminus) filled with about 400 microtubules arrayed longitudinally in pentagonal and hexagonal groupings. In these clusters the microtubules form the corners and the latter are connected by microfilaments. A neurofilament is often present at the center of each polygonal group. ×4,960

proven far easier to prepare. The corneal lens of the fly is an excellent example of the "thicker is easier" maxim. At low magnification a number of individual facets are noted, each with its helicoidal-appearing arrangements of chitin microfibrils (Fig. 5). This section is thick enough so that the base of an interfacetal hair is visible. The latter is a tiny one-celled mechanoreceptor which possesses a tubular body at its distal (dendritic) terminus. The hexagonal and pentagonal arrangement of the 400 microtubules with connecting microfilaments in this sensory process are easily resolved (Fig. 6).

The Future of HVEM

With over 50 microscopes of 1 MV capability (or greater) in the world it is probable that a peak of HVEM scopes has been reached. Many of these instruments came into being through government or industrial sponsorship whose resources are nowadays more limited. However, even if no new HVEMs are built, qualitatively the existing instruments enable technological advances in many fields of electron microscopy: electron optics, generators and accelerators, specimen preparation and wet cells. In the effort to obtain highest resolution it seems that this may be achieved at accelerating voltages near the megavolt range. On the practical level King et al. (1980) pointed out that 4−5 Å resolution is usually sufficient for most biological work. In neuroanatomy, this range is certainly adequate for observations of synaptic membranes and postsynaptic receptors.

Acknowledgements. The research on HVEM was supported by a grant from the National Institutes of Health, EYO-1868, and a Hatch Grant No. 2100. Professor Stanley D. Carlson, Department of Entomology, is thanked for reviewing earlier drafts of this manuscript. Special thanks are extended to Professor Hans Ris, Department of Zoology for permitting the writer to use the UW-Madison-HVEM. Dr. James Pawley, Physicist in Charge and Dr. Damian Newberger, HVEM Specialist, also assisted this research in the HVEM Laboratory.

References

Adams JC (1977) Technical consideration of the use of horseradish peroxidase as a neuronal marker. Neuroscience 2:141−146

Adams ME, O'Shea M (1981) Vacuolation of an identified peptidergic (proctolin-containing) neuron. Brain Res 230:439−444

Adams ME, O'Shea M (1983) Peptide cotransmitter at a neuromuscular junction. Science 221:286−288

Alheid GF, Edwards SB, Kitai ST, Park MR, Switzer RC (1981) Methods for delivering tracers. In: Heimer L, Robards MJ (eds) Neuroanatomical tract-tracing methods. Plenum New York, pp 91−116

Altman JS, Tyrer NM (1980) Filling selected neurons with cobalt through cut axons. In: Strausfeld NJ, Miller TA (eds) Neuroanatomical Techniques: Insect Nervous System. Springer, Berlin Heidelberg New York, pp 374−402

Altman JS, Shaw MK, Tyrer NM (1979) Visualisation of synapses of physiologically identified cobalt-filled neurones in the locust. J Physiol 296:2−3 P

Anderson H, Edwards JS, Palka J (1980) Developmental neurobiology of invertebrates. Ann Rev Neurosci 3:97−139

Anderson PN, Medlen AR, Mitchell J, Mayor D (1981) The uptake and retrograde transport of horseradish peroxidase − polylysine conjugate by ligated postganglionic sympathetic nerves in vitro. J Neurocytol 10:19−26

Anderson WA (1972) The use of myoglobin as an ultrastructural tracer. Reabsorption and translocation of protein by renal tubuli. J Histochem Cytochem 20:672−684

Avrameas S (1969) Coupling of enzymes to proteins with glutaraldehyde. Use of the conjugates for the detection of antigens and antibodies. Immunochemistry 6:43−52

Avrameas S, Ternynck T (1969) The cross-linking of proteins with glutaraldehyde and its use for the preparation of immunoadsorbents. Immunochemistry 6:53−66

Avrameas S, Ternynck T (1971) Peroxidase labelled antibody and Fab conjugates with enhanced intracellular penetration. Immunochemistry 8:1175−1179

Avrameas S, Ternynck T, Guesdon JL (1978) Coupling of enzymes to antibodies and antigens. Scand J Immunol 8:Suppl 7:7−23

Axelsson S, Björklund A, Falck B, Lindvall O, Svensson L (1973) Glyoxylic acid condensation: A new fluorescence method for the histochemical demonstration of biogenic monoamines. Acta Physiol Scand 87:57−62

Bacon J, Strausfeld NJ (1980) Nonrandom resolution of neuron arrangements. In: Strausfeld NJ, Miller TA (eds) Neuroanatomical techniques: Insect nervous system. Springer: Berlin Heidelberg New York, pp 359−372

Bacon JP, Altman JS (1977) A silver intensification method for cobalt-filled neurones in whole-mount preparations. Brain Res 138:359−363

Baker JR (1958) Principles of biological microtechniques. Wiley, New York, p 290

Barnstable CJ (1980) Monoclonal antibodies which recognize different cell types in the rat retina. Nature 286:231−235

Basinger SF, Gordon WC, Lam DMK (1979) Differential labelling of retinal neurones by ³H-2-deoxyglucose. Nature 280:682−684

Bassemir UK, Strausfeld NJ (1983) The cytology of cobalt-filled neurons in flies: Cobalt deposits at pre- and postsynaptic sites, mitochondria and the cytoskeleton. J Neurocytol (in press)

Bassiri RM, Utiger RD (1972) The preparation and specificity of antibody to thyrotropin releasing hormone. Endocrinology 90:722−727

Bate CM (1978) Development of sensory systems in arthropods. In: Jacobson M (ed) Handbook of sensory physiology. IX: Development of sensory systems. Springer, Berlin Heidelberg New York, pp 1−53

Baudry N (1968) The histological study of neurosecretion in the ventral nervous chain of *Rhodnius prolixus* (Hemiptera). C R Hebd Sceances Acad Sci Ser D 267:2356−2359

Bauer PS, Stacey TR (1977) The use of PIPES buffer in the fixation of mammalian and marine tissues for electron microscopy. J Microscopy 109:315−327

Beattie TM (1971) Histology, histochemistry and ultrastructure of neurosecretory cells in the optic lobe of the cockroach, *Periplaneta americana*. J Insect Physiol 17:1843−1855

Beckstead RM (1976) Convergent thalamic and mesencephalic projections to the anterior medial cortex in the rat. J Comp Neurol 156:403−416

Beltz BS, Kravitz EA (1983) Mapping of serotonin-like immunoreactivity in the lobster nervous system. J Neurosci 3:585−602

Benedetti EL (1973) In: Favard P (ed) Freeze-Etching Techniques and Applications. Société Française de Microscopie Electronique, Paris

Benefield DJ, Weinstein RS (1975) Image analysis of fractured biomembranes using computer simulation techniques. In: Proc 33rd Ann Mtg Electron Microsc Soc Am, Claiton's Pub Div, Baton Rouge, pp 214−215

Bennet MVL (1973a) Function of electronic junctions in embryonic and adult tissues. Fed Proc 32:35−75

Bennet MVL (1973b) Permeability and structure of electronic junctions and intercellular movements of tracers. In: Kater SB, Nicholson C (eds) Intracellular staining in neurobiology. Springer, Berlin Heidelberg New York pp 115−134

Bentley D, Keshishian H (1982) Pathfinding by peripheral neurons in grasshoppers. Science 218:1082−1088

Berger B, Glowinski J (1978) Dopamine uptake into serotonergic terminals *in vitro:* A valuable tool for histochemical differentiation of catecholaminergic and serotoninergic terminals in rat cerebral structures. Brain Res 147:29−45

Berger B, Nguyen-Legros J, Thierry AM (1978) Demonstration of horseradish peroxidase and fluorescent catecholamines in the same neuron. Neurosci Lett 9:297−302

Bergroth V, Reitamo S, Konttinen YT, Lalla M (1980) Sensitivity and nonspecific staining of various immunoperoxidase techniques. Histochemistry 68:17−22

Berlind A (1981) Mobilization of a peptide neurohormone for release during a physiological secretion cycle. Gen Comp Endocrinol 44:444−453

Bernard J, Guillet JC, Coillot JP (1980) Evidence for a barrier between blood and sensory terminal in an insect mechanoreceptor. Comp Biochem Physiol 67A:573−579

Berns MW (1971) A simple and versatile argon laser microbeam. Exp Cell Res 65:470−473

Berns MW (1972) Partial cell irradiation with a tunable organic dye laser. Nature (Lond) 240:483−485

Berns MW, Aist J, Edwards J, Strahs K, Girton J, McNeill P, Rattner JB, Kitzes M, Hammer-Wilson M, Liaw L-H, Siemans A, Koontz M, Peterson S, Brenner S, Burt J, Walter R, Bryant PJ, Van Dyk D, Coulombe J, Cahill T, Berns GS (1981) Laser microsurgery in cell and developmental biology. Science 213:505−513

Berod A, Hartman BK, Pujol JF (1981) Importance of fixation in immunohistochemistry: use of formaldehyde solutions of variable pH for the localization of tyrosin hydroxylase. J Histochem Cytochem 29:844−850

388   References

Bessis M, Gires F, Mayer G, Nomarksi B (1962) Irradiation des organites cellulaires à l'aide d'un laser à rubis. CR Acad Sci 255:1010
Bishop AE, Polak JM, Bloom SR, Pearse AGE (1978) A new universal technique for the immunocytochemical localization of peptidergic innervation. Proc Soc Endocrinol 77:25P–26P
Bishop CA, Bishop LG (1981) Vertical motion detectors and their synaptic relations in the third optic lobe of the fly. J Neurobiol 12:281–296
Bishop CA, O'Shea M (1982) Neuropeptide proctolin (H-Arg-Tyr-Leu-Pro-Thr-OH): Immunocytochemical mapping of neurons in the central nervous system of the cockroach. J Comp Neurol 207:223–238
Bishop CA, O'Shea M (1983) Serotonin immunoreactive neurons in the central nervous system of an insect (*Periplaneta americana*). J Neurobiol 14:251–269
Bishop CA, O'Shea M, Miller RJ (1981) Neuropeptide proctolin (H-Arg-Tyr-Leu-Pro-Thr-OH): Immunological detection and neuronal localization in insect central nervous system. Proc Nat Acad Sci US 78:5899–5902
Björklund A, Skagerberg G (1979) Simultaneous use of retrograde fluorescent tracers and fluorescence histochemistry for convenient precise mapping of monoaminergic projections and collateral arrangements in the CNS. J Neurosci Methods 1:261–277
Björklund A, Falck B, Klemm N (1970) Microspectrofluorometric and chemical investigation of catecholamine-containing structures in the thoracic ganglia of Trichoptera. J Insect Physiol 16:1147–1154
Blackstad TW (1965) Mapping of experimental axon degeneration for the study of neuronal relations. Z Zellforsch 67:819–834
Blackstad TW (1970) Electron microscopy of Golgi preparations for the study of neuronal relations. In: Nauta WJH, Ebbesson SOF (eds), Contemporary Research Methods in Neuroanatomy. Springer, Berlin Heidelberg New York, pp 186–216
Blackstad TW (1975a) Golgi preparations for electron microscopy: controlled reduction of the silver chromate by ultraviolet illumination. In: Santini M (ed) Golgi Centennial Proceedings, Perspectives in Neurobiology. Raven, New York, pp 123–132
Blackstad TW (1975b) Electron microscopy for experimental axonal degeneration in photochemically modified Golgi preparations: a procedure for precise mapping of nervous connections. Brain Res 95:191–210
Blessing WW, Furness JB, Costa M, Chalmers JP (1978) Localization of catecholamine fluorescence and retrogradely transported horseradish peroxidase within the same nerve cell. Neurosci Lett 9:311–315
Blomquist A, Westman J (1975) Combined HRP and Fink-Heimer staining applied on the gracile nucleus in the cat. Brain Res 99:339–342
Bodian D (1937) A new method for staining nerve fibres and nerve endings in mounted paraffin sections. Anat Rec 69:153–162
Boer HH, Schot LPC, Roubos EW, Maat A ter, Lodder JC, Reichelt D, Swaab DF (1979) ACTH-like immunoreactivity in two electrotonically coupled giant neurons in the pond snail *Lymnaea stagnalis*. Cell Tissue Res 202:231–240
Boer HH, Schot LPC, Veenstra JA, Reichelt D (1980) Immunocytochemical identification of neural elements in the central nervous systems of a snail, some insects, a fish, and a mammal with an antiserum to the molluscan cardio-excitatory tetrapeptide FMRFamide. Cell Tissue Res 213:21–27
Bolstad G, Kalland T, Srebro B, Stente-Larsen G (1979) Modifications of the glyoxylic acid method for the visualization of catecholamines in vertebrate and invertebrate species. Comp Biochem Physiol 62C:61–65
Boorsma DM (1977) Immunoperoxidase. Technical aspects and some applications in dermatology. Dissertation, Vrije Universität te Amsterdam
Boorsma DM, Kalsbeek GL (1975) A comparative study of horseradish peroxidase conjugates prepared with one-step and two-step method. J Histochem Cytochem 23:200–207

Boorsma DM, Streefkerk JG (1979) Periodate or glutaraldehyde for preparing peroxidase conjugates? J Immunol Methods 30:245–255

Boschek CB (1971) On the fine structure of the peripheral retina and the lamina of the fly, *Musca domestica.* Z Zellforsch 110:336–349

Boschek CB, Hamdorf K (1976) Rhodopsin particles in the photoreceptor membrane of an insect. Z Naturforsch 31:763

Bosman FT, Cramer-Kuijnenburg G (1980) A simplified method for the rapid preparation of peroxidase-anti-peroxidase (PAP) complexes. Histochemistry 67:243–248

Bouteille M (1976) The "LIGOP" method for routine ultrastructural autoradiography. J Microsc Biol Cell 27:121–127

Bowker RM, Steinbusch HWM, Coulter JD (1981) Serotonergic and peptidergic projections to the spinal cord demonstrated by a combined retrograde HRP histochemical and immunocytochemical staining method. Brain Res 211:412–417

Braak H, Braak E (1982) Simple procedure for electron microscopy of Golgi impregnated nerve cells. Neurosci Lett 32:1–4

Branton D (1966) Fracture faces of frozen membranes. Proc Natl Acad Sci USA 55:1048–1055

Branton D, Bullivant S, Gilula NB, Karnarsky MJ, Moor H, Muhlethaler K, Northcote DH, Packer L, Satir B, Satir P, Speth V, Staehelin LA, Steere RL, Weinstein RS (1975) Freeze-etching nomenclature. Science 190:54–56

Broadwell RD, Brightman MW (1976) Entry of peroxidase into neurons of the central and peripheral nervous systems from extracerebral and cerebral blood. J Comp Neurol 166:257–283

Broadwell RD, Brightman MW (1979) Cytochemistry of undamaged neurons transporting exogenous protein in vivo. J Comp Neurol 185:31–74

Brown BE (1975) Proctolin: A peptide transmitter candidate in insects. Life Sci 17:1241–1252

Brown BE (1977) Occurrence of proctolin in six orders of insects. J Insect Physiol 23:861–864

Brown BE, Starrat AN (1975) Isolation of proctolin, a myotropic peptide, from *Periplaneta americana.* J Insect Physiol 21:1879–1881

Brozman M (1978) Immunohistochemical analysis of formaldehyde- and trypsin- or pepsin-treated material. Acta Histochem 63:251–260

Brushart TM, Mesulam MM (1980) Transganglionic demonstration of central sensory projections from skin and muscle with HRP-lectin conjugates. Neurosci Lett 17:1–6

Buchner E (1976) Elementary movement detectors in an insect visual system. Biol Cybern 24:85–101

Buchner E, Buchner S (1979) 2-deoxy-D-glucose maps movement-specific nervous activity in the second visual ganglion of *Drosophila.* Science 205:687–688

Buchner E, Buchner S (1980) Mapping stimulus-induced nervous activity in small brains by ^{3}H-2-deoxy-D-glucose. Cell Tissue Res 211:51–64

Buchner E, Buchner S (1983) Neuroanatomical mapping of visually induced nervous activity in insects by ^{3}H-deoxyglucose. In: Ali MA (ed) Photoreception and vision in invertebrates. Plenum, New York, pp 623–634

Buchner S, Buchner E (1982) Functional neuroanatomical mapping in insects by ^{3}H-2-deoxy-D-glucose at electron microscopical resolution. Neurosci Letters 28:235–240

Buffa R, Solcia E, Fiocca R, Crivelli O, Pera A (1979) Complement mediated binding of immunoglobulins to some endocrine cells of the pancreas and gut. J Histochem Cytochem 27:1279–1280

Buijs RM, Swaab DF (1979) Immuno-electron microscopical demonstration of vasopressin and oxytocin synapses in the limbic system of the rat. Cell Tissue Res 204:355–365

Bullivant S (1973) Freeze-etching and freeze-fracturing. In: Koehler JK (ed) Advanced Techniques in Biological Electron Microscopy. Springer, Berlin Heidelberg New York pp 67–112

Bullivant S, Ames A (1966) A simple freeze-fracture replication method for electron microscopy. J Cell Biol 29:435−337

Bülthoff H, Poggio T, Wehrhahn C (1980) 3-D analysis of the flight trajectories of flies (*Drosophila melanogaster*). Z Naturforsch 35c:811−815

Bunt AH, Haschke RH (1978) Features of foreign proteins affecting their retrograde transport in axons of the visual system. J Neurocytol 7:656−678

Bunt AH, Haschke RH, Lund RD, Calkins DF (1976) Factors affecting retrograde axonal transport of horseradish peroxidase in the visual system. Brain Res 102:152−155

Burkhardt W, Braitenberg V (1976) Some peculiar synaptic complexes in the first visual ganglion of the fly, *Musca domestica*. Cell Tissue Res 173:287−308

Burrows M (1980) The tracheal supply to the central nervous system of the locust. Proc R Soc London 207B:63−78

Cajal SR, Castro F de (1933) Elementos de téchnica micrográfica del sistemo nervioso. Tipografican Artistica Alameda, Madrid

Campbell DH, Garvey JS, Cremer NE, Sussdorf DH (1964) Methods in immunology. Benjamin, New York, Amsterdam

Campos-Ortega JA, Strausfeld NJ (1973) Synaptic connections of intrinsic cells and basket arborizations in the external plexiform layer of the fly's eye. Brain Res 59:119−136

Carlsson A (1966) Drugs which block the storage of 5-hydroxytryptamine and related amines. Handb Exp Pharmacol 19:527−592

Carson KA, Lucas WJ, Gregg JM, Hanker JS (1980) Facilitated ultracytochemical demonstration of retrograde axonal transport of horseradish peroxidase in peripheral nerve. Histochem 67:113−124

Carson KA, Mesulam MM (1981) Tracing neuronal connections by electron microscopy using horseradish peroxidase: a comparison of eight methods. Amer Soc Neurosci 7:490

Case R (1957) Differentiation of the effects of pH and CO_2 on the spiracular function of insects. J Cell Comp Physiol 49:103−113

Caspersson TG, Hillarp NA, Ritzen M (1966) Fluorescence microspectrophotometry of cellular catecholamines and 5-hydroxytryptamine. Exp Cell Res 42:415−428

Chan KY, Haschke RH (1981) Differential uptake of horseradish peroxidase isoenzymes by cultured neuroblastoma cells. J Neurochem 36:339−342

Chan KY, Bunt AH, Haschke RH (1980) Endocytosis and compartmentation of peroxidases and cationized ferritin in neuroblastoma cells. J Neurocytol 9:381−403

Chan-Palay V, Palay SL (1972) High-voltage electron microscopy of rapid Golgi preparations. Neurons and their processes in the cerebellar cortex of monkey and rat. Z Anat Entwicklungsgesch 137:125−152

Chandler JA (1978) X-ray microanalysis in the electron microscope. In: Glauert AM (ed) Practical methods in electron microscopy. North Holland, Amsterdam New York Oxford

Chi C, Carlson SD (1976a) High voltage electron microscopy of the optic neuropile of the housefly, *Musca domestica*. Cell Tissue Res 167:537−545

Chi C, Carlson SD (1976b) The housefly interfacetal hair. Cell Tissue Res 166:353−363

Chi C, Carlson SD (1979) Ordered membrane particles in rhabdomeric microvilli of the housefly (*Musca domestica* L). J Morph 161:309−321

Chi C, Carlson SD (1980a) Membrane specializations in the first optic neuropil of the housefly (*Musca domestica* L). J Neurocytol 9:429−449

Chi C, Carlson SD (1980b) Membrane specialization in the first optic neuropil of the housefly. II. Junctions between glial cells. J Neurocytol 9:451−469

Chi C, Carlson SD (1981) Lanthanum and freeze fracture studies on the retinular cell junction in the compound eye of the housefly. Cell Tissue Res 214:541

Chi C, Carlson SD, St Marie RL (1979) Membrane specializations in the peripheral retina of the housefly *Musca domestica* L. Cell Tissue Res 198:501−520

Clark DG, Macmurchie DD, Elliott E, Wolcott RG, Landel AM, Raftery MA (1972) Elapid neurotoxins. Purification, characterization, and immunochemical studies of α-bungarotoxin. Biochemistry 11:1663−1668

Clark G (1973) Neurological staining methods: Bodian's protargol method. In: Clark G (ed) Staining procedures used by the biological stain commission. Williams and Wilkins, Baltimore pp 98−100

Clayton CJ, McNeill TH, Sladek JR Jr (1981) A comparison of neuropeptide immunocytochemistry in fluid-fixed and freeze-dried brains. Cell Tissue Res 220:223−230

Coggshall JC (1978) Neurons associated with the dorsal longitudinal flight muscles of *Drosophila melanogaster*. J Comp Neurol 177:707−720

Cohen J, Selvendran SY (1981) A neural cell-surface antigen is formed in the CNS but not in peripheral neurones. Nature 291:421−423

Colwell SA (1975) Thalamocortical-corticothalamic reciprocity: a combined anterograde-retrograde tracer technique. Brain Res 92:443−449

Coons AH (1956) Histochemistry with labeled antibody. Int Rev Cytol 5:1−23

Coons AH (1971) The developments in immunochemistry. Ann Acad Sci, NY 177:5−9

Coons AH, Creech HJ, Jones RN (1941) Immunological properties of an antibody containing a fluorescent group. Proc Soc Exp Biol, NY 47:200−202

Coons AH, Leduc EH, Connolly JM (1955) Studies on antibody production. I. A method for the histochemical demonstration of specific antibody and its application to a study of the hyperimmune rabbit. J Exp Med 102:49−60

Costa M, Buffa R, Furness JB, Solcia E (1980) Immunohistochemical localization of polypeptides in peripheral autonomic nerves using whole mount preparations. Histochemistry 65:157−165

Cottrell GA, Schot LPC, Dockray GJ (1983) Identification and probable role of a single neurone containing the neuropeptide *Helix* FMRF amide. Nature 304:638−640

Cowan WM, Cuenod M (1975) The use of axonal transport for the study of neural connections: A retrospective survey. In: Cowan WM, Cuenod M (eds) The use of axonal transport for studies of neuronal connectivity. Elsevier, Amsterdam, pp 2−24

Cremer C, Zorn C, Cremer T (1974) An ultraviolet laser microbeam for 257 nm. Microsc Acta 75:331−337

Crone C (1978) D-glucose − fuel for the brain. Trends Neurosci 1:120−122

Crow T, Heldman E, Hacopian V, Enos R, Alkon DL (1979) Ultrastructure of photoreceptors in the eye of *Hermissenda* labeled with intracellular injection of horseradish peroxidase. J Neurocytol 8:181−195

Cullheim S, Kellerth JO (1976) Combined light and electron microscopic tracing of neurons, including axons and synaptic terminals, after intracellular injection of horseradish peroxidase. Neurosci Lett 2:307−313

Curran RC, Gregory J (1977) The unmasking of antigens in paraffin sections of tissue by trypsin. Experientia 33:1400−1407

Dedmon RE, Holmes AW, Deinhardt F (1965) Preparation of fluorescein isothiocyanate-labeled γ-globulin by dialysis, gel filtration, and ion-exchange chromatography in combination. J Bact 89:734−739

Delcomyn F (1981) Nickel chloride for intracellular staining of neurons in insects. J Neurobiol 12:623−627

Denk H, Radaszkiewicz T, Weirich E (1977) Pronase pretreatment of tissue sections enhances sensitivity of unlabeled antibody enzyme (PAP) technique. J Immunol Methods 15:163−167

DeOlmos J, Heimer L (1980) Double and triple labeling of neurons with fluorescent substances; the study of collateral pathways in the ascending Raphe system. Neurosci Lett 19:7−12

DeRiemer SA, Macagno ER (1981) Light microscopic analysis of contacts between pairs of identified leech neurons with combined use of horseradish peroxidase and Lucifer yellow. J Neurosci 1:650−657

Detwiler PB, Sarthy PV (1981) Selective uptake of Lucifer yellow by bipolar cells in turtle retina. Neurosci Lett 22:227−232

Dismukes RK (1979) New concepts of molecular communication among neurons. Behav Brain Sci 2:409–448

Dockray GJ (1980) Cholecystokinins in rat cerebral cortex: Identification, purification and characterization by immunochemical methods. Brain Res 188:155–165

Dockray GJ, Duve H, Thorpe A (1981) Immunochemical characterization of gastrin/cholecystokinin-like peptides in the brain of the blowfly, *Calliphora vomitoria*. Gen Comp Endocrinol 45:491–496

Doerr-Schott J, Joly L, Dubois MP (1978) Sur l'existence dans la pars intercerebralis d'un insecte (*Locusta migratoria* R. et F.) des cellules neurosécrétrices fixant un antisérum antisomatostatine. CR Acad Sci Ser D 286:93–95

Dubois MP (1976) Mise en évidence par immunofluorescence d'un décapeptide, le LH-RF, dans l'éminence médiane. Ann Histochem 21:269–278

Dubois-Dalq M, McFarland D, McFarlin M (1977) Protein A-peroxidase: a valuable tool for the localization of antigens. J Histochem Cytochem 25:1201–1206

Dunker RO, Harris AB, Jenkins DP (1976) Kinetics of horseradish peroxidase migration through cerebral cortex. Brain Res 118:199–217

Dupouy G, Perrier F (1962) Principaux résultats obtenus avec un microscope électronique fonctionnant sous un million volts. In: Bree Jr SS (ed) 5th International Congress for Electron Microscopy 1: pp 1–3

Duve H (1978) The presence of a hypoglycemic and hypotrehalocemic hormone in the neurosecretory system of the blowfly *Calliphora erythrocephala*. Gen Comp Endocrinol 36:102–110

Duve H, Thorpe A (1979) Immunofluorescent localization of insulin-like material in the median neurosecretory cells of the blowfly, *Calliphora vomitoria* (Diptera). Cell Tissue Res 200:187–191

Duve H, Thorpe A (1980) Localization of pancreatic polypeptide (PP)-like immunoreactive material in neurones of the brain of the blowfly, *Calliphora erythrocephala* (Diptera). Cell Tissue Res 210:101–109

Duve H, Thorpe A (1981) Gastrin/cholecystokinin (CCK)-like immunoreactive neurones in the brain of the blowfly, *Calliphora erythrocephala* (Diptera). Gen Comp Endocrinol 43:381–391

Duve H, Thorpe A (1982) The distribution of pancreatic polypeptide in the nervous system and gut of the blowfly, *Calliphora vomitoria* (Diptera). Cell Tissue Res 227:67–77

Duve H, Thorpe A (1983) Immunocytochemical identification of α-endorphin-like material in neurones of the brain and corpus cardiacum of the blowfly, *Calliphora vomitoria* (Diptera). Cell Tissue Res 233:415–426

Duve H, Thorpe A, Lazarus NR (1979) Isolation of material displaying insulin-like immunological and biological activity from the brain of the blowfly *Calliphora vomitoria*. Biochem J 184:221–227

Duve H, Thorpe A, Lazarus NR, Lowry PJ (1981a) A neuropeptide of the blowfly *Calliphora vomitoria*. Biochem J 197:767–770

Duve H, Thorpe A, Lazarus NR, Lowry PJ (1982) A neuropeptide of the blowfly *Calliphora vomitoria* with an amino acid composition homologous with vertebrate pancreatic polypeptide. Biochem J 201:429–432

Duve H, Thorpe A, Neville R, Lazarus NR (1981b) Isolation and partial characterization of pancreatic polypeptide-like material in the brain of the blowfly *Calliphora vomitoria*. Biochem J 197:767–770

Duve H, Thorpe A, Strausfeld NJ (1983) Cobalt-immunocytochemical identification of peptidergic neurons in *Calliphora* innervating central and peripheral targets. J Neurocytology 12:847–861

Dvorak DR, Bishop LG, Eckert HE (1975) On the identification of movement detectors in the fly optic lobe. J Comp Physiol 100:5–23

Dymond GR, Evans PD (1979) Biogenic amines in the nervous system of the cockroach, *Periplaneta americana*: Association of octopamine with mushroom bodies and dorsal unpaired median (DUM) neurons. Insect Biochem 9:535–545

Eckert HE (1979) Elektrophysiologie und funktionelle Bedeutung bewegungssensitiver Neurone in der Sehbahn von Dipteren (*Phaenicia*). Habilitationsschrift, Ruhr-Universität Bochum

Eckert HE, Bishop LG (1978) Anatomical and physiological properties of the vertical cells in the third optic ganglion of *Phaenicia sericata* (Diptera, Calliphoridae). J Comp Physiol 126:57–86

Eckert HE, Boschek CB (1980) The use of horseradish peroxidase as a marker in the arthropod central nervous system. In: Strausfeld NJ, Miller TA (eds) Neuroanatomical techniques: Insect nervous system. Springer, New York Heidelberg Berlin, pp 325–339

Eckert M (1973) Immunologische Untersuchungen des neuroendokrinen Systems von Insekten. III. Immunhistochemische Markierung des Neuroendokrinen Systems von *Periplaneta americana* durch Fraktionierung von Retrocerebralkomplexextrakten gewonnenen Anti-Seren. Zool Jb Physiol 77:50–59

Eckert M (1977) Immunologische Untersuchungen des neuroendokrinen Systems von Insekten. IV. Differenzierte immunhistochemische Darstellung von Neurosekreten des Gehirns und der Corpora cardiaca bei der Schabe *Periplaneta americana*. Zool Jb Physiol 81:25–41

Eckert M (1980) Enzymmarkierte Antikörper in der Immunhistochemie. Acta Histochem Suppl XXII:55–65

Eckert M (1981 a) Immunhistochemie neurogener Peptide. In: Gersch M, Richter K (eds) Das peptiderge Neuron. Jena, pp 67–96

Eckert M (1981 b) Immunological differentiation of the neurosecretory system in insect brain. In: Sehnal F, Zabża A, Menn JJ, Cymborowski B (eds) Regulation of insect development and behaviour. Wroclaw Technical University Press, Wroclaw, pp 57–66

Eckert M, Gersch M (1978) Immunological investigations on the neuroendocrine system of the cockroach, *Periplaneta americana* L. In: Bargmann W, Oksche A, Polenov A, Scharrer B (eds) Neurosecretion and neuroendocrine activity. Evolution, structure and function. Springer, Berlin Heidelberg New York, pp 365–369

Eckert M, Agricula H, Penzlin H (1981) Immunocytochemical identification of proctolin-like immunoreactivity in the terminal ganglion and hindgut of the cockroach *Periplaneta americana* (L). Cell Tissue Res 217:633–645

Eckert M, Gersch M, Wagner M (1971) Immunologische Untersuchungen des neuroendokrinen Systems von Insekten. II. Nachweis von Gewebeantigenen des Gehirns und der Corpora cardiaca von *Periplaneta americana* mit fluorescein- und peroxydasemarkierten Antikörpern. Zool Jb Physiol 76:29–35

Edwards J, Chen S, Berns MW (1981) Cercal sensory development following laser microlesions of embryonic apical cells in *Acheta domesticus*. J Neurosci 1:250–258

Ellisman MH, Staehelin LA (1979) Electronically interlocked electron gun shutter for preparing improved replicas of free-fracture specimens. In: Rash JE, Hudson CS (eds) Freeze Fracture: Methods, Artifacts and Interpretations. Raven, New York, pp 123–125

El-Salhy M (1981) Immunohistochemical localization of pancreatic polypeptide (PP) in the brain of the larval instar of the hoverfly, *Eristalis aeneus*. Experientia 37:1009–1010

El-Salhy M, Abou-El-Ela R, Falkmer S, Grimelius I, Wilander E (1980) Immunohistochemical evidence of gastro-entero-pancreatic neurohormonal peptides of vertebrate type in the nervous system of the larva of a dipteran insect, the hoverfly, *Eristalis aeneus*. Regul Pept 1:187–204

Endo Y, Iwanaga T, Fujita T, Nishiitsutsuji U (1982) Localization of pancreatic polypeptide (PP)-like immunoreactivity in the central and visceral nervous systems of the cockroach *Periplaneta*. Cell Tissue Res 227:1–9

Essner D (1974) Hemeproteins. In: Hayat MA (ed) Electron microscopy of enzymes. Principles and methods. Vol 2, Van Nostrand, New York, pp 1–33

Evans PD (1980) Biogenic amines in the insect nervous system. Adv Insect Physiol 15:317–473

Evans PD, O'Shea M (1978) The identification of an octopaminergic neuron and the modulation of a myogenic rhythm in the locust. J Exp Biol 73:235−260

Fairén A, Peters A, Saldanha S (1977) A new procedure for examining Golgi impregnated neurons by light and electron microscopy. J Neurocytol 6:311−337

Fairén A, Valverde F (1980) A specialized type of neuron in the visual cortex of the cat: A Golgi and electron microscopy study of chandelier cells. J Comp Neurol 194:761−779

Feder N (1970) A heme-peptide as an ultrastructural tracer. J Histochem Cytochem 18:911−913

Feigl F (1956) Spot tests in organic analysis. Elsevier, Amsterdam London New York Princeton

Fischbach KF, Heisenberg M (1981) Structural brain mutant of *Drosophila melanogaster* with reduced cell number in the medulla cortex and with normal optomotor yaw response. Proc Natl Acad Sci USA 78:1105−1109

Fischel R (1910) Der histochemische Nachweis der Peroxydase. Klin Woche Wien 23:1557

Flagg-Newton J, Simpson I, Loewenstein WR (1979) Permeability of the cell-to-cell membrane channels in mammalian cell junction. Science 205:405−407

Flanagan TRJ (1982) Morphology of a small set of catecholaminergic interganglionic interneurons within the ventral ganglia of *Rhodnius prolixus*. (in press)

Folkersen J, Teisner B, Ahrous S, Svehag SE (1978) Affinity chromatographic purification of the pregnancy zone protein. J Immunol Methods 23:117−125

Frank E, Harris WA, Kennedy MB (1980) Lysophosphatidyl choline facilitates labeling of CNS projections with horseradish peroxidase. J Neurosci Methods 2:183−189

Fredman SM, Jahan-Parwar B (1980) Cobalt-mapping of the nervous system: evidence that cobalt can cross a neuronal membrane. J Neurobiol 11:209−214

Frens G (1973) Controlled nucleation for the regulation of the particle size in monodisperse gold solutions. Nature Phys Sci 241:20−22

Friedel T, Loughton BG, Andrew RD (1980) A neurosecretory protein from *Locusta migratoria*. Gen Comp Endocrinol 41:487−498

Frigero NA, Shaw MJ (1969) A simple method for determination of glutaraldehyde. J Histochem Cytochem 17:176−181

Frisby JP (1979) Seeing: illusion brain and mind. Oxford University Press, Oxford New York Toronto Melbourne

Fritsch HAR, Noorden S van, Pearse AGE (1979) Localisation of somatostatin-, substance P- and calcitonin-like immunoreactivity in the neural ganglion of *Ciona intestinalis* (L). Ascidiaceae. Cell Tissue Res 202:263−274

Frontali N (1968) Histochemical localization of catecholamines in the brain of normal and drug-treated cockroaches. J Insect Physiol 14:881−886

Furness JB, Costa M (1975) The use of glyoxylic acid for the fluorescence histochemical demonstration of peripheral stores of noradrenaline and 5-hydroxytryptamine in whole mounts. Histochemie 41:335−352

Fuxe K, Johnsson G (1973) The histochemical fluorescence method for the demonstration of catecholamines: theory, practice and application. J Histochem Cytochem 21:293−311

Gallyas F (1971) A principle for silver staining of tissue elements by physical development. Acta Morphol Acad Sci Hung 19:57−91

Gallyas F, Görcs T, Merchenthaler I (1982) High grade intensification of the end product of the diaminobenzidine reaction for peroxidase histochemistry. J Histochem Cytochem 30:183−184

Gaunt WA, Gaunt ON (1978) Three-dimensional reconstruction in biology. University Park Press, Baltimore

Geiger G, Nässel DR (1981) Visual orientation behaviour of flies after selective laser beam ablation of interneurons. Nature 293:398−399

Geiger G, Nässel DR (1982) Visual processing of moving single objects and wide-field patterns in flies: behavioural analysis after laser-surgical removal of interneurons. Biol Cybern 44:141−149

Geiger G, Boulin C, Bücher R (1981) How the eyes add together: monocular properties of the visually guided orientation behaviour of flies. Biol Cybern 41:71–78

Geisert EE (1976) The use of the horseradish peroxidase for defining pathways: A new application. Brain Res 117:130–135

Gerfen CR, O'Leary DDM, Cowan WM (1982) A note on the transneuronal transport of wheat germ agglutinin-conjugated horseradish peroxidase in the avian and rodent visual systems. Exp Brain Res 48:443–448

Gersch M, Fischer F, Unger H, Kabitzka W (1961) Vorkommen von Serotonin im Nervensystem von *Periplaneta americana* L. (Insecta). Z Naturforsch 16 b:351–352

Gerschenfeld HN, Paupardin-Tritsch D, Detrre P (1981) Neuronal response to serotonin: a second view. In: Jacobs BL, Gelperin A (eds) Serotonin Neurotransmission and Behavior. MIT, Cambridge London, pp 105–130

Gershorn M (1977) Biochemistry and physiology of serotonergic transmission. Handbook of physiology, Section I, vol 1. Am Physiol Soc, Bethesda, pp 573–605

Ghysen A (1978) Sensory neurons recognize defined pathways in *Drosophila* central nervous system. Nature, London, 274:869–872

Ghysen A (1980) The projection of sensory neurons in the central nervous system of *Drosophila:* Choice of the appropriate pathway. Dev Biol 78:521–541

Gilbert CD, Wiesel TN (1979) Morphology and intracortical projections of functionally characterized neurons in the cat visual cortex. Nature, London 280:120–125

Gilbert LI, Bollenbacher WE, Agui N, Granger NA, Sedlak BJ, Gibbs D, Buys CM (1981) The prothoracicotropes: Source of the prothoracicotropic hormone. Am Zool 21:641–653

Gillett R, Goll K (1972) Glutaraldehyde, its purity and stability. Histochem 30:162–167

Gilmour D (1965) The metabolism of insects. Oliver and Boyd, Edinburgh

Giloh H, Sedat JW (1982) Fluorescence microscopy: reduced photobleaching of rhodamine and fluorescein protein conjugates by n-propyl gallate. Science 217:1252–1255

Giorgi PP, Zahnd J (1978) Anterograde and retrograde transport of horseradish peroxidase isoenzymes in the retinotectal fibres of *Xenopus* larvae. Neurosci Lett 10:109–114

Glantz RM, Kirk MD (1980) Intercellular dye migration and electronic coupling within neuronal networks of the crayfish brain. J Comp Physiol 140:121–133

Glauert AM (1975) Fixation, dehydration and embedding of biological specimens. In: Glauert AM (ed) Practical methods in electron microscopy. North Holland, Elsevier, Amsterdam Oxford New York, 3/1

Glenn LL, Burke RE (1981) A simple and inexpensive method for three-dimensional visualisation of neurons reconstructed from serial sections. J Neurosci Methods 4:127–134

Goldfischer S, Novikoff AB, Albala A, Biempica L (1970) Haemoglobin uptake by rat hepatocytes and its breakdown within lysosomes. J Cell Biol 44:513

Goldstein RS, Weiss KR, Schwartz JH (1982) Intraneuronal injection of horseradish peroxidase labels glial cells associated with the axons of the giant metacerebral neuron of *Aplysia.* J Neurosci 2:1567–1577

Golgi C (1873) Sulla struttura della sostanza grigia dell cervello. Gazz Med Lombarda 33:244–246

Gomez-Ramos P, Rodriguez-Echandia EL (1981) Retrograde axonal transport and transneuronal transference of horseradish peroxidase in the rat ciliary ganglion. Experentia 37:1337–1339

Gonatas NK, Harper C, Mizutani T, Gonatas JO (1979) Superior sensitivity of conjugates of horseradish peroxidase with wheat germ agglutinin for studies of retrograde axonal transport. J Histochem Cytochem 27:728–734

Goodman CS (1982) Embryonic development of identified neurons in the grasshopper. In: Spitzer NC (ed) Neuronal development. Plenum, New York London, pp 171–212

Goodman CS, Spitzer NC (1979) Embryonic development of identified neurones: differentiation from neuroblast to neurone. Nature 280:208–214

Görcs T, Antal M, Olah E, Szekely G (1979) An improved cobalt labelling technique with complex compounds. Acta Biol Acad Sci Hung 30 (1–2):79–88

Goude B, Antoine JC, Gonata NK, Stieber A, Avrameas (1981) A comparative study of fluid-phase and adsorptive endocytosis of horseradish peroxidases in lymphoid cells. Exp Cell Res 132:375–386

Graham GS (1918) Benzidine as a peroxidase reagent for blood smears and tissues. J Med Res 89:15

Graham RC, Karnovsky MJ (1966) The early stages of absorption of injected horseradish peroxidase in the proximal tubules of the mouse kidney: ultrastructural cytochemistry by a new technique. J Histochem Cytochem 14:291–302

Graham RC, Kellermayer RW (1968) Bovine lactoperoxidase as a cytochemical protein tracer for electron microscopy. J Histochem Cytochem 16:275

Graham RC, Lundholm U, Karnovsky MJ (1965) Cytochemical demonstration of peroxidase activity with 3-amino-9-ethylcarbozole. J Histochem Cytochem 13:150

Graphics Standards Planning Committee (1979) Status report part III. Computer graphics 14:II 1–II 179

Graybiel AM, Devor M (1974) A microelectrophoretic delivery technique for use with horseradish peroxidase. Brain Res 68:167–173

Green SH (1981) Segment-specific organization of leg motorneurons is transformed in bithorax mutants of *Drosophila*. Nature, London, 292:152–154

Greenspan R, Finn J, Hall JC (1980) Acetylcholinesterase mutants in *Drosophila* and their effects on the structure and function of the central nervous system. J Comp Neurol 189:741–774

Gregory GE (1980) The Bodian protargol technique. In: Strausfeld NJ, Miller TA (eds) Neuroanatomical techniques: Insect nervous system. Springer, New York Heidelberg Berlin, pp 75–95

Griffin G, Watkins LR, Mayer DJ (1979) HRP pellets and slow-release gels: two new techniques for greater localization and sensitivity. Brain Res 168:595–601

Griffiths GW (1979) Transport of glial cell acid phosphatase by endoplasmic reticulum into damaged axons. J Cell Sci 36:361–389

Griffiths GW, Boschek CB (1976) Rapid degeneration of visual fibres following retinal lesions in the dipteran compound eye. Neurosci Lett 3:253–258

Grimmelikhuijzen CJP, Sundler F, Rehfeld JF (1980) Gastrin/CCK-like immunoreactivity in the nervous system of *Coelenterates*. Histochemistry 69:61–68

Grimmelikhuijzen CJP, Balfe A, Emson PC, Powell D, Sundler F (1981) Substance P-like immunoreactivity in the nervous system of *Hydra*. Histochemistry 71:325–333

Grimmelikhuijzen CJP, Carraway RE, Rökaeus Å, Sundler F (1981) Neurotensin-like immunoreactivity in the nervous system of *Hydra*. Histochemistry 72:199–209

Grimmelikhuijzen CJP, Dockray GJ, Schot LPC (1982) FMRF amidelike immunoreactivity in the nervous system of *Hydra*. Histochemistry 73:499–508

Gros C, Lafon-Cazal M, Dray F (1978) Présence de substances immunoréactivement apparentées aux enképhalines chez un Insecte, *Locusta migratoria*. CR Acad Sci Paris, Ser D 287:647–650

Gross H (1979) Advances in ultrahigh vacuum freeze fracturing at very low specimen temperature. In: Rash JE, Hudson CJ (eds) Freeze fracture methods, artifacts and interpretations. Raven, New York, pp 127–139

Gross H, Bas E, Moor H (1978) Freeze fracturing in ultrahigh vacuum (UHV) at −196 °C. J Cell Biol 76:712–728

Grube D (1980) Immunoreactivities of gastrin (G) cells. II. Nonspecific binding of immunoglobulins to G-cells by ionic interactions. Histochemistry 66:149–167

Grube D, Weber E (1980) Immunoreactivities of gastrin (G-) cells. I. Dilution-dependent staining of G-cells by antisera and nonimmune sera. Histochemistry 65:223–237

Gu J, DeMey D, Moeremans M, Polak JM (1981) Sequential use of the PAP and immunogold staining methods for the light microscopical double staining of tissue antigens. Its application to the study of regulatory peptides in the gut. Regul Pept 1:365

Guesdon JL, Ternynck T, Avrameas S (1979) The use of avidin-biotin interaction in immunoenzymatic techniques. J Histochem Cytochem 27:1131–1139

Hackney CM, Altman JS (1982) Cobalt mapping of the nervous system: how to avoid artefacts. J Neurobiol 13:403−411

Hall CE (1950) A low temperature replica method for electron microscopy. Appl Physics 21:60−62

Hall TA (1971) The microprobe assay of chemical elements. In: Oster G (ed), Physical techniques in biological research. Academic Press, New York London, pp 157−275

Hall TA (1972) Preparation and examination of biological samples. Proc 7th Natn Conf Electron Probe Analysis Society of America, paper No. 40

Hall TA, Clarke-Andersen H, Appleton T (1973) The use of thin specimens for X-ray microanalysis in biology. J Microsc 99:177−182

Hanker JS, Ellis LC, Rustioni A, Carson KA, Reiner A, Eldred W, Karten HJ (1981) The ultrastructural demonstration of the retrograde axonal transport of horseradish peroxidase in nervous tissues by transmission and high voltage electron microscopy. In: Johnson JE (ed) Current trends in morphological techniques: CRC Press, Boca Raton Florida, 55−91

Hanker JS, Kasler F, Bloom MG, Copeland JS, Seligman AM (1967) Coordination polymers of osmium: The nature of osmium black. Science 156:1737

Hanker JS, Norden JJ, Diamond IT (1976) Horseradish peroxidase tracers with fluorescent reporter groups. J Histochem Cytochem 24:609

Hanker JS, Yates PE, Metz CB, Rustioni A (1977) A new specific, sensitive and non-carcinogenic reagent for the demonstration of horseradish peroxidase. J Histochem 9:789−792

Hanson ED (1977) Origin and early evolution of animals. Wesleyan Univ Press, Middletown CT, pp 65

Harary F (1969) Graph Theory. Addison-Wesley, Reading MA.

Harreveld AJ van, Crowell J (1964) Electron microscopy after rapid freezing on a metal surface and substitution fixation. Anat Rec 149:381−386

Harris WA, Ready DF, Lipson ED, Hudspeth AJ, Stark WS (1977) Vitamin A deprivation and *Drosophila* photo pigments. Nature 266:648−650

Hartman BK (1973) Immunfluorescence of dopamine-β-hydroxylase. Application of improved methodology to the localization of the peripheral and central noradrenergic nervous system. J Histochem Cytochem 21:312−332

Hartman BK, Udenfriend S (1969) A method for immediate visualization of proteins in acrylamide gels and its use for preparation of antibodies to enzymes. Anal Biochem 30:391−394

Hausen K (1976a) Struktur, Funktion und Konnektivität bewegungsempfindlicher Interneurone im dritten optischen Neuropil der Schmeißfliege *Calliphora erythrocephala*. Thesis, Eberhard-Karls-Universität Tübingen

Hausen K (1976b) Functional characterization and anatomical identification of motion sensitive neurons in the lobula plate of the blowfly *Calliphora erythrocephala*. Z Naturforsch 31 c:629−633

Hausen K (1981) Monokulare und binokulare Bewegungsauswertung in der Lobula Plate der Fliege. Verh Dtsch Zool Ges 1980. Fischer, Stuttgart, pp 49−70

Hausen K, Strausfeld NJ (1980) Sexually dimorphic interneuron arrangements in the fly visual system. Proc R Soc Lond, B 208:57−71

Hausen K, Wolburg-Buchholz K, Ribi WA (1980) The synaptic organisation of visual interneurons in the lobula complex of flies. A light and electron microscopical study using silver-intensified cobalt-impregnations. Cell Tissue Res 208:371−387

Hautzer NW, Wittkuhn JF, McCaughey WTE (1980) Trypsin digestion in immunoperoxidase staining. J Histochem Cytochem 28:52−53

Hawkins R, Hass WK, Ransohoff J (1979) Measurement of regional brain glucose utilization *in vivo* using 2-^{14}C-glucose. Stroke 10:690−703

Hayat MA (1970) Principles and techniques of electron microscopy. Vol I. Biological applications. Van Nostrand, Reinhold, New York

Hayat MA (1973) Specimen preparation in electron microscopy of enzymes. In: Hayat MA (ed), Principles and methods, Vol I. Van Nostrand, Reinhold, New York, pp 1−35

Hayes NL, Rustioni A (1979) Dual projections of single neurons are visualized simultaneously: use of enzymatically inactive (^{3}H)HRP. Brain Res 165:321−326

Hazlett DTG (1977) A sensitive new immunoadsorbent technique for the detection of antibodies to an animal virus. Immunochemistry 14:473−476

Heisenberg M, Buchner E (1977) The role of retinula cell types in visual behavior of *Drosophila melanogaster.* J Comp Physiol 117:127−162

Heisenberg M, Wonnenberger R, Wolf R (1978) Optomotorblind H^{31} − a *Drosophila* mutant of the lobula plate giant neurons. J Comp Physiol 124:287−296

Hendrickson AE (1969) Electron microscopic radioautography: identification of origin of synaptic terminals in normal nervous tissue. Science 165:194−196

Hengstenberg R (1977) Spike response of a non-spiking visual interneurone. Nature 270:338−340

Hengstenberg R (1981) Visuelle Drehreaktionen von Vertikalzellen in der Lobula Plate von *Calliphora.* Verh Dtsch Zool Ges 1981, Fischer, Stuttgart, p 180

Hengstenberg R, Hengstenberg B (1980) Intracellular staining of insect neurons with Procion yellow. In: Strausfeld NJ, Miller TA (eds) Neuroanatomical techniques: Insect nervous system. Springer, New York Heidelberg Berlin, pp 307−324

Hengstenberg R, Sandemann DC (1982) Kompensatorische Kopf-Roll-Bewegungen von Fliegen. Verh Dtsch Zool Ges 1982, Fischer, Stuttgart, pp 313

Hengstenberg R, Hausen K, Hengstenberg B (1982) The number and structure of giant vertical cells (VS) in the lobula plate of the blowfly *Calliphora erythrocephala.* J Comp Physiol 149:163−177

Herbert GA (1976) Unproved salt fractionation of animal serums for immunofluorescence studies. J Dent Res 55, Spec Issue A: A 33−A 37

Herzog V, Miller F (1972) The localization of endogenous peroxidase in the lacrymal gland of the rat during postnatal development: Electron microscope, cytochemical and biochemical studies. J Cell Biol 53:662

Heuser JE, Kirschner MW (1980) Filament organization revealed in platinum replicas of freeze-dried cytoskeletons. J Cell Biol 86:212−234

Heuser JE, Miledi R (1971) Effect of lanthanum ions on function and structure of frog neuromuscular junctions. Proc R Soc Lond, B 179:247−260

Heuser JE, Reese TS (1973) Evidence for recycling of synaptic vesicle membrane during transmitter release at the frog neuromuscular junction. J Cell Biol 57:315−344

Heuser JE, Reese RS (1977) Structure of the synapse. In: Brookhart JM, Mountcastle VB, Kandel ER, Geiger SR (eds) Handbook of physiology, section 1: The nervous system, vol 1: Cellular biology of neurons. Bethesda, MD: Am Physiol Soc, pp 261−294

Heuser JE, Reese TS (1979) Synaptic-vesicle exocytoses captured by quick freezing. In: Schmitt FO, Worden FG (eds) The Neurosciences. MIT Press, Cambridge MA, Fourth Study Program. pp 573−600

Heuser JE, Reese TS (1981) Structural changes after transmitter release at the frog neuromuscular junction. J Cell Biol 88:564−580

Heuser JE, Reese TS, Landis DMD (1974) Functional changes in frog neuromuscular junctions studied with freeze-fracture. J Neurocytol 3:109−131

Heuser JE, Reese TS, Landis DMD (1975) Preservation of synaptic structure by rapid freezing. Cold Spring Harb Symp Quant Biol XI:17−24

Heuser JE, Reese TS, Dennis MJ, Jan Y, Jan L, Evans L (1979) Synaptic vesicle exocytosis captured by quick freezing and correlated with quantal transmitter release. J Cell Biol 81:275−300

Heyderman E (1979) Immunoperoxidase technique in histopathology: application, methods, and controls. J Clin Pathol 32:971−978

Hirose G, Jacobson M (1979) Clonal organization of the central nervous system of the frog. I Clones stemming from individual blastomeres of the 16-cell and earlier stages. Dev Biol 71:191−202

Hirsch PB (1974) High voltage electron microscopy in the U.K. In: Swann PR, Humphreys CJ, Goringe MJ (eds) High voltage electron microscopy. Academic Press, New York, 1—8

His W (1887) Über die Methoden der plastischen Rekonstruktion und über deren Bedeutung für Anatomie und Entwicklungsgeschichte. Anat Anz 2:382—394

Hoffmann GE, Knigge KM, Moynihan JA, Melnyk V, Arimura A (1978) Neuronal fields containing luteinizing hormone releasing hormone (LHRH) in mouse brain. Neuroscience 3:219—231

Hökfelt T, Ljungdahl A (1972) A modification of the Falck-Hillarp formaldehyde fluorescence method using the vibratome: Simple, rapid and sensitive localization of catecholamines in sections of unfixed brain tissue. Histochemie 29:325—339

Hökfelt T, Elde R, Fuxe K, Johansson O, Ljungdahl A, Goldstein M, Luft R, Efendic S, Wilsson G, Terenius L, Ganten D, Jeffcoate SL, Rehfeld J, Said S, Perez de la Mora M, Possani L, Tapia R, Teran L, Palacios R (1978) Aminergic and peptidergic pathways in the nervous system with special reference to the hypothalamus. In: Reichlin S, Baldessarinin RJ, Martin JB (eds) The hypothalamus. Raven, New York, pp 69—135

Hökfelt T, Johansson O, Ljungdahl Å, Lundberg JM, Schultzberg M (1980a) Peptidergic neurones. Nature 284:515—521

Hökfelt T, Lundberg JM, Schultzberg M, Johansson O, Ljungdahl Å, Rehfeld J (1980b) Coexistence of peptides and putative transmitters in neurons. In: Costa E, Trabucchi M (eds) Neural peptides and neuronal communication. Raven, New York, pp 1—23

Holtzmann E (1977) The origin and fate of secretory packages, especially synaptic vesicles. Neuroscience 2:327—355

Holtzmann E, Peterson ER (1969) Uptake of protein by mammalian neurons. J Cell Biol 40:863—869

Holtzmann E, Freeman AR, Kashner LA (1971) Stimulation dependent alterations in peroxidase uptake at lobster neuromuscular junctions. Science NY 173:733—736

Hongo T, Kudo N, Yamashita M, Ishizuka N, Mannen H (1981) Transneuronal passage of intraaxonally injected horseradish peroxidase (HRP) from group Ib and II fibers into the secondary neurons in the dorsal horn of the cat spinal cord. Biomed Res 2:722—727

Horn DHS, Wilkie JS, Sage BA, O'Connor JD (1976) A high affinity antiserum specific for the ecdysone nucleus. J Insect Physiol 22:901—905

Hoyle G, O'Shea M (1974) Intrinsic rhythmic contractions in insect skeletal muscle. J Exp Zool 189:407—412

Hoyle G, Dagan D, Moberly B, Colquhoun W (1974) Dorsal unpaired median insect neurons make neurosecretory endings on skeletal muscle. J Exp Zool 187:159—165

Hsu SM, Raine L (1981) Protein A and biotin in immunohistochemistry. J Histochem Cytochem 29:1349—1353

Hsu SM, Raine L, Fanger H (1981) Use of avidin-biotin-peroxidase techniques: A comparison between ABC and unlabeled antibody (PAP) procedures. J Histochem Cytochem 29:577—580

Hudson B, Makin MJ (1970) The optimum tilt angle for electron stereo-microscopy. J Physics E 3:311

Hudson CS, Rash JE, Graham WF (1979) Introduction to sample preparation for freeze fracture. In: Rash JE, Hudsen CS (eds) Freeze fracture: Methods, artifacts and interpretations. Raven, New York, pp 1—10

Isenberg G, Bielser W, Meier-Ruge W, Remy E (1976) Cell surgery by laser micro-dissection: A preparative method. J Microsc 107:19—24

Isobe Y, Nakane PK, Brown WR (1977) Studies on translocation of immunoglobulin across intestinal epithelium. I. Improvements in the peroxidase-labeled antibody method for application to study of human intestinal mucosa. Acta Histochem Cytochem 10:161—171

Itaya SK, VanHoesen GW (1982) WGA-HRP as a transneuronal marker in the visual pathways of monkey and rat. Brain Res 236:199—204

Jacobson M, Hirose G (1981) Clonal organization of the central nervous system of the frog. II. Clones stemming from individual blastomeres of the 32- and 64- cell stages. J Neurosci 1:271–284

Jankowska E, Rastad J, Westman J (1976) Intracellular application of horseradish peroxidase and its light and electron microscopical appearance in spinocervical tract cells. Brain Res 105:557–562

Jensen K, Wirth N (1978) "Pascal": user manual and report. Springer, Berlin Heidelberg New York

Jirmanova I, Libelius R, Lundquist I, Thesleff S (1977) Protamine induced intracellular uptake of horseradish peroxidase and vacuolation in mouse skeletal muscle in vitro. Cell Tissue Res 176:463–473

Johnson GM, Gloria C, Noghureira A (1981) A simple method of reducing the fading of immunofluorescence during microscopy. J Immunol Methods 43:349

Julesz B (1971) Foundations of cyclopean perception. University of Chicago Press, Chicago London, pp 406

Karnovsky MJ (1965) A formaldehyde-glutaraldehyde fixation of high osmolarity for use in electron microscopy. J Cell Biol 27:137a

Karnovsky MJ, Rice DF (1969) Exogenous cytochrome as an ultrastructural tracer. J Histochem Cytochem 17:751–753

Kater SB, Nicholson C (1973) Intracellular staining in neurobiology. Springer, Berlin Heidelberg New York

Kater S, Nicholson C, Davis WJ (1973) Guide to intracellular staining techniques. In: Kater SB, Nicholson C (eds) Intracellular staining in neurobiology. Springer, Berlin Heidelberg New York, pp 307–325

Kato M, Fujimori B, Hirata Y (1968) An electron microscopic study of intracellularly stained neurons. Brain Res 9:390–393

Kawarai Y, Nakane PK (1970) Localization of tissue antigens of the ultrathin sections with peroxidase-labeled antibody method. J Histochem Cytochem 18:161–166

Keefer DA (1978) Horseradish peroxidase as a retrogradely transported, detailed dendritic marker. Brain Res 140:15–32

Kerkut GA, Walker FJ (1962) Marking individual nerve cells through electrophoresis of ferrocyanide from a microelectrode. Stain Technol 37:217–219

Kerkut GA, Sedden CB, Walker RJ (1967) Uptake of DOPA and 5-hydroxytryptophane by monoamine-forming neurons in the brain of *Helix aspersa*. Comp Biochem Physiol 23:159–162

Kern M, Zimmermann A, Wegener G (1980) Anpassung im Energiestoffwechsel der Cerebralganglien von Insekten. Verh Dtsch Zool Ges 1980:332

Kernighan BW, Ritchie DM (1978) The C programming language. Prentice-Hall, Englewood Cliffs

King DG, Wyman RJ (1980) Anatomy of the giant fibre pathway in *Drosophila*. I. Three thoracic components of the pathway. J Neurocytol 9:753–770

King MV, Parsons DF, Turner JN, Chang BB, Ratkowski J (1980) Progress in applying the high-voltage electron microscope to biomedical research. Cell Biophys 2:1–95

Kirkham JB, Goodman LJ, Chappell KL (1975) Identification of cobalt in processes of stained neurones using X-ray energy spectra in the electron microscope. Brain Res 85:33–37

Kitai ST, Bishop GA (1981) Horseradish peroxidase. Intracellular staining of neurons. In: Heimer L, Robards MJ (eds) Neuroanatomical tract-tracing methods. Plenum New York 263–277

Klemm N (1974) Vergleichend-histochemische Untersuchungen über die Verteilung monoamin-haltiger Strukturen im Oberschlundganglion von Angehörigen verschiedener Insektenordnungen. Entomol Germ 1:21–49

Klemm N (1976) Histochemistry of putative transmitter substances in the insect brain. Prog Neurobiol 7:99–169

Klemm N (1980) Histochemical demonstration of biogenic monoamines (Falck-Hillarp method) in the insect nervous system. In: Strausfeld NJ, Miller TA (eds) Neuroanatomical techniques: Insect nervous system. Springer, New York Heidelberg Berlin, pp 51–73

Klemm N, Axelsson S (1973) Determination of dopamine, noradrenaline and 5-hydroxy-tryptamine in the cerebral ganglion of the desert locust, *Schistocerca gregaria* Forsk. (Insecta, Orthoptera). Brain Res 57:289–298

Klemm N, Schneider L (1975) Selective uptake of indolamine into nervous fibres in the brain of the desert locust, *Schistocerca gregaria* Forskal (Insecta). A fluorescence and electron microscopic investigation. Comp Biochem Physiol 50 C:177–182

Klemm N, Sundler F (1983) The organization of catecholamine-containing and serotonin immunoreactive fibres in the corpora pedunculata of the desert locust, *Schistocerca gregaria* Forsk. Neurosci Lett 36:13–14

Klemm N, Steinbusch HWM, Sundler F (1983) Serotonin immunoreactive neurons and their projection in the brain of the cockroach, *Periplaneta americana* (L.). Brain Res (in press)

Kolb H (1970) Organization of the outer plexiform layer of the primate retina: electron microscopy of Golgi impregnated cells. Phil Trans R Soc Lond B 258:261–283

Konttinen YT, Reitamo S (1979) Effect of fixation on the antigenicity of human lactoferrin in paraffin-embedded tissues and cytocentrifuged cell smears. Histochemistry 62:55–64

Kramer KJ (1980) Insulin-like and glucagon-like hormones in insects. In: Miller TA (ed) Springer Series in Experimental Entomology: Neurohormonal techniques in insects. Springer, Berlin Heidelberg New York pp 116–136

Kramer KJ, Speirs RD, Childs CN (1977a) Immunochemical evidence for a gastrin-like peptide in insect neuroendocrine system. Gen Comp Endocrinol 32:423–426

Kramer KJ, Tager HS, Childs CN, Speirs RD (1977b) Insulin-like hypoglycemic and immunological activities in honeybee royal jelly. J Insect Physiol 23:293–295

Kramer KJ, Tager HS, Childs CN (1980) Insulin-like and glucagon-like peptides in insect hemolymph. Insect Biochem 10:179–182

Kristensson K (1970) Transport of fluorescent protein tracer in peripheral nerves. Acta Neuropathol 16:293–300

Kristensson K (1975) Retrograde axonal transport of protein tracers. In: Cowan WM, Cuenod M (eds), The use of axonal transport for studies of neuronal connectivity. Elsevier, Amsterdam, pp 71–82

Kristensson K (1978) Retrograde transport of macromolecules in axons. Ann Rev Pharmacol Toxicol 18:97–110

Kristensson K, Olsson J (1973) Diffusion pathways and retrograde axonal transport of protein in peripheral nerves. Prog Neurobiol 1:85–109

Kristensson K, Olsson Y (1971) Retrograde axonal transport protein. Brain Res 29:363–365

Kristensson K, Olsson Y (1976) Retrograde transport of horseradish peroxidase in transected axons. 3. Entry into injured axons and subsequent localization in pericaryon. Brain Res 115:201–213

Kristensson K, Olsson T (1978) Uptake and retrograde axonal transport of horseradish peroxidase in botulinum-intoxicated mice. Brain Res 155:118–123

Kristensson K, Strömberg E, Elofsson R, Olsson Y (1972) Distribution of protein tracers in the nervous system of the crayfish (*Astacus astacus* L.) following systemic and local application. J Neurocytol 1:35–48

Kübler O, Gross H (1978) UHV freeze-fracturing and image processing applied to the purple membrane. Proc Int Congr Electron Microsc 9th 2:143

Kübler O, Gross H, Moor H (1978) Complementary structures of membrane fracture faces obtained by ultrahigh vacuum freeze-fracturing at −196 °C and digital image processing. Ultramicroscopy 3:161–168

Kuhlmann WD (1977) Ultrastructural immunoperoxidase cytochemistry. Prog Histochem Cytochem 10:1–57

Kuypers HGJM, Bentivoglio M, Kooy D, Catsmann-Berrevoets CE (1979) Retrograde transport of bisbenzimide and propridium iodide through axons to their parent cell bodies. Neurosci Lett 12:1–7

Lane NJ (1974) The organization of insect nervous systems. In: Treherne JE (ed), Insect neurobiology. Elsevier, New York, pp 1–71

Lane NJ (1979a) Freeze-fracture and tracer studies on the intercellular junctions of insect rectal tissue. Tissue and Cell 11:481–506

Lane NJ (1979b) A new kind of tight junction-like structure in insect tissues. J Cell Biol 83:82A

Lane NJ (1981) Vertebrate-like tight junctions in the insect eye. Exp Cell Res 132:482–488

Lane NJ, Skaer H le B (1980) Intercellular junctions in insect tissues. In: Berridge MJ, Treherne JE, Wigglesworth VB (eds) Advances in insect physiology. Academic Press, London New York, pp 35–213

Lane NJ, Swales LS (1978a) Changes in the blood-brain barrier of the central nervous system in the blowfly during development, with special reference to the formation and disaggregation of gap and tight junctions. I. Larval development. Dev Biol 62:389–414

Lane NJ, Swales LS (1978b) Changes in the blood-brain barrier of the central nervous system in the blowfly during development, with special reference to the formation and disaggregation of gap and tight junctions. II. Pupal development and adult flies. Dev Biol 62:415–431

Lane NJ, Swales LS (1979) Intercellular junctions and the development of the blood-brain barrier in *Manduca sexta.* Brain Res 168:227–245

Lane NJ, Skaer H le B, Swales LS (1977) Intercellular junctions in the central nervous system of insects. J Cell Sci 26:175–199

Larsson LI (1980) Problems and pitfalls in immunocytochemistry of gut peptides.In: Glass GBJ (ed) Gastrointestinal hormones. Raven, New York

Larsson LI (1981) Peptide immunocytochemistry. Prog Histochem Cytochem 13:1–85

Larsson LI, Rehfeld JF (1977a) Evidence for a common evolutionary origin of gastrin and cholecystokinin. Nature, 269:335–338

Larsson LI, Rehfeld JF (1977b) Characterization of antral gastrin cells with region-specific antisera. J Histochem Cytochem 25:1317–1321

Larsson LI, Rehfeld JF (1979) A peptide resembling COOH-terminal tetrapeptide amide of gastrin from a new gastrointestinal endocrine cell type. Nature 277:575–578

Larsson LI, Schwartz TS (1977) Radioimmunocytochemistry – A novel immunocytochemical principle. J Histochem Cytochem 25:1140–1148

Laughlin SB (1974) Neural integration in the first optic neuropile of dragonflies. II. Receptor signal interactions in the lamina. J Comp Physiol 92:357–375

Laughlin SB (1981) Neuronal principles in the peripheral visual systems of invertebrates. In: Autrum H (ed) Handbook of sensory physiology, Vol VII/6B. Springer, Berlin Heidelberg New York, pp 133–280

LaVail JH (1975) Retrograde cell degeneration and retrograde transport techniques. In: Cowan WM, Cuenod M (eds) The use of axonal transport for studies of neuronal connectivity. Elsevier, Amsterdam, pp 217–248

LaVail JH (1978) A review of the retrograde transport techniques. In: Robertson RT (ed), Neuroanatomical research techniques. Academic Press, New York, pp 355–384

LaVail JH, LaVail MM (1972) Retrograde axonal transport in the central nervous system. Science 176:1416–1417

LaVail JH, LaVail MM (1974) The retrograde intraaxonal transport of horseradish peroxidase in the chick visual system: a light and electron microscopic study. J Comp Neurol 157:303–358

Lechago J, Sun NCJ, Weinstein WM (1979) Simultaneous visualization of two antigens in the same tissue section by combining immunoperoxidase with immunofluorescence techniques. J Histochem Cytochem 27:1221–1225

Lechan RM, Nestler JL, Jacobson S (1981) Immunohistochemical localization of ret-rogradely and anterogradely transported wheat germ agglutinin (WGA) within the central nervous system of the rat: Application to immunostaining of a second antigen within the same neuron. J Histochem Cytochem 29:1255—1262

Leduc EH, Scott GB, Avrameas S (1969) Ultrastructural localization of intracellular immune globulins in plasma cells and lymphoblasts by enzyme-labeled antibodies. J Histochem Cytochem 17:211—224

Lee MT (1982) Regeneration and functional reconnection of an identified vertebrate central neuron. J Neurosci 2:1793—8111

Lent CM (1981) Morphology of neurons containing monoamines within leech segmental ganglia. J Exp Zool 216:311—316

Lent CM (1982) Fluorescent properties of monoamine neurone following glyoxylic acid treatment of intact leech ganglia. Histochem (in press)

LeRoith D, Lesniak M, Roth J (1981) Insulin in insects and annelids. Diabetes 30:70—76

LeVay S (1973) Synaptic patterns in the visual cortex of the cat and monkey: Electron microscopy of Golgi preparations. J Comp Neurol 150:53—86

Levinthal C, Ware R (1972) Three-dimensional reconstruction from serial sections. Nature 236:207—210

Lewis PR (1977) Other cytochemical methods for enzymes. In: Glauert AM (ed), Practical methods in electron microscopy, Vol 5. North Holland, Amsterdam, pp 225—287

Lewis PR, Henderson Z (1980) Tracing putative cholinergic pathways by a dual cytochemical technique. Brain Res 196:489—493

Lillie RD (1969) HJ Conn's biological stains. Williams and Wilkins, Baltimore, p 269

Lindsay RD (1977) Computer analysis of neuronal structures. Plenum, New York

Lindvall O, Björklund A (1974) The glyoxylic acid fluorescence histochemical method: A detailed account of the methodology for the visualization of central catecholamine neurons. Histochem 39:97—127

Liposits Z, Görcs T, Gallyas F, Kosaras B, Setalo G (1982) Improvement of the electron microscopic detection of peroxidase activity by means of silver intensification of the diaminobenzidine reaction in the rat nervous system. Neurosci Lett 31:7—11

Lipp HP, Schwegler H (1980) Improved transport of horseradish peroxidase after incubation with a non-ionic detergent (Nonidet P-40) into mouse cortex and observations on the relationship between spread at the injection site and amount of transported label. Neurosci Lett 20:49—54

Lison L (1936) Histochemie animale. Gautier-Villars, Paris

Litchy WJ (1973) Uptake and retrograde transport of horseradish peroxidase in frog sartorius nerve in vitro. Brain Res 56:377—381

Livett BC (1978) Immunohistochemical localization of nervous system — specific proteins and peptides. In: Bourne GH, Danielli JF (eds) Intern Rev Cytol Suppl 7. Academic Press, New York, pp 53—237

Livingstone MS, Hubel DH (1983) Specificity of cortico-cortical connections in monkey visual system. Nature 304:531—539

Ljungdahl A, Hökfelt T, Goldstein M, Park D (1975) Retrograde peroxidase tracing of neurons combined with transmitter histochemistry. Brain Res 84:313—319

Lohs-Schardin M, Cremer C, Nusslein-Volhard C (1979) A fate map for the larval epidermis of *Drosophila melanogaster:* Localized cuticle defects following irradiation of the blastoderm with an ultraviolet laser microbeam. Dev Biol 73:239—255

Lompré AM, Bouveret P, Leger J, Schwartz K (1979) Detection of antibodies specific to sodium dodecyl sulfate-treated proteins. J Immunol Methods 28:143—148

Loren I, Björklund A, Falck B, Lindvall O (1980) The aluminiumformaldehyde (ALFA) histofluorescence method for improved visualization of catecholamines and indoleamines. 1. A detailed account of the methodology for central nervous tissue using paraffin, cryostat or vibratome sections. J Neurosci Methods 2:277—300

Lubinska L (1975) On axoplasmic flow. Int Rev Neurobiol 17:241—296

Lucy JA (1970) The fusion of biological membranes. Nature, 227:815—817

Lundquist I, Josefsson J-O (1971) Sensitive method for determining of peroxidase activity in tissue by means of coupled oxidation reaction. Anal Biochem 41:567−577

Lux DH, Globus A (1968) Effects on IPSPs of cat motoneurons due to intra- and extracellular iontophoresis of CuSO$_3$. Brain Res 9:377−380

Macagno ER (1977) Abnormal connectivity following UV-induced cell death during *Daphnia* development. In: Lash JW, Burges MM (eds) Cell and tissue interactions. Raven, New York

Macagno ER (1978) Mechanisms for the formation of synaptic projections in the arthropod visual system. Nature 275:318−320

Macagno ER, Levinthal C, Sobel I (1979) Three-dimensional computer reconstruction of neurons and neuronal assemblies. Annu Rev Biophys Bioeng 8:323−351

Macagno ER, Muller KJ, Kristan BW, DeRiemer SA, Stewart R, Granzow B (1981) Mapping of neuronal contacts with intracellular injection of horseradish peroxidase and Lucifer yellow in combination. Brain Res 217:143−149

Maddrell SHP, Gee JD (1974) Potassium-induced reverse of the diuretic hormones of *Rhodnius prolixus* and *Glossina austeni:* Ca^{2+}-dependence, time course and localization of neurohemal areas. J Exp Biol 61:155−171

Malmgren LT, Brink JJ (1975) Permeability barriers to cytochrome c in nerves of adult and immature rats. Anat Rec 181:755−766

Malmgren LT, Olsson Y (1978) A sensitive method for histochemical localization of horseradish peroxidase in neurons following retrograde axonal transport. Brain Res 148:279−294

Malmgren LT, Olsson Y, Olsson T, Kritensson K (1978) Uptake and retrograde axonal transport of various exogenous macromolecules in normal and crushed hypoglossal nerves. Brain Res 153:477−493

Maranto AR (1982) Neuronal mapping: a photooxidation reaction makes Lucifer yellow useful for electron microscopy. Science 217:953−955

Marr D, Poggio T (1976) Cooperative computation of stereo disparity. Science 194:283−287

Marr D, Poggio T (1979) A computational theory of human stereo vision. Proc R Soc Lond B 204:301−328

Mason DY, Sammons R (1978) Rapid preparation of PAP for immunocytochemical use. J Immunol Methods 20:317−324

Mata M, Fink DJ, Gainer H, Smith CB, Davidsen L, Savaki H, Schwartz WJ, Sokoloff L (1980) Activity dependent energy metabolism in rat posterior pituitary primarily reflects sodium pump activity. J Neurochem 34:213−215

Maul GG (1979) Temperature dependent changes in intramembrane particle distribution. In: Rash JE, Hudson CS (eds) Freeze fracture methods, artifacts and interpretations. Raven, New York, pp 37−42

Mazur P (1970) Cryobiology, the freezing of biological systems. Science 168:939−949

McIntyre JA, Gilula NB, Karnovsky MJ (1974) Cryoprotectant induced redistribution of intramembranous particles in mouse lymphocytes. J Cell Biol 60:192−203

McLean JD, Nakane PK (1974) Periodate-lysine-paraformaldehyde fixative: A new fixative for immunoelectron microscopy. J Histochem Cytochem 22:1077−1082

McLean JD, Singer SJ (1970) A general method for the specific staining of intracellular antigens with ferritin-antibody conjugates. Proc Nat Acad Sci 65:122−128

McNeill TH, Sladek JR Jr (1980) Simultaneous monoamine histofluorescence and neuropeptide immunocytochemistry: V. A methodology for examining correlative monoamine-neuropeptide neuroanatomy. Brain Res Bull 5:599−608

McNeill TH, Scott DE, Sladek JR Jr (1980) Simultaneous monoamine histofluorescence and neuropeptide immunocytochemistry: I. Localization of catecholamines and gonadotropin-releasing hormone in the rat median eminence. Peptides 1:59−68

Meek GA (1976) Practical electron microscopy for biologists. 2nd ed, Wiley, New York, pp 528

Meinertzhagen IA (1973) Development of the compound eye and optic lobe of insects. In: Young D (ed) Developmental neurobiology of arthropods. Cambridge University Press, pp 51–104

Mensah PL, Cascio A, Thompson RF, Glanzman F, Glanzman D (1980) Vesicular transport of horseradish peroxidase by ependymal cells of the medulla oblongata. Brain Res 196:483–488

Mepham BL, Frater W, Mitchell BS (1979) The use of proteolytic enzymes to improve immunoglobulin staining by the PAP technique. J Histochem 11:345–357

Merryman HT (1950) Replication of frozen liquids by vacuum evaporation. J Appl Physics 21:68

Mesulam MM (1976) The blue reaction product in horseradish peroxidase neurohistochemistry: Incubation parameters and visibility. J Histochem Cytochem 24:1273–1280

Mesulam MM (1978) Tetramethyl benzidine for horseradish peroxidase neurohistochemistry: A non-carcinogenic blue reaction product with superior sensitivity for visualizing neural afferents and efferents. J Histochem Cytochem 26:106–117

Mesulam MM (1981) Enzyme histochemistry of horseradish peroxidase for tracing neuronal connections with the light microscope. In: Johnson JE (ed) Current trends in morphological techniques, Vol I. CRC Press Inc, Boca Raton Florida, 1–54

Mesulam MM (ed) (1982) Tracing neural connections with horseradish peroxidase. John Wiley, Sussex England

Mesulam MM, Rosene DL (1979) Sensitivity in horseradish peroxidase neurohistochemistry: A comparative and quantitative study of nine methods. J Histochem Cytochem 27:763–773

Metzler DE (1977) The chemical reactions of living cells. Academic Press, New York

Meyer EP (1983) The identification of mass impregnated visual cells in the ant *Cataglyphis bicolor*. Experientia (in press)

Micheel B (1978) The use of trypan blue counterstaining in the sepharose bead immunofluorescence test. The application of the test for the demonstration of primate retrovirus specific antibodies and antigens. Acta Biol Med Ger 27:K 19

Miller JP, Selverston AI (1979) Rapid killing of single neurons by irradiation of intracellularly injected dyes. Science 206:702–704

Millonig G (1961) Advantages of a phosphate buffer for OsO$_4$ solutions in fixation. J Appl Phys 32:1637

Millonig G, Marinozzi V (1968) Fixation and embedding in electron microscopy. In: Barer R, Coslett VE (eds) Advances in optical and electron microscopy, Vol 2. Academic Press, New York, p 251

Mobbs PG (1976) Golgi staining of material containing cobalt-filled profiles in the insect CNS. Brain Res 105:563–566

Molinoff PB, Landsberg L, Axelrod J (1969) An enzymatic assay for octopamine and other β-hydroxylated phenylamines. J Pharmacol Exp Ther 170:253–261

Monsell EM (1980) Cobalt and horseradish peroxidase tracer studies in the stellate ganglion of *Octopus*. Brain Res 184:1–9

Moor H (1959) Platin-Kohle-Abdruck-Technik angewandt auf den Feinbau der Milchröhren. J Ultrastruct Res 2:393–422

Moor H (1971) Recent progress in the freeze-etching technique. Phil Trans R Soc Lond, B 261:121–131

Moor H (1973) Evaporation and electron guns. In: Benedetti EL, Favard P (eds) Freeze-etching techniques and applications. Soc Française de Microscopie Electronique, Paris, pp 27–30

Moor H (1977) Limitations and prospects of freeze-fixation and freeze etching. In: Bailey GW (ed) Proc 35th Annual Meeting Electron Microscopy Soc Am Boston, Mass, Claitor's Div Baton Rouge, pp 334–337

Moor H, Mühlethaler K, Waldner H, Frey-Wyssling A (1961) A new freezing-ultramicrotome. J Biophys Biochem Cytol 10:1–13

Moor H, Pfenninger K, Akert K (1969) Synaptic vesicles in electron micrographs of freeze-etched nerve terminals. Science 164:1405–1407

Moore RY (1980) Fluorescence histochemical methods. In: Heimer L, Robards MJ (eds) Neuroanatomical tract-tracing methods. Plenum New York, pp 441–482

Moore RY, Loy R (1978) Fluorescence histochemistry. In: Robertson RT (ed) Neuroanatomical research techniques. Academic Press, New York, pp 115–139

Moran DT, Rowley III JC (1975) High voltage and scanning electron microscopy. J Ultrastruct Res 50:38–46

Moreau R, Raoelison C, Sutter BC Jr (1981) An intestinal insulinlike molecule in *Apis mellifica* L. (Hymenoptera). Comp Biochem Physiol 69:79–83

Morell JI, Greenberger LM, Pfaff DW (1981) Comparison of horseradish peroxidase visualization methods: Quantitative results and further technical specifics. J Histochem Cytochem 29:903–916

Moriarty GC (1976) Immunocytochemistry of the pituitary glycoprotein hormones. J Histochem Cytochem 24:846–863

Muller KJ (1979) Synapses between neurones in the central nervous system of the leech. Biol Rev 54:99–134

Muller KJ, McMahan UJ (1976) The shapes of sensory and motor neurons and the distribution of their synapses in ganglia of the leech: a study using intracellular injection of horseradish peroxidase. Proc R Soc B 194:481–499

Muller LL, Jacks TJ (1975) A rapid chemical dehydration of samples for electron microscopic examination. J Histochem Cytochem 21:107

Murphey RK (1973) Characterization of an insect neuron which cannot be visualized in situ. In: Kater SB, Nicholson B (eds) Intracellular staining in neurobiology. Springer, Berlin Heidelberg New York, pp 135–150

Myhrberg HE, Elofsson R, Aramant R, Klemm N, Laxmyr L (1979) Selective uptake of exogenous catecholamines into nerve fibres in crustaceans. A fluorescence histochemical investigation. Comp Biochem Physiol 62C:141–150

Nairn RC (1976) Fluorescent protein tracing. 4th ed, Churchill Livingstone, New York

Nakane PK (1968) Simultaneous localization of multiple tissue antigens using peroxidase-labeled antibody method: A study on pituitary glands of the rat. J Histochem Cytochem 16:560–577

Nakane PK (1971) Application of peroxidase-labeled antibodies to intracellular localization of hormones. Acta Endocrinol (Kbh) Suppl 153:190–204

Nakane PK (1975) Localisation of hormones with the peroxidase-labeled antibody method In: O'Malley BW, Hardman JG (eds) Methods Enzymol 37:133–144

Nakane PK, Kawaoi A (1974) Peroxidase-labeled antibodies: A new method of conjugation. J Histochem Cytochem 22:1084–1091

Nakane PK, Pierce GB Jr (1966) Enzyme-labeled antibody: preparation and application for the localization of antigens. J Histochem Cytochem 14:929–931

Nässel DR (1981) Transneuronal labeling with horseradish peroxidase in the visual system of the house fly. Brain Res 206:431–438

Nässel DR (1982) Transneuronal uptake of horseradish peroxidase in the central nervous system of dipterous insects. Cell Tissue Res 225:639–662

Nässel DR (1983) Extensive labeling of injured neuron with seven different heme peptides. Histochem (in press)

Nässel DR, Geiger G (1983) Neuronal organization in fly optic lobes altered by laser ablations in development of by mutations of the eye. J comp Neurol 218:1–17

Nässel DR, Berriman JA, Seyan HS (1981) Cytochrome c as a high resolution marker of neurons for light and electron microscopy. Brain Res 206:439–445

Nässel DR, Klemm N (1983) Serotonin-like immunoreactivity in the optic lobes of three insect species. Cell Tissue Res 232:129–140

Nässel DR, Sivasubramanian P (1983) Neural differentiation in fly CNS transplants cultured in vivo. J Exp Zool 225:301–310

Nässel DR, Strausfeld NJ (1982) A pair of descending neurons with dendrites in the optic lobes projecting directly to thoracic ganglia of dipterous insects. Cell Tissue Res 226:355—362

Nässel DR, Geiger G, Seyan HS (1983) Differentiation of fly visual interneurons after laser ablation of their central targets early in development. J comp Neurol 217:8—15

Nässel DR, Berriman JA, Seyan HS (1981) Cytochrome c as a high resolution marker of neurons for light and electron microscopy. Brain Res 206:439—445

Nauta WJH, Gygax PA (1951) Silver impregnation of degenerating axon terminals in the central nervous system. Stain Technol 26:5—11

Newman WM, Sproull RF (1979) Principles of interactive computer graphics. (2nd Ed) McGraw-Hill, New York

Nicholson C, Kater SB (1973) The development of intracellular staining. In: Kater SB, Nicholson C (eds) Intracellular staining in neurobiology. Springer, Berlin Heidelberg New York, pp 1—9

Nicod I (1983) Visual mutants of *Drosophila melanogaster*. Functional neuroanatomical mapping of nervous activity by ^{3}H-deoxyglucose. Doctoral Thesis, University of Lausanne

Nicol D, Meinertzhagen IA (1982) An analysis of the number and composition of the synaptic populations formed by photoreceptors of the fly. J Comp Neurol 207:29—44

Noorden S van, Fritsch HAR, Grillo TAI, Polak JM, Pearse AGE (1980) Immunocytochemical staining for vertebrate peptides in the nervous system of a gastropod mollusc. Gen Comp Endocrinol 40:375—376

Norris DM (1979) Chemoreceptor proteins. In: Narahashi T (ed) Neurotoxicology of insecticides and pheromones. Plenum, New York, pp 59—77

Notani GW, Parsons JA, Erlandsen SL (1979) Versatility of *Staphylococcus aureus* protein A in immunocytochemistry. Use in unlabeled antibody enzyme system and fluorescent methods. J Histochem Cytochem 27:1438—1444

Nowakowski RS, LaVail JH, Rakic P (1975) The correlation of the time of origin of neurons with their axonal projection: the combined use of (^{3}H) thymidine autoradiography and horseradish peroxidase histochemistry. Brain Res 99:343—348

Nowotny A (1969) Basic exercises in immunochemistry. A laboratory manual. Springer, Berlin Heidelberg New York, pp 157—158

Okoshi T (1976) Three-dimensional imaging techniques. Academic Press, New York San Francisco London, pp 403

Olsson T, Kristensson K (1978) A simple histological method for double labeling of neurons by retrograde axonal transport. Neurosci Lett 8:265—268

Olsson Y, Hossman KA (1970) Fine structural localization of exudated protein tracers in the brain. Acta Neuropathol 16:103—116

Orchard I, Loughton BG (1980) A hypolipaemic factor from the corpus cardiacum of locusts. Nature 286:494—496

Orchard I, Webb RA (1980) The projections of neurosecretory cells in the brain of the North American medicinal leech *Macrobdella decora* using intracellular injection of horseradish peroxidase. J Neurobiol 11:229—242

Orci L, Perrelet A (1975) Freeze etch histology. Springer, Berlin Heidelberg New York

Ormanns W, Pfeifer U (1981) A simple method for incubation of tissue sections in immunohistochemistry. Histochemistry 72:315—319

Ornstein L (1968) Benzidine analogues used with α-naphthol in new variants of oxidase and peroxidase cytochemical reactions. 19th Ann Meeting, Histochem Soc New Orleans, p 6

Ørstavik TB, Brandtzaeg P, Nustad K, Pierce JV (1981) Effects of different tissue processing methods on the immunohistochemical localization of kallikrein in the pancreas. J Histochem Cytochem 29:985—988

Osborne MP (1980) Electron-microscopic methods for nervous tissue. In: Strausfeld NJ, Miller TA (eds) Neuroanatomical techniques: Insect nervous system. Springer, New York Heidelberg Berlin, pp 205—239

O'Shea M (1982) Peptide neurobiology: An identified neuron approach with special reference to proctolin. Trends Neurosci 5:69–73

O'Shea M, Adams ME (1981) Pentapeptide (proctolin) associated with an identified neuron. Science 213:567–569

O'Shea M, Bishop CA (1983) Neuropeptide proctolin associated with an identified skeletal motoneuron. J Neurosci 2:1242–1251

O'Shea M, Evans PD (1979) Potentiation of neuromuscular transmission by an octopaminergic neurone in the locust. J Exp Biol 79:169–190

Palay SL, Chan-Palay V (1974) Cerebellar cortex cytology and organization. Springer, Berlin Heidelberg New York

Palka J, Schubiger M (1980) Formation of central patterns by receptor cell axons in *Drosophila*. In: Siddiqi O, Babu P, Hall LM, Hall JC (eds) Development and neurobiology of *Drosophila*. Plenum, New York, pp 223–246

Palka J, Lawrence PA, Hart HS (1979) Neuronal projection patterns from homeotic tissue of *Drosophila* studied in bithorax mutants and mosaics. Dev Biol 69:549–575

Panov AA (1980) Demonstration of neurosecretory cells in the insect central nervous system. In: Strausfeld NJ, Miller TA (eds) Neuroanatomical techniques: Insect nervous system. Springer, New York Heidelberg Berlin, pp 25–50

Parnavelas JG, Sullivan K, Lieberman AR, Webster KE (1977) Neurons and their synaptic organization in the visual cortex of the rat. Electron microscopy of Golgi preparations. Cell Tissue Res 183:499–517

Pasik P, Pasik T, Saavedra JP (1982) Immunocytochemical localization of serotonin at the ultrastructural level. J Histochem Cytochem 30:760–764

Pawley J (1980) Recent improvements in the EM-7 at the Madison Bio-Technology Resource. 38th Ann Proc Electron Microscopy Soc Am (Bailey GW ed), San Francisco CA

Pearse AGE (1960) Histochemistry: Theoretical and applied. Churchill, London, Table 52, p 691

Pearse AGE (1972) Histochemistry: theoretical and applied. Vol 2, Churchill, London

Pearse AGE (1980) In: Histochemistry theoretical and applied, Vol 1. Preparative and optical technology. Churchill, Livingstone, 4th Ed

Pearse AGE, Polak JM (1975) Bifunctional reagents as vapour- and liquid-phase fixatives for immunohistochemistry. J Histochem 7:179–186

Peetoom F (1967) Antiserum production. Acta Histochem Suppl 7:43–54

Peskar B, Spector S (1973) Serotonin: radioimmunoassay. Science 179:1340–1341

Peters A, Fairén A (1978) Smooth and sparsely-spined stellated cells in the visual cortex of the rat: a study using a combined Golgi-electron microscopy technique. J Comp Neurol 181:129–172

Peters A, Proskauer CC, Feldman ML, Kimerer L (1979) The projection of the lateral geniculate nucleus to area 17 of the rat cerebral cortex. V. Degenerating axon terminals synapsing with Golgi impregnated neurons. J Neurocytol 8:331–357

Petrusz P, Sar M, Ordronneau P, DiMeo P (1976) Specificity in immunocytochemical staining. J Histochem Cytochem 24:1110–1115

Phillips CE (1980) Intracellularly injected cobaltous ions accumulate at synaptic densities. Science 207:1477–1479

Pichon Y (1974) The pharmacology of the insect nervous system. In: Rockstein M (ed), Physiology in insects, volume 4. Academic Press, New York, pp 101–174

Pickel V (1979) Immunocytochemical localization of neuronal antigens: Tyrosine hydroxylase, substance P, Met[5]-encephalin. Fed Proc 38:2374–2380

Pickel VM, Tong HJ, Reis DJ, Leeman SE, Miller RJ (1979) Electron microscopic localization of substance P and encephalin in axon terminals related to dendrites of catecholaminergic neurons. Brain Res 160:387–400

Piek T, Mantel P (1977) Myogenic contractions in locust muscle induced by proctolin and by wasp, *Philanthus triangulum* venom. J Insect Physiol 23:321–325

Pierantoni R (1974) An observation on the giant fibre posterior optic tract of the fly. Biokybernetik 5:157–163

Pierantoni R (1976) A look into the cock-pit of the fly: the architecture of the lobula plate. Cell Tissue Res 171:101−122

Pinching AJ, Brook RNL (1973) Electron microscopy of single cells in the olfactory bulb using Golgi impregnation. J Neurocytol 2:157−170

Pique L, Cesselin F, Strauch G, Valcke JC, Bricaire H (1978) Specificity of anti-LH-RH antisera induced by different immunogens. Immunochemistry 15:55−60

Pitman RM (1979) Block intensification of neurons stained with cobalt sulphide: a method for destaining and enhanced contrast. J Exp Biol 78:295−297

Pitman RM, Tweedle CD, Cohen MJ (1972) Branching of central neurons: Intracellular cobalt injection for light and electron microscopy. Science 176:412−414

Pitman RM, Tweedle CD, Cohen MJ (1973) The form of nerve cells: determination by cobalt impregnation. In: Kater SB, Nicholson C (eds) Intracellular staining in neurobiology. Springer, Berlin Heidelberg New York, pp 83−97

Plapinger RES, Linus SL, Kawashima T, Deb C, Seligman AM (1968) Preparation and structure-activity-relationship of reagents for cytochrome oxidase activity: Potential for light and electron microscopy. Histochem 14:1−6

Poggio T, Reichardt W (1981) Visual fixation and tracking by flies: mathematical properties of simple control systems. Biol Cybern 40:101−112

Polge C, Smith AU, Parkes AS (1947) Revival of spermatozoa after vitrification and dehydration at low temperatures. Nature 164:666

Politoff A, Pappas GD, Bennet MVL (1974) Cobalt ions can cross an electronic synapse if the concentration is low. Brain Res 76:343−346

Poole AR, Howell JI, Lucy JA (1970) Lysolecithin and cell fusion. Nature 227:810−814

Porstmann T, Porstmann B, Schmechta H, Nugel E, Seifert R, Grunow R (1981) Effect of IgG-horseradish peroxidase conjugates purified on ConA-Sepharose upon sensitivity of enzyme immunoassay. Acta Biol Med Germ 40:849−859

Porter KR, Tucker JB (1981) The ground substance of the living cell. Sci Am 244:57−67

Power ME (1943) The effect of reduction in numbers of ommatidia upon the brain of Drosophila melanogaster. J Exp Zool 94:33−72

Power ME (1948) The thoracico-abdominal nervous system of an adult insect, Drosophila melanogaster. J Comp Neurol 88:347−409

Priestly JV, Somogyi P, Cuello AC (1981) Neurotransmitter-specific projection neurons revealed by combining PAP immunohistochemistry with retrograde transport of HRP. Brain Res 220:231−240

Qualman SJ, Keren DF (1979) Immunofluorescence of deparaffinized, trypsin-treated tissues: Preservation of antigens as an adjunct to diagnosis of disease. Lab Invest 41:483−489

Quicke DLJ, Brace RC (1979) Differential staining of cobalt- and nickel-filled neurons using rubeanic acid. J Microsc 115:1−4

Quicke DLJ, Brace RC, Kirby P (1980) Intensification of nickel- and cobalt-filled neurone profiles following differential staining by rubeanic acid. J Microsc 119:267−272

Rademakers LHPM (1977) Identification of a secretomotor centre in the brain of Locusta migratoria, controlling the secretory activity of the adipokinetic hormone producing cells of the corpus cardiacum. Cell Tissue Res 184:381−395

Ramón-Moliner E, Ferrari J (1972) Electron microscopy of previously identified cells and processes within the central nervous system. J Neurocytol 1:85−100

Ramón-Moliner E, Ferrari J (1976) Electron microscopy of Golgi-stained material following lead chromate substitution. Brain Res 103:339−344

Rash JE, Hudson CS (1979) Freeze fracture methods, artifacts and interpretations. Raven, New York

Reading M (1977) A digestion for the reduction of background staining in the immunoperoxidase method. J Clin Pathol 30:88−90

Ready JF (1978) Industrial applications of lasers. Academic Press, New York

Rebhun LI (1972) Freeze-substitution and freeze-drying. In: Hayat MA (ed) Principles and techniques of electron microscopy, Vol 2. Nostrand van, New York

Reese TS, Bennett MVL, Feder N(1971) Cell-to-cell movement of peroxidases injected into the septate axon of crayfish. Anat Rec 169:409

Rehfeld JF (1980) Cholecystokinin. Trends Neurosci 3:65−67

Reichardt LF, Mathew WD (1982) Monoclonal antibodies. Applications to studies on the chemical synapse. Trends Neurosci 5:24−31

Rémy C, Dubois MP (1981) Immunohistological evidence of methionine encephalin-like material in the brain of the migratory locust. Cell Tissue Res 218:271−278

Rémy C, Girardie J (1980) Anatomical organization of two vasopressin-neurophysin-like neurosecretory cells throughout the central nervous system of the migratory locust. Gen Comp Endocrinol 40:27−35

Rémy C, Giradie J, Dubois MP (1977) Exploration immunocytologique des ganglions cérébroides et sous-oesophagien due phasme *Clitumnus extradentatus:* existence d'une neuroscrétion apparentée à la vasopressine-neurophysine. CR Acad Sci 285: D 1495−1497

Rémy C, Girardie J, Dubois MP (1978) Présence dans le ganglion sous-oesophagien de la chenille processionnaire du Pin (*Thaumetopoea pityocampa* Schiff) de cellules révélées en immunofluorescence par un anticorps anti-a-endorphine. CR Acad Sci Ser D 286:651−653

Rémy C, Girardie J, Dubois MP (1979) Vertebrate neuropeptide-like substances in the suboesophageal ganglion of two insects: *Locusta migratoria* R and F (Orthoptera) and *Bombyx mori* L (Lepidoptera) immunocytological investigation. Gen Comp Endocrinol 37:93−100

Reynolds ES (1963) The use of lead citrate at high pH as an electron opaque stain in electron microscopy. J Cell Biol 17:208−212

Ribi WA (1975) The first optic ganglion of the bee. I. Correlation between visual cell types and their terminals in the lamina and medulla. Cell Tissue Res 165:103−111

Ribi WA (1976a) A Golgi-electron microscope method for insect nervous tissue. Stain Technol. 51:13−16

Ribi WA (1976b) The first optic ganglion of the bee. II. Topographical relationship of second order neurons within a cartridge and to groups of cartridges. Cell Tissue Res 171:359−373

Ribi WA (1978) A unique hymenopteran compound eye. The retina fine structure of the digger wasp *Sephex cognatus* Smith (Hymenoptera, Sphecidae). Zool Jb Anat 100:299−342

Ribi WA, Berg GJ (1980) Light and electron microscopic structure of Golgi-stained neurons in the vertebrate brain (new rapid Golgi procedure). Cell Tissue Res 205:1−10

Richmann RA, Kopf GS, Hamat P, Johnson RA (1980) Preparation of cyclic nucleotide antisera with thyroglobulin-cyclic nucleotide conjugates. J Cycl Nucl Res 6:461−468

Ris H, Allen C (1975) Embedding in negative stain as a method to preserve and contrast isolated cell organelles for HVEM. Proc 4th Int Congr Microscopic Electronique à Haute Tension, Toulouse, pp 365−368

Ritzen M (1967) Cytological identification and quantitation of biogenic monoamines. A microspectrofluorimetric and autoradiographic study. Thesis, Stockholm

Robertson HA, Juorio AV (1976) Octopamine and some related non-catecholic amines in invertebrate nervous systems. Int Rev Neurosci 19:173−224

Robinson AL (1981) Electron microscope center opens at Berkeley. Science 211:1407−1410

Rogers AW (1973) Techniques of autoradiography. Elsevier, Amsterdam

Rogers DF (1976) Mathematical elements for computer graphics. McGraw-Hill, New York

Rooijen N van (1980) Six methods for separate detection of two different antigens in the same tissue section. J Histochem Cytochem 28:716

Rooijen N van (1981) The application of autoradiography and other immunohistochemical techniques to the same tissue section or cell smear. J Immunol Methods 40:247−252

Roth J (1982) The protein A-gold (pAG) technique; qualitative and quantitative approach for antigen localization on thin sections. In: Bullock SR, Petrusz P (eds) Techniques in Immunocytochemistry. Vol 1. Academic Press, London, pp 107−133

Roth J, Bendayan M, Orci L (1978) Ultrastructural localization of intracellular antigens by the use of protein A-gold complex. J Histochem Cytochem 26:1074—1081

Roth J, Bendayan M, Orci L (1980) FITC-protein A-gold complex for light and electron microscopic immunohistochemistry. J Histochem Cytochem 28:55—57

Roth RL, Sokolove PG (1976) Histological evidence for direct connection between the optic lobes of the cockroach *Leucophaea maderae*. Brain Res 87:23—39

Rowell FCH (1963) A general method for silvering invertebrate central nervous systems. Quart J Microsc Sci 104:81—87

Ruda M, Coulter JD (1982) Axonal and transneuronal transport of wheat germ agglutinin demonstrated by immunocytochemistry. Brain Res 249:237—246

Saavedra JM, Brownstein M, Axelrod J (1973) A specific and sensitive enzymatic-isotope microassay for serotin in tissue. J Pharmac Exp Ther 186:508—515

Saint Marie RL (1981) A thin section and freeze fracture study of intercellular junctions and synaptic vesicle activity in the first optic neuropil of the housefly compound eye. Ph D thesis in Neurosciences. University of Wisconsin, Madison, pp 138

Saint Marie RL, Carlson SD (1982) Synaptic vesicle activity in stimulated and unstimulated photoreceptor axons in the housefly. A freeze-fracture study. J Neurocytol 11:141—161

Saint Marie RL, Carlson SD (1983a) The fine structure of neuroglia in the lamina of the housefly *Musca domestica* L. J Neurocytol 12:213—241

Saint Marie RL, Carlson SD (1983b) Glial membrane specializations and the compartmentalization of the lamina ganglionaris of the housefly compound eye. J Neurocytol 12:243—275

Sakai M, Yamaguchi T (1983) Differential staining of insect neurons with nickel and cobalt. J Insect Physiol 29:393—397

Salema R, Brandao I (1973) The use of PIPES buffer in the fixation of plant cell for electron microscopy. J Submicr Cytol 5:79

Sandemann DC, Markl H (1980) Head movements in flies (*Calliphora*) produced by deflexion of the halteres. J Exp Biol 85:43—60

Satir BH, Satir P (1979) Partitioning of intramembrane particles during the freeze-fracture procedure. In: Rash JE, Hudson CS (eds), Freeze fracture: Methods, artifacts and interpretations. Raven, New York, pp 43—49

Satir P, Gilula NB (1973) The fine structure of membranes and intercellular communication in insects. Ann Rev Entomol 18:143—166

Saunders BC, Holmes-Siedle AG, Stark BP (1964) Peroxidase: The properties and users of a versatile enzyme and related catalysts. Butterworth, London

Sawchenko PE, Swanson LW (1981) A method for tracing biochemically defined pathways in the central nervous system using combined fluorescence retrograde transport and immunohistochemical techniques. Brain Res 210:31—51

Schacher SM, Holtzmann E, Hook DC (1974) Uptake of horseradish peroxidase by frog photoreceptor synapses in the dark and the light. Nature 249:261—263

Schenk EA, Churukian CJ (1974) Immunofluorescence counterstains. J Histochem Cytochem 22:962—966

Schiff RI, Gennaro JF Jr (1979) The role of the buffer in the fixation of biological specimens for transmission and scanning electron microscopy. Scanning 2:135—148

Schipper J, Tilders FJH (1983) A new technique for studying specificity of immunocytochemical producers: specificity of serotonin immunostaining. J Histochem Cytochem 31:12—18

Schipper J, Steinbusch HWM, Verhofstad AAJ, Tilders FJH (1981) Quantitative immunofluorescence of serotonin. Proc 22nd Dutch Fed Meeting, p 390

Schulze WH (1909) Verh Dt Path Ges, 13 and Beitr Path Anat, 45. Quoted by Lison (1936) and Pearse (1972)

Schürmann FW (1980) Experimental anterograde degeneration of nerve fibres: A tool for combined light- and electron microscopic studies of the insect nervous system. In:

Strausfeld NJ, Miller TA (eds) Neuroanatomical techniques: Insect nervous system. Springer, New York Heidelberg Berlin, pp 263–281

Schwaab ME (1977) Ultrastructural localization of nerve growth factor-horseradish peroxidase (NGF-HRP) coupling product after retrograde axonal transport in adrenergic neurons. Brain Res 130:190–196

Schwartz WJ, Smith CB, Davidsen L, Savaki H, Sokoloff L, Mata M, Fink DJ, Gainer H (1979) Metabolic mapping of functional activity in the hypothalamo-neurohypophysical system of the rat. Science 205:723–725

Scott GL, Guillery RW (1974) Studies with the high voltage electron microscope of normal, degenerating and Golgi impregnated neuronal processes. J Neurocytol 3:567–590

Sejnowski TJ, Reingold SC, Kelley DB, Gelperin A (1980) Localization of ^{3}H-2-deoxyglucose in single molluscan neurones. Nature 287:449–451

Seligman AM, Karnovsky MJ, Wasserkrug HL, Hanker JS (1968) Non-droplet ultrastructural demonstration of cytochrome oxidase activity with a polymerizing osmophilic reagent diaminobenzidine (DAB). J Cell Biol 38:1

Seligman AM, Seito T, Plapinger RE (1970) Some cytochemical correlations between oxidase activity (cytochrome and peroxidase) and chemical structure of bis-(phenylenediamines). Histochem 23:63

Séláló G, Flerkó B (1978) Brain cells as producers of releasing and inhibiting hormones. In: Bourne GH, Danielli JF (eds) Int Rev Cytol, Suppl 7. Academic Press, New York, pp 1–52

Shaw SR (1975) Retinal resistance barriers and electrical lateral inhibition. Nature 255:480–483

Shaw SR (1979) Signal transmission by graded slow potentials in the arthropod visual system. In: Schmitt FO, Worden RG (eds) The Neurosciences: Fourth study program. MIT Press, Cambridge MA, pp 275–295

Shaw SR (1981) Anatomy and physiology of identified non-spiking cells in the photoreceptor-lamina complex of the compound eye of insects, especially Diptera. In: Roberts A, Bush BMH (eds) Neurons without impulses. Cambridge University Press, Cambridge pp 61–116

Shaw SR, Stowe S (1982) Freeze-fracture evidence for gap junctions connecting the axon terminals of dipteran photoreceptors. J Cell Sci 53:115–141

Shelton PMJ, Horridge GA, Meinertzhagen IA (1971) Reconstruction of synaptic geometry and neural connections from serial thick sections examined by the medium high voltage electron microscope. Brain Res 29:374–377

Shivers RR (1976) Trans-glial channel-facilitated translocation of tracer protein across ventral nerve root sheaths of crayfish. Brain Res 108:47–58

Sigee DC (1976) A resin slide technique to select fixed embedded cells for transmission electron microscopy. J Microsc 108:325–329

Sigee DC, Kearns LP (1980) Detection of nickel in the chromatin of dinoflagellates by X-ray microanalysis. In: Brown, Sunderman (eds) Nickel toxicology. Academic Press, London, pp 163–166

Silva PP da (1979) Interpretation of freeze-fracture and freeze-etch images: Morphology and realism. In: Rash JE, Hudson CS (eds), Freeze-fracture: Methods, artifacts and interpretations. Raven, New York, pp 185–193

Simmons PJ (1981) Synaptic transmission between second- and third-order neurones of a locust ocellus. J Comp Physiol 145:265–276

Singer W, Hollander H, Vanegas H (1977) Decreased peroxidase labeling of lateral geniculate neurons following deafferentation. Brain Res 120:133–137

Skaer H Le B, Lane NJ (1974) Junctional complexes, perineurial and gliaaxonal relationship and the ensheathing structures of the insect nervous system. A comparative study using conventional and freeze-cleaving techniques. Tissue and Cell 6:695–718

Skowsky MR, Fisher DA (1972) The use of thyroglobulin to induce antigenicity to small molecules. J Lab Clin Med 80:134–144

Slade TC, Mills J, Winlow W (1981) The neuronal organization of the paired pedal ganglia of *Lymnaea staghalis.*Comp Biochem Physiol 69A:789−803

Slemmon JR, Salvaterra PM, Soaito K (1980) Preparation and characterization of peroxidase-anti-peroxidase Fab complex. J Histochem Cytochem 28:10−15

Sleytr UB, Messner P (1978) Freeze fracturing in normal vacuum reveals ringlike yeast plasmalemma structures. J Cell Biol 79:276−280

Sleytr UB, Robards AW (1977) Freeze-fracturing: A review of methods and results. J Microsc 111:77−100

Smith RE, Farquhar MG (1966) Lysosome function in the regulation of the secretory process in cells of the anterior pituitary gland. J Cell Biol 31:319−347

Smyth TR (1977) Analytical biochemistry of insect neurotransmitters and their enzymes. In: Turner RB (ed) Analytical biochemistry of insects. Elsevier, New York, pp 289−312

Snow PJ, Rose PK, Brown A (1976) Tracing axons and axon collaterals of spinal neurons using intracellular injection of horseradish peroxidase. Science 191:312−313

Sobel I, Levinthal C, Macagno ER (1980) Special techniques for the automatic computer reconstruction of neural structures. Ann Rev Biophys 9:347−362

Sofroniew MV, Madler M, Müller OA, Sxriba PC (1978) A method for the consistent production of high quality antisera to small peptide hormones. Fresenius Z Anal Chem 290:163

Sokoloff L (1978) Mapping cerebral functional activity with radioactive deoxyglucose. Trends Neurosci 1:75−79

Sokoloff L, Reivich M, Kennedy C, Des Rosiers MH, Patlak CS, Pettigrew KD, Sakurada O, Shinohara M (1977) The ^{14}C-deoxyglucose method for the measurement of local cerebral glucose utilization: theory, procedure and normal values in the conscious and anesthetized albino rat. J Neurochem 28:897−916

Somogyi P, Hodgson A, Smith A (1979) An approach to tracing neuron networks in the cerebral cortex and basal ganglia. Combination of Golgi staining, retrograde transport of horseradish peroxidase and anterograde degeneration of synaptic boutons in the same material. Neuroscience 4:1805−1852

Speck PT (1981a) NEU Program Reference Manual. EMBL Internal Report, Heidelberg

Speck PT (1981b) NEU Users Manual. EMBL Internal Report, Heidelberg

Speck PT (1981c) Automatische Darstellung und Interpretation von Linien- und Kantenstrukturen in Digitalbildern. In: Radig B (ed) Modelle und Strukturen. DAGM Symposium, Hamburg, Oktober 1981. Springer, Berlin Heidelberg New York

Spencer HJ, Lynch G, Jones RK (1978) The use of somatofugal transport of horseradish peroxidase for tract tracing and cell labelling. In: Robertson RT (ed) Neuroanatomical research techniques. Academic Press, New York, pp 291−316

Spurr AR (1969) A low viscosity epoxyresin embedding medium for electron microscopy. J Ultrastruc Res 26:31−43

Staehelin LA (1973) Analysis and critical evaluation of the information contained in freeze-etch micrographs. In: Benedetti EL, Favard P (eds) Freeze-etching techniques and applications. Soc Française de Microscopie Electronique, Paris, pp 113−134

Staehelin LA, Bertaud W (1971) Temperature and contamination dependent freeze-etch images of frozen water and glycerol solutions. J Ultrastruct Res 37:146−168

Staines WA, Kimura H, Fibiger HC, McGeer EG (1980) Peroxidase-labeled lectin as a neuroanatomical tracer: evaluation in a CNS pathway. Brain Res 197:485−490

Starrat AN, Brown BE (1975) Structure of the pentapeptide proctolin, a proposed neurotransmitter in insects. Life Sci 17:1253−1256

Steere RL (1957) Electron microscopy of structural detail in frozen biological specimens. J Biophys Biochem Cytol 3:45−60

Steere RL (1973) Preparation of high association freeze etch, freeze fracture, frozen surface and freeze dried replicas in a single freeze-etch module, and the use of stereo electron microscopy to obtain maximum information from them. In: Benedetti EL, Favard P (eds) Freeze-etching techniques and applications. Soc Française de Microscopie Electronique, Paris, pp 223−255

Steere RL, Rash JE (1979) Use of double tilt device (Goniometer) to obtain optimum contrast in freeze-fracture replicas. In: Rash JE, Hudson CS (eds) Freeze Fracture: Methods, artifacts and interpretations. Raven, New York, pp 161–167

Steere RL, Erbe EF, Moseley JM (1979) Controlled contamination of freeze fractured specimens. In: Rash JE, Hudson CS (eds) Freeze fracture: Methods, artifacts and interpretations. Raven, New York, pp 99–109

Steere RL, Erbe EF, Moseley JM (1980) Prefracture and cold fracture images of yeast plasma membranes. J Cell Biol 86:113–122

Stefani M, Martino C de, Zamboni C (1967) Fixation of ejaculated spermatozoa for electron microscopy. Nature 216:173–174

Stein O, Stein Y (1971) Light and electron microscopic radioautography of lipids: techniques and biological applications. Adv Lipid Res 9:1

Steinbusch HWM, Verhofstad AAJ (1978) Immunofluorescent staining of serotonin in the central nervous system. Adv Pharmacol Ther 2:151–160

Steinbusch HWM, Verhofstad AAJ, Joosten HWJ (1978) Localization of serotonin in the central nervous system by immunohistochemistry: description of a specific and sensitive technique and some applications. Neuroscience 3:811–819

Steinbusch HWM, Verhofstad AAJ, Joosten HWJ (1982) Antibodies to serotonin for neuroimmunocytochemical studies. J Histochem Cytochem 30:756–759

Steinbusch HWM, Verhofstad AAJ, Joosten HWJ (1983) Antibodies to serotonin for neuroimmunocytochemical studies on the central nervous system. Methodological aspects and applications. In: Cuello C (ed) IBRO Handbook: Neuroimmunocytochemistry. Wiley, Chichester

Steiner AL, Parker CW, Kipnis DM (1970) The measurement of cyclic nucleotides by radioimmunoassay. In: Greengard P, Costa E (eds) Role of cyclic AMP in cell function. Adv Biochem Psychopharmacol 3:Raven, New York, pp 89–111

Stell WK (1964) Correlated light and electron microscopy observations on Golgi preparations of goldfish retina. J Cell Biol 23:89 a

Stent GS (1981) Strength and weakness of the genetic approach to the development of the nervous system. Annul Rev Neurosci 4:163–194

Sternberger LA (1974) Immunocytochemistry. In: Oster AG, Weiss L (eds), Foundations of Immunology. Prentice-Hall, Englewood Cliffs, NJ

Sternberger LA (1979) Immunocytochemistry. 2nd ed. Wiley, New York Chichester Brisbane Toronto, pp 354

Sternberger LA, Hardy PH, Cuculis JJ, Meyer HG (1970) The unlabeled antibody enzyme method of immunohistochemistry. Preparation and properties of soluble antigen-antibody complex (horseradish peroxidase–antihorseradish peroxidase) and its use in identification of spirochetes. J Histochem Cytochem 18:315–333

Stevens JK, Davis TL, Friedman N, Sterling P (1980) A systematic approach to reconstructing microcircuitry by electron microscopy of serial sections. Brain Res Rev 2:265–293

Steward O (1983) Horseradish peroxidase and fluorescent substances and their combination with other techniques. In: Heimer L, Robards MJ (eds) Neuroanatomical tract-tracing methods . Plenum New York 279–310

Stewart WW (1978) Functional connections between cells, as revealed by dye-coupling with a highly fluorescent naphthalimide tracer. Cell 14:741–759

Stewart WW (1981) Lucifer dyes – highly fluorescent dyes for biological tracing. Nature 292:17–21

Stocker M, Hilgenfeldt U, Gross F (1979) Production of antibodies against bradykinin. Experienta 35:1113–1114

Stolinski C, Breathnach AS (1975) Freeze-fracture replication of biological tissues. Academic Press, New York

Stone JV, Mordue W, Betley KE, Morris HR (1976) Structure of locust adipokinetic hormone, a neurohormone that regulates lipid utilisation during flight. Nature 265:207–211

Stone JV, Mordue W, Broomfield CE, Hardy PM (1978) Structure-activity relationship for the lipid-mobilising action of locust adipokinetic hormone. Synthesis and activity of a series of hormone analogues. Eur J Biochem 89:195−202

Strambi C, Rougon-Rapuzzi G, Cupo A, Martin N, Strambi A (1979) Mise en évidence immunocytologique d'un composé apparenté à la vasopressine dans le système nervoux du Grillon *Acheta domesticus*. C. R. Acad Sci Paris, Ser D 288:131−133

Straus N (1971) Inhibition of peroxidase by methanol and by methanolnitroferricyanate for use in immunoperoxidase procedures. J Histochem Cytochem 19:682−688

Strausfeld NJ (1973) Golgi method, invertebrate applications. In: Gray P (ed), The encyclopedia of microscopy and microtechnique. Van Nostrand London Melbourne, pp 225−229

Strausfeld NJ (1976) Atlas of an insect brain. Springer, Berlin Heidelberg New York

Strausfeld NJ (1980) The Golgi method, its application to the insect nervous system and the phenomenon of stochastic impregnation. In: Strausfeld NJ, Miller TA (eds) Neuroanatomical techniques: Insect nervous system. Springer, Berlin Heidelberg New York, pp 131−190

Strausfeld NJ (1983) Functional neuroanatomy of the blowfly's visual system. In: Ali MA (ed) Photoreception and vision in invertebrates. Plenum, New York, pp 483−521

Strausfeld NJ, Bassemir UK (1983) Cobalt-coupled neurons of a giant fibre system in Diptera. J Neurocytol 12:(in press)

Strausfeld NJ, Campos-Ortega JA (1973) The L4 monopolar neurone: a substrate for lateral interaction in the visual system of the fly *Musca domestica*. Brain Res 59:97−117

Strausfeld NJ, Campos-Ortega JA (1977) Vision in insects: Pathways possibly underlying neural adaptation and lateral inhibition. Science 195:894−897

Strausfeld NJ, Bacon JP (1983) Multimodal convergence in the central nervous system of dipterous insects. In: Horn E (ed) Multimodal convergence of sensory systems. Fortschritt der Zoologie 27. Gustav Fischer, Stuttgart New York

Strausfeld NJ, Hausen K (1977) The resolution of neuronal assemblies after cobalt injection into neuropil. Proc R Soc Lond B 199:463−476

Strausfeld NJ, Miller TA (1980) Neuroanatomical techniques: Insect nervous system. Springer Series in Experimental Entomology (Miller TA (ed)). Springer, New York Heidelberg Berlin

Strausfeld NJ, Nässel DR (1980) Neuroarchitectures serving compound eyes of crustacea and insects. In: Autrum H (ed) Handbook of sensory physiology VII/6B. Springer, Berlin Heidelberg New York, pp 1−132

Strausfeld NJ, Obermayer M (1976) Resolution of intraneuronal and transsynaptic migration of cobalt in the insect visual and nervous systems. J Comp Physiol 110:1−12

Strausfeld NJ, Singh RN (1980) Peripheral and central nervous system projections in normal and mutant (bithorax) *Drosophila melanogaster*. In: Siddiqi O, Babu P, Hall LM, Hall JC (eds) Development and neurobiology of *Drosophila*. Plenum New York, pp 267−291

Strauss W (1964) Factors affecting the cytochemical reaction of peroxidase with benzidine and the stability of the blue reaction product. J Histochem Cytochem 12:462−469

Strauss W (1980) Factors affecting the sensitivity and specificity of the cytochemical reaction of the anti-horseradish peroxidase antibody in lymph tissue sections. J Histochem Cytochem 28:645−652

Streefkerk JG (1972) Inhibition of erythrocyte pseudoperoxidase activity by treatment with hydrogen peroxidase following methanol. J Histochem Cytochem 20:829−831

Stretton AO, Kravitz EA (1968) Neuronal geometry: determination with a technique of intracellular dye injection. Science 162:132−134

Stretton AO, Kravitz EA (1973) Intracellular dye injection: The selection of procion yellow and its application in preliminary studies of neuronal geometry in the lobster nervous system. In: Kater SB, Nicholson C (eds) Intracellular staining in neurobiology. Springer, Berlin Heidelberg New York, pp 21−40

Stuart AE, Hudspeth AJ, Hall ZW (1974) Vital staining of specific monoamine-containing cells in the leech nervous system. Cell Tissue Res 153:55–61

Stumph WE, Elfin SCR, Hood L (1974) Antibodies to proteins dissolved in sodium dodecyl sulfate. J Immunol 113:1752–1756

Suganuma T, Kowa M, Isaka H, Tsukada Y, Murata F (1980) Immunoradioautographic study of alpha fetoprotein in hepatoma cells. Histochemistry 68:129–132

Sulston JE, White JG (1980) Regulation and cell autonomy during postembryonic development of *Caenorhabditis elegans*. Dev Biol 78:577–597

Sundler F, Håkanson R, Alumets J, Walles B (1977) Neuronal localization of pancreatic polypeptide (PP) and vasoactive intestinal peptide (VIP) immunoreactivity in the earthworm (*Lumbricus terrestris*). Brain Res Bull 2:61–65

Sutherland IE, Sproull RF, Schuhmacher RA (1974) A characterization of ten hidden-surface algorithms. Comput Surv 6:1–55

Swaab DF, Pool CW, Leeuwen FW van (1977) Can specificity ever be proved in immunocytochemical staining? J Histochem Cytochem 25:388–391

Székely G, Kosaras B (1976) Dendro-dendritic contact between frog motoneurons shown with the cobalt labeling technique. Brain Res 108:194–198

Székely G, Kosaras B (1977) Electron microscopic identification of postsynaptic dorsal root terminals: a possible substrate of dorsal root potentials in the frog spinal cord. Exp Brain Res 29:531–593

Taban CH, Cathieni M (1979) Localization of substance P-like immunoreactivity in Hydra. Experientia 35:811–812

Tager HS, Kramer KJ (1980) Insect glucagon-like peptides: evidence for a high molecular weight form in midgut from *Manduca sexta* (L). Insect Biochem 10:617

Tager HS, Markese J, Spiers RD, Kramer KJ (1975) Glucagon-like immunoreactivity in insect corpus cardiacum. Nature 254:707–708

Tager HS, Markese J, Kramer KJ, Spiers RD, Childs CN (1976) Glucagon-like and insulin-like hormones of the insect neurosecretory system. Biochem J 156:515–520

Taghert PH, Bastiani M, Ho RK, Goodman CS (1983) Guidance of pioneer growth cones: filipodial contacts and coupling revealed with an antibody to Lucifer yellow. Dev Biol (in press)

Taylor AC, Weiss P (1965) Demonstration of axonal flow by the movement of tritium-labeled protein in mature optic nerve fibres. Proc Nat Acad Sci Washington 54:1521–1527

Teichberg S, Holzmann E, Crain SM, Peterson ER (1975) Circulation and turnover of synaptic vesicle membrane in cultured fetal mammalian spinal cord neurons. J Cell Biol 67:215–230

Teugels E, Ghysen A (1983) Independence of the numbers of legs and leg ganglia in *Drosophila* bithorax mutants. Nature 304:440–442

Thomas RC, Wilson VJ (1966) Precise localization of Renshaw cells with a new marking technique. Nature 206:211–213

Thompson EB, Bailey CH (1979) Two different and compatible intraneuronal labels for ultrastructural study of synaptically related cells. Brain Res 173:201–208

Thomsen M (1965) The neurosecretory system of the adult *Calliphora erythrocephala*. II. Histology of the neurosecretory cells of the brain and some related structures. Z Zellforsch 67:693–717

Timm F (1958a) Zur Histochemie der Schwermetalle. Das Sulfid-Silber-Verfahren. Dtsch Z Ges Gerichtl Med 46:706–711

Timm F (1958b) Zur Histochemie des Ammonshorngebietes. Z Zellforsch Mikrosk Anat 48:548–555

Torre JC de la (1980) An improved approach to histofluorescence using the SPG method for tissue monoamines. J Neurosci Methods 3:1–5

Torre JC de la, Surgeon JW (1976) A methodological approach to rapid and sensitive monoamine histofluorescence using a modified glyoxylic acid technique: The SPG method. Histochem 49:81–93

Townes CH (1962) Optical masers and their possible applications to biology. Biophys J 2:325–329

Tramu G, Pillez A, Leonardelli J (1978) An efficient method of antibody elution for the successive or simultaneous localization of two antigens by immunocytochemistry. J Histochem Cytochem 26:322–324

Treherne JE (1974) The environment and function of insect nerve cells. In: Treherne JE (ed) Insect neurobiology. North Holland, Amsterdam, pp 187–244

Treherne JE, Pichon Y (1972) The insect blood-brain barrier. Adv Insect Physiol 9:257–313

Treilhou-Lahille F, Crescent M, Taboluet J, Moukhtar MS, Milhaus G (1981) Influences of fixatives on the immunodetection of calcitonin in mouse "C" cells during pre- and postnatal development. J Histochem Cytochem 29:1157–1163

Trelstad RL (1969) The effect of pH on the stability of purified glutaraldehyde. J Histochem Cytochem 17:756–757

Triller A, Korn H (1981) Interneuronal transfer of horseradish peroxidase associated with exo/endocytotic activity in adjacent membranes. Exp Brain Res 43:233–236

Trisler A (1982) Are molecular markers of cell position involved in the formation of neuronal circuits? Trends Neurosci 5:306–310

Trujillo-Cenóz O (1965) Some aspects of the structural organization of the arthropod eye. Cold Spring Harbor Symp Quant Biol 30:371–382

Trujillo-Cenóz O, Melamed J (1970) Light and electron microscope study of one of the systems of centrifugal fibres found in the lamina of muscoid flies. Z Zellforsch 110:336–349

Tubbs RR, Sheibani K (1981) Chromogens for immunochemistry. J Histochem Cytochem 29:684

Turner PT, Harris AB (1973) Ultrastructure of synaptic vesicle formation in cerebral cortex. Nature 242:57–58

Turner PT, Harris AB (1974) Ultrastructure of exogenous peroxidase in cerebral cortex. Brain Res 74:305–326

Twarog BM, Page IH (1953) Serotonin content of some mammalian tissues and urine. Am J Physiol 175:157–161

Tweedle CD (1978) Single-cell staining techniques. In: Robertson RT (ed) Neuroanatomical research techniques. Academic Press, New York, pp 141–174

Tyrer NM, Bell EM (1974) The intensification of cobalt-filled neurone profiles using a modification of Timm's sulphide-silver method. Brain Res 73:151–155

Tyrer NM, Shaw MK, Altman JS (1980) Intensification of cobalt-filled neurons in sections (light and electron microscopy). In: Strausfeld NJ, Miller TA (eds) Neuroanatomical techniques: Insect nervous system. Springer, New York Heidelberg Berlin, pp 426–446

Valnes K, Brandtzaeg P (1981) Unlabeled antibody peroxidase-antiperoxidase method combined with direct immunofluorescence. J Histochem Cytochem 29:703–711

Vandesande F (1979) A critical review of immunocytochemical methods for light microscopy. J Neurosci Methods 1:3–23

Vandesande F, Dierickx K (1975) Identification of the vasopressin producing and of the oxytocin producing neurons in the hypothalamic magnocellar neurosecretory system of the rat. Cell Tissue Res 164:153–162

Vanegas H, Hollander H, Distel H (1978) Early stages of uptake and transport of horseradish peroxidase by cortical structures, and its use for the study of local neurons and their processes. J Comp Neurol 177:193–212

Venable JH, Coggeshall R (1965) A simplified lead citrate stain for use in electron microscopy, J Cell Biol 25:407–408

Venkatachalam MA, Fahimi HD (1969) The use of beef liver catalase as a protein tracer for electron microscopy. J Cell Biol 42:480

Visser TJ, Klootwijk W, Doctor R, Hennemann G (1977) A new radioimmunoassay of thyrotropin-releasing hormone. FEBS Lett 83:37–40

Wagner H (1981) Fliegen beginnen mit der Landung, wenn die „relative retinale Expansionsgeschwindigkeit" eines Landeobjekts einen kritischen Wert überschreitet. Verh Dtsch Zool Ges, Fischer, Stuttgart, p 279

Wagner M (1967) Fluoreszierende Antikörper und ihre Anwendung in der Mikrobiologie. In: Bieling R, Kathe J, Mayr A (eds) Infektionskrankheiten und ihre Erreger. Vol 5, Jena

Walker RJ, Kerkut GA (1978) The first family (adrenaline, noradrenaline, dopamine, octopamine, tyramine, phenylethanolamine and phenylethylamine). Comp Physiol 61 C:261−266

Walsh C (1979) Enzymatic reaction mechanisms. Freeman, San Francisco

Ware RW, LoPresti V (1975) Three-dimensional reconstruction from serial sections. Int Rev Cytol 40:325−440

Warr WB, de Olmos JS, Heimer L (1981) Horseradish Peroxidase. The basic procedure. In: Heimer L, Robards MJ (eds) Neuroanatomical tract-tracing methods. Plenum New York, pp 207−262

Wässle H, Hausen K (1981) Extracellular marking and retrograde labelling of neurons. In: Heym C, Forssman WG (eds) Techniques in neuroanatomical research. Springer, Berlin Heidelberg New York, pp 317−338

Watson AHD, Burrows M (1981) Input and output synapses on identified motor neurons of a locust revealed by the intracellular injection of horseradish peroxidase. Cell Tissue Res 215:325−332

Weber E, Evans CJ, Samuelsson SJ, Barchas JD (1981) Novel peptide neuronal system in rat brain and pituitary. Science 214:1248−1250

Wehrhahn C, Poggio T, Bülthoff H (1982) Tracking and chasing in houseflies (Musca). An analysis of 3-D flight trajectories. Biol Cybern 45:123−130

Weinstein RS, Benefiel DJ, Pauli BV (1979) Use of computers in the analysis of intramembrane particles. In: Rash JE, Hudson CS (eds) Freeze fracture: Methods, artifacts and interpretations. Raven, New York

Weiss P, Hiscoe HB (1948) Experiments on the mechanism of nerve growth. J Exp Zool 107:315−395

Weisblatt DA, Sawyer RT, Stent GS (1978) Cell lineage analysis by intracellular injection of a tracer enzyme. Science 202:1295−1298

Welinder KG, Massa G (1977) Amino acid sequences of heme-linked, histidine-containing peptides of five peroxidases from horseradish and turnip. Eur J Biochem 23:353−358

West LS (1951) The housefly. Comstock, Ithaca New York

West RW (1976) Light and electron microscopy of the ground squirrel retina: functional considerations. J Comp Neurol 168:355−378

Whaley WM, Govindachari TR (1951) The Picet-Spengler synthesis of tetrahydroisoquinones and related compounds. In: Adams R (ed) Organic reactions Vol 6. Wiley, New York, pp 151−190

Wilczynski W, Zakon H (1982) Transcellular transfer of HRP in the amphibian visual system. Brain Res 239:29−40

Williams RJP, Moore GR, Wright PE (1977) Oxidation-reduction properties of cytochromes and peroxidases. In: Addison AW, Cullen WR, Dolphin D, James BR (eds) Biological aspects of inorganic chemistry. Wiley, New York, pp 369−401

Wilson JA (1981) Unique, identifiable local nonspiking interneurons in the locust mesothoracic ganglion. J Neurobiol 12:353−366

Winer JA (1977) A review of the status of the horseradish peroxidase method in neuroanatomy. Biobehav Rev 1:45−54

Wirth N (1976) Algorithms + data structures = programs. Prentice-Hall, Englewood Cliffs

Wohlers DW, Huber F (1982) Processing of sound signals by six types of neurons in the prothoracic ganglion of the cricket, Gryllus campestris L. J Comp Physiol 146:161−173

Wolf KV, Stockem W, Wohlfarth-Botterman KE, Moor H (1981) Cytoplasmic actomyosin fibrils after preservation with high pressure freezing. Cell Tissue Res 217:479−495

Wulliamý T, Ratray S, Mirsky R (1981) Cell surface antigen distinguishes sensory and autonomic peripheral neurons from central neurons. Nature 291:418−420

Wunderlich F, Hozel DF, Speth V, Fischer H (1974) Differential effects of temperature on the nuclear and plasma membranes of lymphoid cells. Biochem. Biophys. Acta 373:34−43

Wyatt GR (1967) The biochemistry of sugars and polysaccharides in insects. In: Beament JWL, Treherne JE, Wigglesworth VB (eds) Adv Insect Physiol 4:287−360

Wyckoff RWG (1949) Freeze-dried preparations for the electron microscope. Science 104:36−37

Yasuda K, Yamamoto N, Yamashita S (1981) Use of peroxidase-avidin conjugates for the demonstration of intracellular antigen. Experientia 37:306−308

Yen EH (1981) A graphics glossary. Computer Graphics 15:208−229

Yezierski RP, Bowker RM (1981) A retrograde double label tracing technique using horseradish peroxidase and the fluorescent dye 4,6-diamino-2-phenylindole 2 HCl (DAPI). J Neurosci Methods, 4:53−62

Young WG, Deutsch JA (1980) Effects of blood-glucose levels on ^{14}C-2-deoxyglucose uptake in rat-brain tissue. Neurosci Lett 20:89−93

Yu WA (1980) Uptake sites of horseradish peroxidase after injections into peritoneal structures: defining some pitfalls. J Neurosci Methods 2:123−133

Yui R, Fujita T, Ito S (1980) Insulin-, gastrin-, pancreatic polypeptide-like immunoreactive neurons in the brain of the silkworm *Bombyx mori* Biomed Res 1:42−46

Zacks SI, Saito A (1969) Uptake of exogenous horseradish peroxidase by coated vesicles in mouse neuromuscular junctions. J Histochem Cytochem 17:161−170

Zimmerman EA, Krupp L, Hoffman DL, Mathew E, Nilavar G (1980) Exploration of peptidergic pathways in brain by immunocytochemistry: a ten year perspective. Peptides 1 Suppl 1:3−10

Zimmermann H (1979) Vesicle recycling and transmitter release. Neuroscience 4:1773−1804

Zipser B (1982) Complete distribution patterns of neurons with characteristic antigens in the leech central nervous system. J Neurosci 2:1453−1469

Zipser B, McKay R (1981a) Monoclonal antibodies specific for identifiable leech (*Haemopis marmorata*) neurons. In: Mackay R, Raff MC, Reichardt F (eds) Cold Spring Harbour reports of the neurosciences. 2. Monoclonal antibodies to neural antigens. Cold Spring Harbour, New York, pp 91−100

Zipser B, McKay R (1981b) Monoclonal antibodies distinguish identifiable neurons in the leech. Nature 289:549−554

Subject Index

Ablation, laser 206
  behavioral correlates of 219–220
  histology of 212–216, 219
  neurosecretory cells 273
  targets 216–218
Acoustic interneurons 139, 145
Activity staining 225–238 (*see also* ³H-2-De-oxy-D-glucose)
Affinity chromatography method 278 (*see also* Hapten isolation)
Aldehyde fluorescence 239, 302
  monoamine loading for 319
  reserpine/vinblastine treatment for 320
  specificity of 321
Aldehydes, blocking of 285
3-Amino-9-ethylcarbazole (*see* Peroxidase activity, demonstration)
Ammonium sulphide carrier, cacodylate 21, 119
  ethanol 113
  PIPES buffer 21
Anaesthesia 3
Antennal nerves 42, 56–57, 117, 127 (*see also Musca domestica, Calliphora*)
Antibodies, chromatography of 279
  isolation of 278–281
  monoclonal 134, 301
  optimal ratio 285
  precipitation of 279
  production of 275–277
Antibody conjugation, to fluorochromes 283–284
  to HRP 284
Antibody generation, by hapten-protein conjugates 272
  by succinylation 275

Antigen, preparation 271–275
  separation by polyacrylamide gel electrophoresis 272
Antigenic determinant 252, 334
Antisera, region-specific 252
  terminus-specific 334
Antisera adsorption 277–278
  controls 257–259
  solid phase affinity purification 278
Antisera quality, enzyme immunoassay 276
  estimation 276
  Ouchterlony technique 276
  radioimmunoassay (RICH) 250, 255, 276
Apis mellifica, electron microscopy of lamina 10, 11, 13
  Golgi-stained lamina neurons 11, 12
  ocellar L-neurons 137
  serotonin in 307
Aplysia, neuron-glia HRP transfer 90
Autoradiography 229–230

Blowfly (*see Calliphora*)
Bluebottle (*see Calliphora*)
Bodian method (*see* Silver staining)
Bombyx mori, endorphin-like peptide 259
  insulin-like peptides 259
  retrocerebral complex 259
Brain-gut peptides (*see* Peptides)
Brain metabolism 225–226, 231
Buffers, additives for whole-mount 5-HT immunostaining 319 (*see also* Lysine)
  cacodylate 91, 119, 138
  glycine- and ethanolamine-HCl 280
  Millonig's phosphate 15, 137
  phosphate (PBS) for immunocytochemistry 285, 295, 304

PIPES (piperazine) 16, 91
sodium acetate 91
Sorensen's phosphate 15, 91
standard phosphate 17
TES [N-Tris (hydromethyl)methyl-2 amino ethane sulfonic acid] 138
Tris-HCl 91, 313
Veronal-sodium acetate 289

Caenorhabditis elegans, computer graphics of 156
laser ablation of 220
Calliphora, antennal lobes 56
antibody-cobalt labelling 331–337
cholecystokinin in 252
descending neurons 119, 151, 202
α-endorphin in 252
Giant Descending Neuron 60
horizontal cells (HS) 174, 196, 197
lobula 119
lobula plate organization 192–203
median bundle neurons 72
median neurosecretory cells 255, 256, 258, 333–337
medulla neurons 84
motor neurons (DLMs) 63
ocellar interneurons 125, 137
photoreceptor HRP-labelling 82
proctolin immunoreactivity in 268
sex-specific neurons 171
vertical cells (VS) 119, 196–198
visual neuropil structure 9, 82
Carbon-platinum electrodes 344–346
Carbon shadowing 345, 347
Cardiac-recurrent nerve, filling of 334
Carrier protein (*see* Hapten conjugates)
Central body 115
Chemical dehydration 68, 69
4-Chloro-1-naphthol 253, 266, 291 (*see also* Peroxidase activity demonstration)
Cholecystokinin (CCK) 250, 252, 267, 331, 333, 334–336 (*see also* Terminus-specific antisera)
Chromogens 47, 253, 266, 292 (*see also* Peroxidase activity demonstration)
Clitumnus extradentatus, neurophysin-vasopressin-like substance in 259
Cobalt affinities to, (*see also* X-ray microanalysis)
cell membrane 33
extracellular spaces 31, 33
gap junctions 31, 33, 35, 42–43
microtubuli 31
mitochondria 28, 41
pre- and postsynaptic sites 31, 42–43

Cobalt artefacts 37, 41, 97, 99, 101
Cobalt contamination 24, 26, 40, 101, 123
Cobalt filling 20, 21, 97
in conjunction with antibody staining 271, 274, 292, 332–337
parallel 97 (*see also* Cobalt artefacts)
sequential 102
Cobalt intensification, Cajal block method 113–114, 117
degree of 24–27, 114
destaining 94
Gallyas block method 27
glycine for block electron microscopy 22
gold toning 114–117
Görcs block method 123
hydroquinone for block electron microscopy 22, 100
hydroquinone for cobalt-Golgi electron microscopy 120–121
hydroquinone for electron microscopic sections 104
hydroquinone for paraffin 336
osmication 22
pyrogallol method 21, 113
with immunocytochemistry 292, 336
Cobalt leakage 35
Cobalt-lysine complex 334
Cobalt-nickel filling 93, 97
Cobalt precipitation, rubeanic acid 94, 97–100
sulphide 19, 21, 97, 113
Cobalt screening, abnormal development 215, 217
mutants 42
Cobalt-silver, electron microscopy 27–37
light microscopy 27–37
uranyl staining for background 27, 41
with other markers (*see* Double marking)
Cockroach (*see Periplaneta americana*)
Colour neutralization 114, 117, 153, 337
Complete Freund's adjuvant 251, 275, 285
Computer graphics, data base 164–167, 185–186
equipment 158–159, 185
operation 163–164, 185–186
programs 159–161, 165, 182, 188
resolution and field size 187
section alignment 161–163, 186
Computer graphics illustrations, anaglyphs 153, 189
hidden lines 172–173
rotations 167–170
shading 175–177
stereopairs 170, 189–191

stick diagrams 179
  wire-frame figures 172–175
Computer image enhancement 157, 180
Computer reconstructions, brain
  regions 168, 223
  depth interpolation for 194
  flight paths 203–204
  giant neurons 152, 169, 175, 176
  neuron sequence 199–200
  sex-specific neurons 171
  synaptic contact layers 195–198
  synaptic sites 178
  visual cortex 181
  wide-field visual neurons 174, 177, 195
Corpus cardiacum antigens 291
Corpus cardiacum-corpus allatum complex 268, 291
  removal of 268, 271
Cricket (see Gryllus bimaculatus)
Cross-linking, agents for 272
  haptens 272
Cryoprotectants 340
Cytochrome c (see Heme protein)

DAB (see 3,3′-Diaminobenzidine)
DEAE cellulose chromatography 261,
  279–280
³H-2-Deoxy-D-glucose 225
  method 227–237
  transport and metabolism 226
Desheathing (see PAP method on whole
  mounts)
Destaining (see Cobalt intensification)
Development, anti-5-HT staining of neurons in 314
  HRP screening of 62–64
  inference based on laser
    ablation 220–224
  synchronized larvae 210
3,3′-Diaminobenzidine 48, 253 (see also
  Peroxidase activity demonstration)
  safety precautions 73
  solutions 66, 90, 125–127, 130, 266, 290,
    313
2,2-Dimethoxypropane (see Chemical dehydration)
Dimethyl sulfoxide (see Heme protein, additives)
Dithio-oxamide (Rubeanic acid: see Cobalt
  precipitation)
Dixipus morosus, proctolin-like immunoreactivity 268
Dopamine-containing cells, screening of
  (see Neutral red)

Double marking, antibodies and
  cobalt 274–275, 292, 331–337
  antibodies and Lucifer yellow 243,
    245–257
  cobalt and Golgi 43, 117–123
  cobalt and reduced silver 113–117
  cytochrome c and Golgi 125
  HRP and autoradiography 78
  HRP and cobalt 75–77, 127–128
  HRP and degeneration 78
  HRP and fluorescence 77
  HRP and Golgi 73–75, 125–127
  HRP and immunocytochemistry 77–78
  HRP and Lucifer yellow 77, 130–131
  lateral white cells 241
  Lucifer yellow and neutral red 242
Drosophila melanogaster 3, 42
  activity staining in 227, 230–235
  brain activity 230–238
  Giant Descending Neurons 59
  insulin-like peptide 264
  vertical cells-descending neurons 177
  visual mutants 232
DUM cells (dorsal unpaired median
  cells) 240
Dye injection (see Lucifer yellow)

E-face, definition of 351
Electrodes, nickel-filled 93
  blocking 93
Electron microscopic sectioning (see Ultramicrotomy)
Endogenous peroxidase,
  inactivation 290–291, 315
α-Endorphin 251
β-Endorphin 267
Enkephalin 250, 267
Enzyme-labelled antibody, production 284–286
Epon (see Plastic embedding)
Eristalis aeneus, vertebrate-like peptides 259

Falck-Hillarp (see Aldehyde fluorescence)
Fate correlation (see Ablation, targets)
Filled neurons, cobalt-complex 334
  cobalt sulphide 20, 21, 113
  cobalt "rubeneate" 93, 94, 97–101
  cobalt-Golgi 123
  heme protein 44–78
  Lucifer yellow 129, 132–137
  nickel precipitates 93, 97–101
Fluorescein isothiocyanate (FITC) 281,
  287, 304
  conjunction to antibodies 283–284

labelled secondary antisera 283, 309
paraffin section staining 288–290
photography of 310
properties of 309–310
whole-mount staining 293
Fluorescence-labelled antibodies, production of 283–289
Fluorescence microscopy, (*see also* Fluorescein isothiocyanate, Lucifer yellow and Rhodamine)
antioxidant 309
combined illumination 137, 145, 336–337
Fluorescent antibody method (FAB) 252–253
Fluorogenic neurons, distributions of 321–323, 325–329 (*see also* Aldehyde fluorescence)
Freeze drying 227–228, 312
Freeze etching 339, 348–350
carbon-Formvar 349
carbon shadowing 345, 347
platinum replicas 345
Freeze fixation 227, 311
Freeze fracture 339, 341–348
apparatus 339–340, 344–346
cell components 364–371
fixation for 340, 342–343
knives for 341–342
organelles 351–357
replicas 339
technique 342–350
Freezing, rapid 341, 343–344
Frozen sections 72
Functional mapping, (*see also* ³H-2-Deoxy-D-glucose)
of movement-specific activity 231–232
of single neurons 232–237

Gap junctions 31, 35, 42, 134, 355–357
Gastrin 267, 331 (*see also* Cholecystokinin)
Gelatine-chrome alum adhesive 306
Giant Descending Neurons (*see Calliphora, Drosophila, Musca*)
Glucagon 267
Glucose, substitution 226, 231
Glycerol-gelatine adhesive 117, 313
Glyoxylic acid, histofluorescence 318–319
Golgi method, buffers for 3, 15–18
chromation 4, 10
de-impregnation 2, 14, 128
electron microscopy for 17, 123
fixation 3, 16–18
ion substitution 2, 128
osmication 4

re-embedding 5, 6
semithin sectioning, staining of 6
specificity for cell components 10–15
thick sectioning 5
Golgi – Colonnier method 119
Golgi rapid method 4, 17–18, 121–123
Gryllus bimaculatus, nickel chloride filling 97

Hanker – Yates (*see* Peroxidase activity demonstration)
Hapten-specific antibodies, isolation of 288
Haptens, conjugates 277
conjugation of 272
Heme protein, additives 64–65, 85, 125
artefacts and errors 81, 85, 127, 128
cytology 78–81
electron microscopy 71
injection methods 61–62, 80, 130
preparation of 52
sites of application 53–63, 65, 79, 80
types used 51–52
Heme protein histochemistry 65–73, 125, 128
combinations with other markers (*see* Double marking)
counterstaining 73
for frozen sections 72–73
for PAP method 312, 313 (*see also* PAP method)
for paraffin sections 266
for whole brains 66–67
Heterocyclic dyes (*see* Neutral red)
High voltage electron microscopy (HVEM) 14, 375
applications 377–379
facilities 376
kV range 376
methods 378, 379–380
of insect neuropil 378, 380–384
Hirudo medicinalis, Lucifer yellow and HRP filling 129–131
Honeybee (*see Apis mellifica*)
Horseradish peroxidase (HRP), artefacts and errors 81, 85, 127–128
cytochrome c as substitute 81, 125
pellets 53, 62, 90
reaction product intensification 73, 83, 85
transneuronal staining 81, 88, 90, 125
uptake mechanisms 85–88
House fly (*see Musca domestica*)
HPLC (High Pressure Liquid Chromatography), protein-peptide separation 250
proctolin-containing neurons 247
purification of insect peptides 262
HRP (*see* Horseradish peroxidase)

IgG-immunoadsorbants 280
IgG, anti-rabbit 289
  fluorescein-conjugated 253
Immunoadsorption 280 (*see also* Antibodies, isolation of)
  with Sepharose 280
Immunocytochemical techniques 244−249, 252−255, 281−288, 294−299, 304−315, 319−320
  fluorescence 252−253 (*see also* Immunofluorescence)
  immunoenzyme 253−255 (*see also* PAP method)
  multiple staining 291
  screening 243
  unlabelled antibody enzyme method 281, 311−313 (*see also* Rhodamine, Fluorescein isothiocyanate)
Immunocytochemistry, biochemical confirmation 243, 259−264
  colchicine enhancement 244
  direct method 288, 293−299
  electron microscopy 293−300 (*see also* Protein A)
  fixatives for 265, 286−287, 304, 311−312, 314, 319, 334
  freeze fixation 311
  indirect method 288
  specificity (absorption controls) 255−259
Immunoelectrophoresis 276 (*see also* Antisera quality)
Immunofluorescence 304−310
  counterstains for 290
  indirect method 265−266
Immunoperoxidase method (*see* PAP method)
Immunoreactivity, loss and recovery of 288
Insulin-like peptide 259−264
Ion exchange (*see* Peptides, purification)
Ion-exchange chromatography 260−262

Laser ablation unit 207−210
Laser surgery 210
Lasers, types of 206−207
Lateral white neuron characterization 241−242
Lissamine rhodamine B 200 sulfonyl chloride (*see* Rhodamine)
Locust (*see* *Locusta migratoria, Schistocerca gregaria*)
Locusta migratoria, A-cell neurosecretion 271
  neurophysin-vasopressin-like substance 259

proctolin-like immuno-reactivity 247−248, 268
Lucifer yellow 92, 129−131
  antibodies to 134
  artefacts of 154
  backfilling with 135
  dye coupling by 133, 134 (*see also* Gap junctions)
  fading of 149−151
  fixation of 135, 137
  iontophoresis 134−135, 245
  photography of 145−148, 183−184
  photosensitization of 134
  Procion yellow, comparison to 132, 149
  properties of 132−133, 135
  selective staining 132, 135
Lucifer yellow preparations,
  sections 141−143
  storage 151, 153
  uses of 132−134, 181
  viewing of 143, 145
  whole mounts 138−141
Luteinizing hormone releasing hormone (LH-RH) 275
Lysine, for immunofixatives 285, 319
Lysolecithin (*see* Heme protein, additives)

Manduca sexta, prothoracicotropic hormone 271
Median neurosecretory cells, antibody-cobalt staining 275, 336
  axon bundle (median bundle) 72
  dendritic domains 338
  vinblastine, effect of 329
Metal ion identification (*see* X-ray microanalysis)
3-Ò-Methylglucose 227, 231
Molluscan cardio-excitatory tetrapeptide (FMRF-amide) 267
Monoamine staining 240−243
Monoamines, in neurons 239−249, 315−330
  transmitters 241, 315
Mouse cortex 180, 181
  Purkinje neurons 7, 8, 68
  pyramidal neurons 7, 8, 180, 181
Multiple cell markers (*see* Double marking)
Musca domestica, activity staining in single neurons 237
  antennal axons 117
  brain activity 230−238
  central body 115
  descending neurons 137, 140, 149
  Giant Descending Neuron 27, 31, 35, 71, 115, 133, 169, 175
  horizontal cells (HS neurons) 115

HVEM studies 378–382
lamina freeze fracture structure 351–374
lobula 217, 219
vertical cells (VS neurons) 24, 214
visual behaviour 203, 204, 219, 220, 231

Nervus corpus cardiacum (ncc I,
  ncc II) 271
Neurosecretory cells 271, 329, 333
Neutral red 240–243, 320, 329
  associated dyes 240
  peptidergic neuron identification 241,
    242
  staining specificity 241, 326
Nickel chloride 93–95
Nonidet P-40 (see Heme protein, additives)

O.C.T. medium 59, 72
Octopamine 240, 241, 247 (see also DUM
  cells)
Ornectes limosus, proctolin-like immunoreactivity 268
Osmication, cobalt electron microscopy 22
  heme protein electron microscopy 69, 90

P-face, definition of 351
Pancreatic polypeptide (PP) 259–264, 267
PAP (peroxidase-antiperoxidase)
  method 255, 266, 281, 290
  on sections 266, 288–291, 311–313
  on whole mounts 244, 292, 313–315
Paraffin embedding, for immunocytochemistry 334
  for Cajal-cobalt double
    marking 115–117
  for Lucifer yellow 142–143
Peptide extraction, fixation artefact 265
Peptides, amino acid sequence 243–264
  antigenic determinant 252, 334
  brain-gut type 259
  co-localization with transmitters 241, 267
  composition studies 247–248, 262–264
  extraction of 260
  hormonal role 264
  insect, synthesis of 268
  insect, types of 259
  metabolism, localization 351
  neuroregulatory 267
  purification 259–269
  transmitter role 241, 264
  unspecificity 248–249
  vertebrate-like, structural homolo-
    gies 267
  vertebrate-like, distribution 267

Periplaneta americana, antineurosecretory
  antisera 268–271
  cobalt-antibody staining 292
  corpus cardiacum 279
  immunocytology electron microscope
    studies 298
  monoamine staining in 240–243
  motor neurons 95
  neurosecretory cell staining 270
  neutral red staining 241
  "proctolin" cells 287
  serotonin 243–245, 272, 306, 308–311
Peroxidase activity, controls 291
  demonstration 47–51, 266, 291, 292 (see
    also PAP method)
  endogenous, inactivation of 290–291,
    315
Peroxidase-antiperoxidase preparation 285–286, 313
Peroxidase antisera, equivalence zone 285
Perioxidase-IgG conjugates 284
p-Phenylenediamine 50 (see also Peroxidase activity, demonstration)
  with pyrocatechol 67–68
Plastic blocks, softening of 5
  thick sectioning of 5
Plastic embedding, araldite 5, 29
  araldite composition 91, 114
  base (see Sylgard)
  Epon 228, 296
  forms for 23, 142
  Lucifer yellow (see Spurr's)
  soft araldite 23
Plastic section re-embedding 5–6, 23–24,
  71, 103–104
Plastic sections for immunocyto-
  chemistry 296–297
Polyacrylamide gel electrophoresis (PAGE)
  (see Peptides, purification)
Proctolin 243, 244, 247–248, 268
  antibodies against 278
  immunoelectrophoresis of 276–277
  influences of fixatives 287, 288
Protein A-colloidal gold 294
Protein A-conjugates 293, 294
Protein A-Gold method 297–299
Prothoracicotropic hormone 271
Purkinje cells 6–8
Pyramidal cells 6–8, 179–181

Radioimmunocytochemistry (RICH) 250,
  255, 276
Replica interpretation 350
Reserpine (see Aldehyde fluorescence)
Retinula cell junctions 361–364

Retrocerebral complex 268, 292
Retzius cell 240
Rhodamine, labelled antisera 281, 283, 304, 334–336
Rhodnius prolixus, monoaminergic neuron innervation 317–330
  peptidergic neurons 329
Ringer solutions, Grace medium 292
  hypotonic for immunocytochemistry 314
  TES 138
Rubeanic acid 94, 97

Schistocerca gregaria, peripheral nerve fills 98–100
  serotonin 305
Schistocerca nitens, proctolin assay 247
Secretin 267
Sephadex gel filtration (*see* Peptides, purification)
Sepharose 4B-IgG gels 280–281
Serotonin (5-HT) 240, 241, 243, 247
  antisera solutions 309
  FITC labelling 309
  radioenzymatic assay for 247
  specificity of 315–316
Silver affinities, to cobalt sulphide (*see* Cobalt-silver)
  to DAB-H$_2$O$_2$ reaction product 126–128
Silver intensification (*see* Cobalt intensification)
Silver staining, brain 214–219, 222
  larvae 213
Solid-phase affinity purification 278
Somatostatin 267
Spurr's (plastic embedding medium), dehydration into 141
  composition of 142
  for heme protein methods 69
  for Lucifer yellow embedding 141–142
  for whole mounts 138–141
Stereopairs, electron microscopy 375, 378
  generation of (*see* Computer graphics illustrations)
Stereoscopes 189–190

Stereoscopy 188–189, 190, 194
Stripping film (*see* Autoradiography)
Substance P 250, 267
Sylgard (Silgel or silicone rubber) 137, 142, 229, 231
Synapses, selective staining (*see* Cobalt affinities to)
  structure 28–35, 42, 367–371
  vesicle release 368

Tetramethylbenzidine 49
Tetramethyl-rhodamine isothiocyanate (TRITC) (*see* Rhodamine)
Three-dimensional reconstruction (*see* Computer graphics illustrations)
Transsynaptic cobalt migration 31, 35, 42
  (*see also* Gap junctions)

Ultramicrotomy (electron microscopic sectioning) for cobalt-silver sulphide 21–27, 103, 104
  cobalt-Golgi 123
  ^{3}H-2-deoxy-D-glucose 228
  Golgi 6
  heme protein 71–73
  immunocytochemistry 299

Vacuum dialysis 278
Vacuum evaporators 339–340
Vapour fixation, formalin 312
  osmium 298
Vaso-active intestinal peptide (VIP) 250
Vertebrate-type peptides (*see* Peptides, vertebrate-like)
Vinblastine 329
Vital staining (*see* Neutral red)

Whole-mount immunocytochemistry 244, 314, 319
Whole mounts (*see* Cobalt intensification, Lucifer yellow preparations)

X-ray (electron probe) microanalysis 37, 38–39, 105–111

Biophysics

Editors: **W. Hoppe, W. Lohmann, H. Markl, H. Ziegler**
With contributions by numerous experts
1983. 852 figures. XXIV, 941 pages
ISBN 3-540-12083-1

Contents: The Structure of Cells (Prokaryotes, Eukaryotes). – The Chemical Structure of Biologically Important Macromolecules. – Structure Determination of Biomolecules by Physical Methods. – Intra- and Intermolecular Interactions. – Mechanisms of Energy Transfer. – Radiation Biophysics. – Isotope Methods Applied in Biology. – Energetic and Statistical Relations. – Enzymes as Biological Catalysts. – The Biological Function of Nucleic Acids. – Thermodynamics and Kinetics of Self-assembly. – Membranes. – Photobiophysics. – Biomechanics. – Neurobiophysics. – Cybernetic. – Evolution. – Appendix. – Subject Index.

Biophysics is the English translation of the completely revised second German edition of this book. A wealth of information, new sections and chapters, and extra figures make **Biophysics** an indispensable text for advanced students and lecturers of physics, chemistry, biology, and medicine. Indeed every scientist interested in biophysical questions will find much in this book essential for a deeper understanding of modern biophysical research.

From the reviews: "… Most of these contributions are very informative reviews of 10–20 pages written for undergraduate students or scientists who wish to broaden the scope of their knowledge. They provide a convenient source of information for many facets of biophysics and cannot be found in such a concise way in other textbooks…" *Nature*

C. S. Crawford

Biology of Desert Invertebrates

1981. 181 figures. XVI, 314 pages
ISBN 3-540-10807-6

Contents: Deserts and Desert Invertebrates. – Adaptations to Xeric Environments. – Life-History Patterns. – Invertebrate Communities: Composition and Dynamics. – Invertebrates in Desert Ecosystems: Summary Remarks. – References. – Index.

Biology of Desert Invertebrates explores the ways in which invertebrate animals function in and contribute to desert ecosystems. The work is a unique overview of desert biology as a whole, synthesizing the author's own extensive investigations into desert invertebrates with his summary of both classical and more recent studies.
The author opens with an introduction to the evolution, distribution and abiotic and biotic elements of deserts themselves, followed by a summary of the vast array of the invertebrates that inhabit them. The major portion of the book is devoted to invertebrate adaptation patterns, life history and community interaction. A final summary integrates these topics and points out areas where further study is still indicated.
Biology of Desert Invertebrates will prove an ideal reference and stimulating source of ideas for students and active desert biologists alike. In addition, it will serve as a catalyst for others to question, explore ultimately respect the rhythms and pulses peculiar to desert life.

Springer-Verlag Berlin Heidelberg New York Tokyo